The Biology of the Monotremes

The Biology of
the Monotremes

MERVYN GRIFFITHS

Deakin, A. C. T.
2600, Australia

ACADEMIC PRESS New York San Francisco London 1978

A Subsidiary of Harcourt Brace Jovanovich, Publishers

ACADEMIC PRESS, INC.
111 Fifth Avenue, New York, New York 10003

United Kingdom Edition published by
ACADEMIC PRESS, INC. (LONDON) LTD.
24/28 Oval Road, London NW1 7DX

Library of Congress Cataloging in Publication Data

Griffiths, Mervyn.
  The biology of the monotremes.

  Bibliography: p.
  Includes index.
  1. Monotremata. I. Title.
QL737.M7G74   599'.1   78–4818
ISBN 0–12–303850–2

Contents

v

5. Temperature Regulation

6. Endocrine Glands and the Glands of the Immune System

7. Special Senses, Organization of the Neocortex, and Behavior

8. Reproduction and Embryology

9. Lactation, Composition of the Milk, Suckling, and Growth of the Young

10. The Affinities of the Monotremes

Preface

Since their discovery in the late eighteenth century the monotremes—the platypus and the echidna, with their mélange of reptilian, mammalian, and specialized characters—have held a fascination for zoologists, particularly those interested in evolution. The outcome of that interest, until recently, has been a literature concerned for the most part with anatomy and embryology. Over the past twenty years, however, there has been an upsurge of research into the physiology, biochemistry, ecology, ethology, and paleontology of the egg-laying mammals. The catalyst for all this activity is the recent spate of publications from paleontologists on the anatomy of extinct Mesozoic mammals whose cranial and postcranial skeletons have been shown to have many characters in common with those of the living monotremes. This has led some biologists to study the monotremes from the possibly naïve standpoint that an understanding of the biology of Mesozoic mammals may result. Others study the monotremes for their intrinsic interest as a taxon of mammals, probably of Mesozoic origin, which has survived the vicissitudes of millions of years of existence during which their relatives have gone to the wall; they seek answers to the questions of how these survivors cope with their modern environments.

Whatever the standpoint, much has been achieved but the information is available only in specialized journals or in theses housed in university libraries; there has been no coordinated account of monotreme biology. This book is an attempt to give an overall picture of what has been done and what is going on at present; thus it is addressed to established research workers in different disciplines, hopefully to young graduates to encourage them to undertake research on monotremes, and to students and teachers concerned with life processes and evolution of mammals.

As far as possible the information in each chapter has been considered under the headings of *Ornithorhynchus, Tachyglossus,* and *Zaglossus,* the three living genera of the monotremes. Occasionally it has been necessary to abandon that arrangement, in part in Chapter 8 where embryos from *Ornithorhynchus* and

Tachyglossus are used to give a sequential description of "monotreme" development, and completely so in Chapter 10.

My thanks go to numerous authors for permission to use their figures and tables and I am particularly indebted to the following friends and colleagues: Ederic Slater for the photography; Frank Knight for the artwork; Douglas Parsonson for the line drawings and graphs; Gutta Schoefl, John Calaby, Hugh Tyndale-Biscoe, Tom Grant, Roslyn Bohringer, Murray Elliott, and Graeme George for advice and information; Harry Frith for extension of museum, library, and typing facilities at C.S.I.R.O. Division of Wildlife Research; Barbara Staples and her staff for bibliographic research; Carol Todd who typed the manuscript from my quasi-handwriting; Jim Menzies, Fathers May and Sourisseau, and Gilles Telsfer for assistance in taking *Zaglossus* in the mountains of Papua; Ray Leckie, Roy Coles, Leckie MacLean, and Terry Rutzou for skilled technical assistance in the laboratory and in the bush; and to my daughter Mrs. Garry Scott for assistance in checking the references.

Mervyn Griffiths

The Biology of the Monotremes

1

The Discovery and the General Anatomy of the Monotremes

The order Monotremata of the subclass of the Mammalia known as the Prototheria is made up of two families: Ornithorhynchidae (platypuses, two genera *Ornithorhynchus,* and the fossil *Obdurodon*); and Tachyglossidae (echidnas, two genera *Tachyglossus*, and *Zaglossus*).

ORNITHORHYNCHUS

Discovery

This beautiful little amphibious mammal, an inhabitant of freshwater streams, lakes, and lagoons in eastern Australia, was first brought to the notice of biologists by John Hunter, the Governor of the penal colony at Port Jackson in New South Wales, who sent a skin and a drawing of the animal to the Literary and Philosophical Society of Newcastle-upon-Tyne in 1798 (cited in Gilbert Whitley, 1975). Home (1802a) relates just how Governor Hunter came by his platypus:

> The Natives sit upon the banks, with small wooden spears, and watch them everytime they come to the surface, till they get a proper opportunity of striking them. This they do with much dexterity; and frequently succeed in catching them in this way.
>
> Governor Hunter saw a native watch one for above an hour before he attempted to spear it, which he did through the neck and foreleg: when on shore, it used its claws with so much force, that they were obliged to confine it between two pieces of board, while they were cutting off the barbs of the spear, to disengage it.

Hunter's drawing (which is ghastly) was reproduced by David Collins (1802) along with a description of the external characters of the animal. However, the first description published was that by George Shaw (1799) in which he named the animal *Platypus anatinus*. The name was changed to *Ornithorhynchus paradoxus* by Blumenbach (1800), and an anatomical description using the name *O. paradoxus* was published by Home (1802a). However, *anatinus* was deemed to have priority, but not *Platypus*, so the animal is now known as *Ornithorhynchus anatinus*. It should be mentioned here that Home (1802a,b) was the first to realize that platypuses and echidnas were closely related and although the echidna had been discovered long before the platypus, the peculiar nature of the anatomy of the monotremes was first discovered with Home's dissection of the platypus.

External Features

The body of the platypus is streamlined (Fig. 1) and compressed dorsoventrally; the snout is prolonged into a muzzle shaped like the bill of a duck, and at its hind end the platypus sports a tail like that of a beaver. The males are larger than the females: the largest male I have taken weighed 2300 g and was 56 cm long from tip of muzzle to tip of tail; the two largest females weighed 1300 and 1325 g and were 49.5 and 47.5 cm long, respectively. T. Grant (personal communication) caught a male 2350 g in weight. The males are testicond but can be distinguished from adult females by the presence of a prominent sharp spur on the medial side of the ankle in the former. Excretory and reproductive products pass to the exterior through a cloacal aperture situated ventrally at the rostral end of the tail in both sexes. There is no external pinna; the aural opening and the eye are close together and are located dorsolaterally in a groove, the lips of which come together and close off both eyes and ears when the animal is submerged.

Although the muzzle externally resembles the bill of a duck, internally it is dissimilar. In *Ornithorhynchus* the muzzle is formed by two diverging ''rostral crura'' (Wilson and Martin, 1893a), each of which consists of a long slender

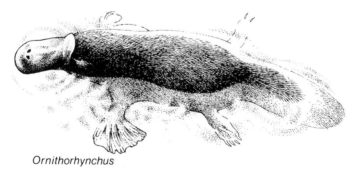

Ornithorhynchus

Figure 1.   *Ornithorhynchus anatinus*. Note backwardly directed frontal shield over forehead and web of manus extending beyond the nails; hindlimbs trail and are used mainly as rudders.

extension of the premaxilla; the crura are inflected sharply anteriorly, forming the limit of the bony rostrum. The crura, however, do not form the margin of the muzzle since they are embedded in a vast plate of cartilage; the crura are separated widely (Fig. 5), the intercrural area being filled-in with cartilage, and their lateral and anterior margins are also invested with cartilage forming the outermost limits of the muzzle. In contradistinction the duck's bill is made of bone and keratin.

The muzzle is covered by a delicate naked skin which at the caudal border is carried back as a fold covering the anterior part of the face almost to the eyes. The nostrils are located near the midline at the rostral end of the muzzle. Just below the external opening of the nostril is a flap of skin which acts as a valve when the animal dives, closing off the nostrils and thus eliminating sense of smell under water. The mandibles are likewise covered with soft skin that is also projected backwards over the throat, but this fold is divided into two at the midline. These dorsal and ventral folds are known as the frontal shields. Both snout and mandibular skin exhibits minute pores; these are the openings of sweat gland ducts and are associated with collections of special sensory organs that are distributed over the entire surface of the muzzle and frontal shields (see Chapter 7 for structure and function).

The platypus feeds in freshwater only and is largely a bottom feeder, swimming along and waggling its muzzle from side to side as it does so. The gape is large and any food item snapped up or otherwise ingested is passed into a capacious cheek-pouch on either side where it is stored in between dives. When the pouches are full the food is transferred to the buccal cavity where it is comminuted by the grinding action of keratinous horny grinding pads on the upper and lower jaws (see also pp. 16, 56, 78). Since the eyes, ears, and olfactory organs are shut off from the outside world when underwater, any discrimination about what is ingested must largely be due to tactile sense of the muzzle (see p. 172).

The margins of the mandibles bear a series of keratinous serrations placed transversely at right angles to the long axis at the posterior end of the mandible but slanting forwards in the mandible's more rostral parts. The tongue is bipartite consisting of a narrow elongated part housed in the hollow between the two mandibles, and a raised oval-shaped portion housed at the posterior end of the buccal cavity. The dorsal surface of the tongue is covered with backwardly directed cuticular papillae and the oval portion bears two large forwardly directed keratinous teeth at its laterorostral borders. These teeth are undoubtedly concerned with transferring and directing cheek-pouch contents to the grinding pads.

The body is covered by a pélage of coarse and fine hairs arranged, as in all mammals, in bundles consisting of a principal hair and varying numbers of neighboring hairs (Haupthaare and Nebenhaare of German authors). The principal hairs are much longer than the fine ones and distally they are expanded into a narrow flattened leaf-shape. These principal hairs point backwards and lie flat on the fine hairs beneath, thereby enhancing the insulative qualities of the pelt. There has been controversy about the numbers of neighboring hairs surrounding each principal hair and the numbers of bundles per unit area of skin in "normal" platypuses (see Spencer and Sweet, 1899). Temple-Smith (1973) and Grant (1976), however, have settled the matter since they found that platypus pelts undergo moult involving even complete loss of principal hairs. This has a bearing on thermoregulation since the pelt plays a major part in regulating heat loss (p. 126). The fur is thickest on the ventral surface of the body, next on the dorsal surface, while that on the dorsal surface of the tail is sparse and coarse; the ventral surface of the tail is practically naked skin. All these facts were known to Home (1802a) who also noted the peculiar structure of the principal hairs. The color of the fur in southeastern Australian platypuses is, in general, dark brown on the back and a mixture of rufous brown and silvery grey on the venter. The platypuses of northeastern Queensland are said to be of a rufous color on the shoulders.

The skin is loosely attached to an extensive panniculus carnosus muscle that confers a fantastic ability of the platypus to change its shape and wriggle out of the firmest grasp.

The forelimbs are short and stoutly built; the digits of the pentadactyl manus bear long nails and are webbed. The web occurs not only between the digits but is carried forward well beyond the nails to form large fan-shaped paddles; these are the main organs of propulsion in water. The hindlimbs are also webbed but they simply trail and act as rudders when the platypus is swimming. The webs and extensions of the forelimbs account for 13.5% of the total surface area of the animal and play important roles in thermoregulation (p. 104). When out of the water the platypus lives in a burrow dug into the bank of a river, pool, or creek. When burrowing the web extensions are folded back and under so that the nails form the most rostral part of the manus. This also takes place when the platypus walks and runs.

Reproductive and Excretory Organs

The female reproductive organs consist of paired ovaries enclosed by paired thin infundibular funnels, paired oviducts, paired uteri each of which communicates separately with a long median unpaired urogenital sinus. At the level of the union of the uteri with the urogenital sinus the ureters from the kidneys enter the sinus opposite the neck of the bladder which depends from its ventral surface. The elongated urogenital sinus communicates with the cloaca; thus eggs, urine, and feces pass out through the one cloacal sphincter, hence the word Monotremata to describe the Order.

Only the left ovary and oviduct are functional; the right ovary exhibits oocytes that never mature; however, the right oviduct while never as well-developed as the left, exhibits uterine glands during the breeding season (p. 219). In the left ovary mature oocytes achieve a diameter of ca. 4.3 mm and, along with their follicular investments, are visible to the naked eye; an ovary with maturing oocytes sticking out all over it gives one the impression of a bunch of grapes.

Home (1802a) discovered and described the essentials of the female reproductive system and realized that it bore no resemblance to the reproductive systems in mammals, nor could he find any resemblance of the platypus uterus to the oviduct of birds. However, he did find a close resemblance of that organ to the oviducts of "ovi-viviparous" lizards; he concluded that the platypus "also is oviviviparous in its mode of generation." In 1884 Caldwell (1884b) finally showed that the platypus is oviparous (p. 209).

The ovaries of *Ornithorhynchus* are attached to the anteroventral faces of the kidneys by a fold of peritoneum. The kidneys are nonlobulated structures, i.e., they are "kidney" shaped and have the typically mammalian blood supply and drainage, consisting of renal artery and renal vein; there is no renal portal system as in birds and reptiles (Hochstetter, 1896).

As mentioned above the males are testicond, the paired testes being suspended posterior and ventral to the kidneys by mesorchial folds of the peritoneum. Seminiferous tubules pass from the anterior end of the testis and join to form a short rete that communicates with the ductuli efferentes (Temple-Smith, 1973). Spermatozoa pass from the testes through the ductuli efferentes into the epididymis, which consists of three regions: caput, corpus, and cauda. From the cauda a short vas deferens conveys the sperm to the unpaired urogenital sinus which is elongated as in the female. The penis depends ventrally from the posterior end of the urogenital sinus and is housed in a thin-walled preputial sac lying ventral to the cloacal chamber. The ventral surface of the latter is perforate so that the erectile penis can pass through the cloaca and out to the exterior through the cloacal aperture. These relationships are illustrated in Fig. 10, actually a photograph of the reproductive system of a male echidna; that of the platypus is practically the same. As Home (1802a) recognized, the urethra of the penis communicates with the lumen of the urogenital sinus and that it conveys sperm only, the urine being passed separately from the urogenital sinus into the cloaca. Just why sperm do not do the same thing is not clear; possibly when the penis is erect it also projects backwards into the sinus and blocks off the passage to the cloaca thus forcing the sperm to enter the seminal urethra. The penis is bifid distally, right and left portions terminating in a bulbous glans, each glans bearing a group of evertible foliate papillae. Branches of the seminal urethra pass to the exterior through these papillae (Temple-Smith, 1973). The penis is about 5–7 cm in length.

Two types of accessory glands are present: paired Cowper's glands which communicate by short ducts with the penile urethra, and a set of disseminate tubular glands which encloses the urogenital sinus (p. 232).

Mammary Glands

Home (1802a) found "There is no appearance, that could be detected of nipples; although the skin on the belly of the female was examined with the utmost accuracy for that purpose." Twenty-four years later Meckelio (1826) found two sets of ventrally located glands in female platypuses, one on each side of the midline about halfway down the body; each set consisted of club-shaped lobules that converged proximally and passed through the corium and skin to the exterior. The sets varied greatly in size from platypus to platypus. Meckel diagnosed these as mammary glands but not everyone was convinced he was right: Geoffroy (cited in Meckel, 1827) thought that the great variation in size ruled out the possibility of them being mammary glands but Meckel stuck to his guns: "Die Grösseverschiedenheit der Drüse bei gleichem Alter und Grösse der Individuen ist offenbar der beste Beweis für die Ansicht, dass diese Drüse zum Zeugungssysteme gehöre, und, in Verbindung mit den übrigen Thatsachen, dass die Brustdrüse sey." A few years later his interpretation was confirmed by Lieutenant the Hon. Lauderdale Maule of the 39th Regiment who communicated his observations to the Zoological Society of London (Maule, 1832):

> During the spring of 1831 being detached in the interior of New South Wales, I was at some pains to discover the truths of the generally accepted belief, namely, that the female *Platypus* lays eggs and suckles its young. By the care of a soldier of the 39th Regiment who was stationed at a post on the Fish River, a mountain stream abounding with *Platypi*, several nests of this shy and extraordinary animal were discovered.

A female and two young were taken from one of these nests; after keeping her and the young for a couple of weeks she was accidentally killed. Upon skinning her while she was still warm it was seen that milk oozed through the fur on her ventral surface. This came from "two teats or canals," both of which contained milk.

Considering the size of the animal the fully lactating glands are enormous. In one platypus examined by Griffiths *et al*. (1973) measuring 43 cm from muzzle to tail, the glands measured 13.5 cm in length stretching from axilla to groin, and around the flank almost to the dorsal surface. Each of the lactating glands is fan-shaped (Fig. 2) since the lobules are expanded distally but converge proximally to pass in the form of ducts through the abdominal musculature and skin to appear at two patches obscured, as Maule noted, by the pélage of the ventral surface. These patches or areolae have a well-defined morphology: the ducts below the skin are distended into lacuna-like spaces reminiscent of storage cisterns (Fig. 3). Peripheral to the distended ducts is a ring of convoluted glands, the

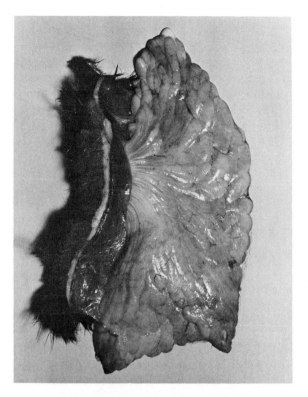

Figure 2.   Mammary gland of lactating platypus. × 0.67 (*From Griffiths et al.*, 1973; *with permission of The Zoological Society of London.*)

ducts of which open to the exterior around the areola. These glands resemble large sweat glands and are known as the Knäueldrüsen of Gegenbaur. The pélage covering the areolae consists of neighboring and principal hairs found all over the body, the follicles being interspersed with sebaceous glands. The ducts of the mammary lobules and the sebaceous glands open to the surface at the base of a principal hair (Gegenbaur, 1886). Apart from the fact that the areola is covered by fur* its structure is remarkably like that of the human nipple and areola. Cowie (1972) has recently compared the different arrangements of the ducts in the mammary system with the four kinds exemplified by those of the rat, ruminant, rabbit, and woman; that of the latter closely resembles the monotreme duct system in that both exhibit main ducts expanded to form sinuses which open to

---

*A lactating platypus taken in the month of February, i.e., near the end of the lactation season (see p. 264), exhibited areolae and neighboring areas of the ventral surface devoid of fur; it is surmised that the nuzzling by two or three young at the areolae over a period of months had worn the fur away. Apparently hair at the areola is not necessary for suckling.

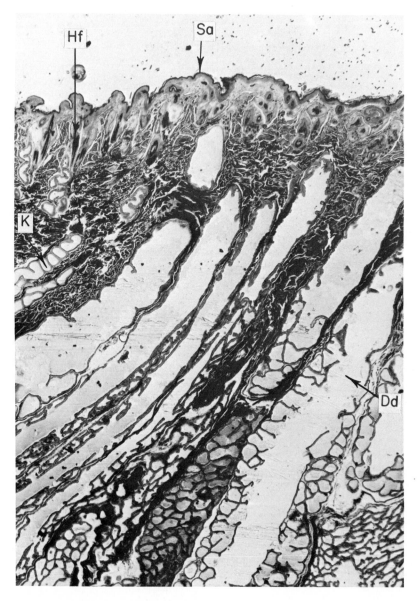

Figure 3. Longitudinal section of areola of platypus mammary gland, × 25. Hf, hair follicle; Sa, surface of areola; Dd, distended duct; K, Knäueldrüsen. (*From Griffiths et al., 1973; with permission of The Zoological Society of London.*)

the exterior of flattened, but raised surfaces of skin. The resemblance is even more striking in the echidna since a pélage of fur is wanting over the areola (Fig. 12). The human nipple also exhibits sebaceous glands and a ring of convoluted glands peripheral to the openings of the ducts on the surface—the glands of Montgomery which Griffiths *et al.* (1973) consider to be homologues of the Knäueldrüsen. There have been reports of the presence of mammary glands in male platypuses but I have never been able to detect them.

Brain

A detailed description of this organ has been given by Hines (1929); consequently only a brief account of structures of the phylogenetic interest and those necessary for an understanding of the aspects of physiology and function of the brain discussed in Chapter 7 will be included here.

The neocortex is enormous (Fig. 4), overshadows the pyriform and hippocampal cortices, and proclaims the brain to be that of a mammal, but for purposes of description we will start at the other end and work up to the forebrain.

The medulla oblongata bears on its ventral surface two prominent elongated protruberances formed by the enormous chief sensory nuclei of the trigeminal system. Bohringer (1977) found that fibers from peripheral receptor complexes of the trigeminal system formed nine major nerve bundles on either side of the muzzle—four mandibular, four maxilliary, and one opthalmic—which unite and

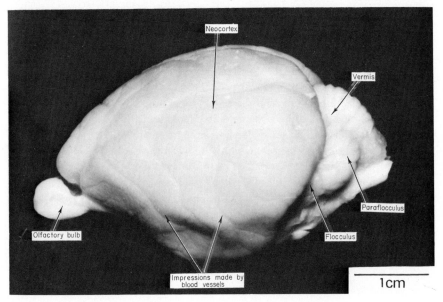

Figure 4. *Ornithorhynchus.* Lateral view of left side of brain. (*From Bohringer and Rowe, 1977.*)

enter the brain as one gigantic trigeminal nerve (V) just anterior to the pons Varolii. It is a mixed nerve consisting of the afferent sensory fibers mentioned above, and efferent motor fibers that take their origin in the trigeminal motor nucleus of the medulla. The sensory fibers terminate in a mesencephalic nucleus and, the majority, in the chief sensory nucleus made up of five large nuclei (Watson *et al.* 1977), three of which (Va, Vb, and Vc) appear to correspond to the caudal, interpolar, and oral parts of the spinal trigeminal tract in primates. Dorsolateral to the rostral half of Vc lie the presumptive homologues of the two portions of the eutherian principal trigeminal nucleus, Vd (parvocellular) and Ve (magnocellular).

Axons from the chief sensory nucleus decussate in the mesencephalon to form, along with fibers from the gracile and cuneate nuclei of the medulla, the medial lemniscus which conveys trigeminal impulses to the nucleus latero ventralis of the thalamus. From here they radiate to the neocortex since mechanical stimulation of receptors in the muzzle gives rise to evoked potentials at a large, but circumscribed, area of the cortex (Bohringer and Rowe, 1977; also see p. 174).

Another ascending tract of sensory fibers is the lateral lemniscus formed by fibers from the dorsal and ventral nuclei of the cochlear division of nerve VIII. En route to the mesencephalon the medial and lateral lemnisci come into contact forming one great ascending tract of sensory fibers.

The cerebellum on the roof of the hindbrain is like that of other mammals in that it consists of a central vermis thrown into various lobes that, according to Dillon (1962), are the lobus centralis, culmen, pyramis, lingula, lobus ventralis (peculiar to monotremes) nodulus, and uvula. The transverse diameter is very wide due to the presence of lateral extensions of the nodulus and pyramis—the flocculus and paraflocculus, respectively. As in all mammals the cerebellum is attached to the brain stem by three peduncles on each side: the corpus restiforme (the largest), the brachium pontis connecting the pons Varolii (found only in mammals) to the cerebellum, and the brachium conjunctivum.

The floor of the midbrain is made up of two longitudinally arranged bundles of fibers, the cerebral peduncles, widely separated anteriorly, but fused together posteriorly into an entity, the tegmentum. Between the peduncles is a ganglion interpedunculare. The posterior part of the roof or tectum of the midbrain is like that in a mammal since there are two distinct domed auditory centers, the inferior colliculi. The anterior portion of the tectum, however, according to Hines is not separated into two distinct domes as in other mammals but resembles a large "flat plateau which merges inperceptibly with the dorsal part of the thalamus laterally . . .".* The medial lemniscus gives off some trigeminal fibers to the superior colliculus but the rest of the medial lemniscus trigeminal fibers pass to the

---

*Campbell and Hayhow (1972) in their study of the primary optic pathways in *Ornithorhynchus* refer to superior colliculi; the latter are certainly present in the tectum of *Tachyglossus* (p. 31).

nucleus ventrolateralis thalami. Auditory fibers from the lateral lemniscus pass to the inferior colliculus and to the medial geniculate body in the thalamus.

The thalamus has a massa intermedia connecting its right and left halves. Apparently this structure is characteristic of mammalian brains since it is absent from those of birds and reptiles but it is said to be present in the turtle (Jollie, 1962). The thalamus is large due to hypertrophy of a lateral group of nuclei found in the pars dorsalis thalami; these nuclei are the dorsolateralis, mediolateralis, and the ventrolateralis. The last is by far the largest in diameter; it ranges from the anterior end of the diencephalon to the hemisphere posteriorly where it pushes into the corpus striatum so that in cross section the nucleus ventrolateralis appears as an island surrounded by a ring of the corpus striatum tissue.

Two other important thalamic way stations for cortical radiations are the medial and lateral geniculate bodies. The medial geniculate receives auditory fibers from the lateral lemniscus and the inferior colliculus; efferent fibers from the medial geniculate pass to the caudal pole of the neocortex—the auditory cortex (p. 177). Concerning the lateral geniculate body, Hines (1929) and Campbell and Hayhow (1972) distinguish ventral and dorsal nuclei which receive fibers from the optic tract. However, Campbell and Hayhow are not prepared to commit themselves to the view that these are homologues of the dorsal and ventral nuclei of the eutherian lateral geniculate body. The optic nerves are small and the decussation at the chiasma is almost complete, only a few fibers passing in the ipselateral tract (Campbell and Hayhow, 1972).

The surface of the neocortex is lissencephalic like that in the opossum *Didelphys virginiana*. There is one fissure, however, which Hines believes to be the cruciate because the largest pyramidal cells in the cortical layer occur near it and because descending pathways from the cortex originate in the vicinity; if this proves to be so, the cortex anterior to the fissure is frontal lobe, and posterior to it the parietal lobe.

The olfactory lobes are joined by short peduncles to the nucleus olfactorius which is divided posteriorly into the tuberculum olfactorium and the long slender pyriform lobe of the cortex. The pyriform cortex is separated from the neopallium by the fissura rhinalis. There is no corpus callosum but there is communication between the two halves of the telencephalon through the anterior commissure and dorsal to this through the hippocampal (dorsal or pallial) commissure.

Skeleton

Chondrocranium

As de Beer and Fell (1936) said ''The chondrocranium of *Ornithorhynchus* is one of the most interesting so far known in all vertebrates, and it owes its interest largely to the fact that it represents a mosaic of characters some of which are purely reptilian and others purely mammalian.'' The development of the chon-drocranium has been described by those authors in three embryos, one hatchling, and one suckling. The sizes of the specimens ranged from 8.5–122 mm.

1. 8.5-mm specimen. The chrondrocranium at this stage consists of two para-chordal cartilages fused together in the midline to form the basal plate. Occipital arches rise from the posterolateral corners of the basal plate; anteriorly this is upturned and from its ventral margin two trabeculae or polar cartilages project forwards, and anterolaterally from the corners of the plate arise two elongated flat cartilages—the pilae antoticae. Just posterior to the trabeculae the basal plate exhibits a lateral process of cartilage on either side, the processus alaris or basitrabecular process. The occipital condyles, even at this early stage, are present as rudimentary bosses at the posterior surface of the occipital arches.

2. 9.0-mm specimen. Anteriorly the trabeculae are prolonged to form the interorbital region, the orbital cartilage on each side taking the form of a plate of cartilage continuous with the distal end of the pila antotica. The processus alaris has extended further laterally and exhibits a distinct condensation of cartilage at its distal end, the anlage of the ala temporalis. The auditory capsules are morphologically more distinct than at the previous stage and a crista parotica is present as a ventrolateral swelling on each capsule. The dorsal portion of each occipital arch is extended anteriorly into a parietal plate that hovers over the auditory capsule but is not yet united to it. The space between the auditory capsule and the occipital arch is the fissura metotica.

Meckel's cartilage, on each side, has appeared and takes the form of a long rod lying beneath the chondrocranium at about an angle of 45° to its long axis, but is not attached to the chondrocranium. At its proximal end Meckel's cartilage exhibits a hook of cartilage—the anlage of the malleus which is homologous with the articular of the jaw of therapsid reptiles. The stylohyal cartilages are present and lie just posterior to Meckel's cartilage but are not yet attached to the cristae paroticae; the two thyrohyal elements are also present and the stylohyals and thyrohyals are interconnected by two basihyals lying transversely and fused together.

3. 9.4-mm specimen. Anteriorly the trabecular plate, as it has become, is narrow and deep dorsoventrally forming an interorbital septum that is continued forwards as a nasal septum. The dorsal margin of the interorbital septum on each side is extended to form a crista galli continuous with the roof of the partially formed nasal capsule that now extends laterally on each side from the dorsal margin of the nasal septum. The anterior part, only, of the nasal capsule is floored and from this floor there extends a cartilaginous plate that is joined anteriorly by its mate of the other side. These plates are the laminae transversales anteriores and they constitute the first indications of the cartilaginous muzzle.

The side wall of the nasal capsule is also very rudimentary and the only indication of it is the paranasal cartilage continuous with the hindmost part of the nasal capsular roof. The sphenethmoid cartilage is not yet apparent.

The anterior edge of the laminae transversales shows a deep notch in the midline and within the notch is housed the anlagen of the prenasal processes of the premaxillae (os carunculae). The egg tooth is also present and is attached to the premaxillae at their symphysis lying immediately below the os carunculae. The orbital cartilages have begun to extend back towards the auditory capsules to which they are united by mesenchyme—the anlage of the orbito-parietal commissure. The auditory capsules are not yet completely chondrified; on the lateral surface of each is the fenestra ovalis blocked by a rudimentary stapes. The crista parotica now exhibits a rostrally directed rodlike projection that has united with the stylohyal cartilage. In between the head (malleus) of Meckel's cartilage and the crista parotica a small blob of cartilage has appeared; this is the incus homologous with the quadrate of the therapsid reptilian jaw suspension. Dorsally the auditory capsule has fused with the parietal plate, which at its anterior margin, now exhibits a projection towards the orbital cartilage. The occipital arch is largely but not completely fused with the hind wall of the auditory capsule so that the fissura metotica is reduced in size. The occipital condyles have become more prominent.

4. 28-mm specimen. This "embryo" is a newly-hatched platypus; the 28 mm refers to overall measurement from tip of snout to tip of rump, but the maximum direct length was only 16.75 mm, a little less than the long diameter of the egg—18 mm. The chondrocranium is little more than a deep saucer with a lot of big holes in it; it is hard to believe that this supports the brain of a suckling mammal. The occipital arches are interconnected dorsally by a narrow band of cartilage, the tectum posterius, thus forming the foramen magnum. The occipito-capsular fissure is obliterated and the jugular foramen (Fig. 5) is the last relic of the foramen metotica.

The auditory capsule is made up of canalicular and cochlear portions anchored to the basal plate by means of the basivestibular commissure attached throughout the length of the cochlear portion, but the cochlear capsule does not reach as far forward as the front of the basal plate.

The incus retains cartilaginous connection with the stapes, the crista parotica, and the malleus (still part of Meckel's cartilage); the fenestra ovalis is filled in by the stapes.

The parietal plate is now connected with the orbital cartilage by means of the orbito-parietal commissure, and the orbital cartilage is attached to the central stem by a preoptic root and the pila antotica. Anteriorly the orbital cartilage is attached to the roof of the nasal capsule by the sphenethmoid commissure.

The processus alaris and its attached ala temporalis project laterally from the

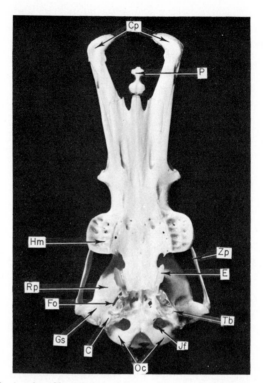

Figure 5. *Ornithorhynchus.* Ventral view of skull × 0.84. Cp, crura of premaxillae; P, prevomer; Zp, zygomatic process of maxilla; E, ectopterygoid; Tb, tympanic bone; Jf, jugular foramen; Oc, occipital condyles; C, crista parotica; Gs, glenoid surface of squamosal; Fo, foramen ovale; Rp, rostral projection of periotic; Hm, housing on maxilla for grinding pad.

central stem, beneath the base of the pila antotica, and form the floor of the cavum epiptericum.* The medial wall is but scantily formed by the pila antotica and at this stage the cavum has no lateral wall.

*The cavum epiptericum in the chondrocranium of reptiles is a space lying between the pilae antoticae and the epipterygoids, and floored by the basitrabecular plate. The cavum persists in this form in the adult since the pilae antoticae ossify to form large pleuro- (=latero) sphenoids forming a major part of the brain case or the medial wall of the cavum; the lateral wall is formed by the epipterygoid. In eutherian and metatherian mammals the pilae antoticae fail to appear and in monotremes they ossify as vestiges so small that they cannot be detected as entities. The medial wall of the cavum now consists of a membrane applied closely to the brain; its outer wall in the eutherians and metatherians is formed by the lamina ascendens (homologue of the epipterygoid) of the ala temporalis and by a membrana-sphenoobturatoria stretching from it which later becomes ossified. In the platypus, however, a lamina ascendens fails to develop and a rostral projection of the periotic ossifies in the membrana-sphenoobturatoria (p. 323). The floor of the cavum epiptericum in all mammals is formed by the ossified ala temporalis.

The nasal cartilages are well chondrified and the primary fenestra narina through which the external nostril passes is formed by the anterior edges of the nasal septum, the roof and side wall of the capsule, and the lamina transversalis which is continued forwards and outwards into the marginal cartilage of the muzzle. Internally the lamina transversalis anterior is projected backwards on each side in the form of a prong of cartilage—the paraseptal cartilage which runs about halfway along the ventrolateral edge of the nasal septum, and then ends freely. Rostrally each paraseptal cartilage forms a ring around Jacobson's organ.

The paranasal cartilage has united to the sphenethmoid commissure and also to the central stem by means of the lamina orbitonasalis which forms a roof over the posterior part of the cavity of the nasal capsule; at this stage there is no hind wall or floor to the capsule.

The foramen olfactorium evehens that leads out of the cranial cavity into the cavity of the supracribrous recess is bounded medially by the crista galli (of the nasal capsule anterodorsally), and ventrally by the anterior edge of the preoptic root of the orbital cartilage.

The stylohyal cartilage is now fused dorsally with the crista parotica; in addition to the thyrohyal, the other elements of the branchial skeleton, the anterior and posterior laryngeal cartilages, have now appeared. These cartilages retain their primitive appearance and persist throughout the life of the platypus (p. 179).

A number of membrane bones are present at this stage. The premaxillae are fused in the midline at a symphysis bearing the egg tooth; the prenasal processes are also fused forming a huge os carunculae. (Figs. 80 and 102, although of a hatchling echidna, serve as illustrations for a hatching platypus.)

The septomaxillae are small, paired bones lying isolated on the dorsal surface of the hinder part of the marginal cartilage. The vomer is an unpaired bone covering the ventral edge of the caudal part of the nasal septum.

The maxillae exhibit alveolar portions, ascending portions that cover the lower part of the paries nasi, palatine processes that extend towards the midline beneath the ventral body of the paries nasi, and zygomatic processes that extend back towards the zygomatic processes of the squamosals—thus the malar arches will be formed.

The palatine is represented on each side by a plate of bone extending back from the hinder part of the ventral edge of the paries nasi to the processus alaris.

The pterygoids are present as lateral wings of the basisphenoid lying above the posterior ends of the palatines and anterior to the processus alaris on each side.

The nasal lies over the roof of the nasal capsule on each side; the tiny frontals lie on the lateral surfaces of the sphenethmoid commissure, and the parietals lie along the dorsal edges of the parietal plates.

The tympanic by now is a sickle-shaped bone adherent to the ventral surface of the auditory capsule supporting the developing tympanic membrane and lying medial to the caudal end of Meckel's cartilage (the malleus); the dentary ensheaths the ventrolateral aspect of the cartilage.

5. At a later stage of development (122-mm suckling) of the chondrocranium three other membrane bones appear: the ectopterygoids, prevomer, and jugals. The latter appears as a small bone in front of and dorsal to the anterior end of the zygomatic process of the squamosal. The ectopterygoid is an ovoid nodule of bone lying posteroventrally to the ala temporalis. De Beer and Fell are of the opinion that it probably represents the nodule of the secondary pterygoid cartilage found in the chondrocraniums of higher mammals. The prevomers appear as two small platelike ossifications lying between the premaxillae and anterior to the unpaired vomer. The two ossifications fuse and form the prevomer or the dumbbell-shaped bone of the adult skull (Fig. 5).

Adult Skull

A peculiarity of the monotreme skulls is that the sutures fuse together at an early age so that they are hard or impossible to detect in mature adults. This has led to misinterpretation of the nature of the bones forming the side wall of the brain case (p. 313).

In general appearance the skull is broad-flattened and exhibits a large brain case for the accommodation of the well-developed forebrain (Fig. 5). Caudally the crura of the premaxillae are wedged in between splintlike rostral prolongations of the nasals and maxillae. The roof of the mouth is a false palate that forms the floor of the two nasopharyngeal passages. This false palate forms in the suckling platypus, behind the original internal nostrils on the true palate by ingrowth of flat extensions from ventral ridges along the palatines and maxillae. The extensions meet ventrally and join a vertical plate dependent from the ventral surface of the vomer. Thus the nasopharyngeal air passages are formed (de Beer and Fell, 1936); they communicate with the pharynx at the posterior end of the false palate. The maxilla has a very strong backwardly directed zygomatic process that overlaps and fuses with the rostral projection of the squamosal. The jugal is a small knob of bone sitting on top of the anterior end of the zygomatic process of the squamosal. There are no teeth in adult jaws but juveniles exhibit true molariform teeth on the posterior parts of the maxilla and dentary. These teeth disappear a month or so after the juvenile leaves the breeding burrow and are replaced by heavily kèratinized pads that serve to grind the food. The long narrow dentary exhibits an indication of a coronoid process and the condyles, elevated relative to the long axis of the bone, articulate with a shallow glenoid surface on the squamosal. Some dentaries exhibit a small inflected angle; the presence or absence of this has nothing to do with age or sex of the platypus.

The roof of the skull looking from front to rear is formed by the paired nasals, very small frontals, very large parietals, and the unpaired supraoccipital. The posterior part of the brain case is formed by the paired exoccipitals which, of course, bear the condyles. The frontals extend ventrally into the orbits where they meet the palatines extending up into the orbit. The lateral wall of the brain case is formed by the periotic, which exhibits a rostral projection replacing the membrana-sphenoobturatoria, by the orbitosphenoid, and by the squamosal. As already mentioned, the ossified ala temporalis (alisphenoid) fails to develop a lamina ascendens (it does, however, do so in *Tachyglossus,* p. 318), so the alisphenoid plays no part in filling the side wall of the brain case. It does play a part, though, in forming the foramen ovale for egress of the mandibular ramus (V3) of the trigeminal nerve. This foramen is bounded anteriorly by the alisphenoid, posteriorly by the true periotic, and laterally by the rostral lamina of the periotic (Figs. 5 and 120).

Between the squamosal and the periotic is a canal or passage open at each end; according to Goodrich (1958) this is supposed to represent a vestige of the post-temporal opening of the well-developed temporal fossa of reptiles. The glenoid cavities for articulation with the condyles of the dentary are located on the ventral aspects of the squamosals.

The floor of the posterior part of the skull is formed by the basioccipital, the basisphenoid complex, by the very small alisphenoids, which lie lateral to the basisphenoid, by the pterygoids, and by the periotic or petrosal bone. Anterior to the pterygoids lie the palatines and the median vomer extending back to the basisphenoid. The anterior end of the brain case is formed by the transverse mesethmoid, a bone made up of ossifications of the trabecular plate, internasal septum, and the nasal capsule. The mesethmoid is perforated by the two olfactory foramina; it is not cribriform as in *Tachyglossus*. A small vertical plate of the mesethmoid unites with the anterior dorsal margins of the orbitosphenoids.

The basisphenoid complex is of particular interest since it exhibits two longitudinal ridges along the sides of the sella turcica, which houses the hypophysis; these are the ossifications of our old friends the pilae antoticae of the chondrocranium, and presumably they are the homologues of the pleuro (= latero) sphenoids of the Reptilia.

The auditory capsule of the chondrocranium has been invested with the periotic or petrosal bone. Posteriorly this exhibits a pars mastoidea and a ventromedial portion, the pars petrosum, which houses the vestibular and acoustic organs. External signs of these can be detected on the ventral surface of the pars petrosum; at its posterolateral corner is a depression—the tympanic cavity, in the roof of which lies the fenestra ovalis leading into the scala media of the cochlea; the middle and inner ear structures are described in Chapter 7.

The two nodular ectopterygoids of the chondrocranium have now assumed the form of two thin horizontally disposed wings of bone projecting laterally from the lateral margins of the alisphenoids. The function of the ectopterygoids is unknown but it would seem likely that they offer a solid resistance to upward movements of the dorsolateral margins of the posterior part of the tongue; this could conceivably facilitate transfer of food to and from the cheek pouches.

Postcranial Skeleton

Monotremes, like all mammals, have seven cervical vertebrae, but unlike those of eutherians and metatherians, they bear cervical ribs as do those of reptiles (Gregory, 1947; Lessertisseur and Sigogneau, 1965). The rest of the vertebral formula in *Ornithorhynchus* includes 17 dorsals, 2 lumbars, 2 sacral, and 21 caudal (Cabrera, 1919).

If one had occasion to read about a pectoral girdle consisting of 2 scapulas (with acromian processes on the anterior borders), 2 coracoids, 2 clavicles, 2 epi- (= pro) coracoids, and a median T-shaped interclavicle, one might be forgiven for thinking it belonged to a therapsid reptile. However, it so happens that the above is a list of the bones of the monotreme pectoral girdle (Fig. 6). The scapula, which is expanded dorsally and has a shallow infraspinous fossa, forms a unit on one side with the ventral coracoid and procoracoid. This unit is tied dorsally to the rib cage and ventrally by the united coracoid and procoracoid to the interclavicle, a T-shaped median bone united posteriorly to the sternum; the cross of the T is clamped anteriorly to the scapulae by paired clavicles which are rigidly united to the intercavicle. The glenoid cavity is a laterally oriented wedge-shaped depression at the union of scapula and coracoid.

Parker (1868; *fide* Cave, 1970) found that the interclavicle consists of two parts: a ventral T-shaped intramembranous ossification (the main mass of bone) and a dorsally located, ovoid, endochondral ossification, the pro-osteon. Cave (1970) demonstrated that the pro-osteon is epiphyseal in nature and that it is invariably demonstrable in the interclavicles of young monotremes.

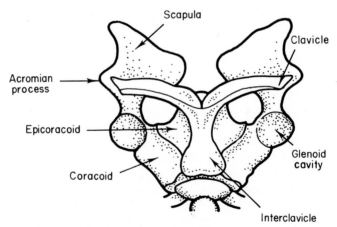

Figure 6.   Pectoral girdle of a monotreme (*Tachyglossus*). (*After Gregory, 1947.*)

The pelvic girdle consists of the usual three elements: dorsal ilium, ventro-posterior ischium, and ventroanterior pubis. The latter two are united ventrally at a symphysis but dorsal to this they are separated by an obturator foramen. The symphysis of the three bones at the acetabulum is incomplete and the latter exhibits a foramen.

The anterior margin of each pubis bears an epipubic hone which projects forward and is movably articulated as it is in the marsupial pelvis. The anterior extremities of the ilia are elongated, narrow, and extend forward and dorsally to unite with the two sacral vertebrae.

As in *Tachyglossus* the hindlimbs are widely everted and the knees are higher than the acetabula when the animal is standing, but the feet point forward, not outward, as in *Tachyglossus*.

TACHYGLOSSUS

Discovery

In London, 1792, George Shaw was presented with a strange quadruped about 18 inches long, with a long naked snout like the beak of a bird, and long sharp spines on its back. This animal had been caught in New Holland "on a large red anthill" and since it exhibited "dentes nulli" and "lingua teres extensilis" he thought it could be related to the South American ant-bear, *Myrmecophaga*, but he was uneasy about assigning it to that genus in view of the presence of spines in the pélage; he suggested it might be a new genus related also to the porcupine: "It is a most striking instance of that beautiful gradation so frequently observed in the animal kingdom by which creatures of one tribe or genus approach to those of a very different one. It forms a connecting link between the very distant genera of *Hystrix* and *Myrmecophaga*; having the external coating and general aspect of the one with the mouth and peculiar generic characters of the other." With reservations he named it *Myrmecophaga aculeata* and published a description and illustration of it (Shaw, 1792), not knowing in the least the true nature of the animal. Ten years later Home (1802b) dissected a specimen and, recognizing its close relationship to his *Ornithorhynchus paradoxus*, published a description of it under the name *Ornithorhynchus hystrix*; he politely acknowledged that "A description and figure of this animal is given by Dr. Shaw in his Zoology under the name of *Myrmecophaga aculeata*." Iredale and Troughton (1934) give, inter alia, an account of the transition from *Ornithorhynchus hystrix* to *Echidna hystrix* to *Echidna aculeata* to the present name *Tachyglossus aculeatus*.

No one knows for certain how Shaw came by his echidna; it has been suggested (Augee, 1975) that it was taken at Adventure Bay in Tasmania in 1792 implying that the type could be *aculeatus setosus* (see p. 59) even though

Shaw's illustration shows emphatically that it is not a Tasmanian echidna. The story behind the adventure Bay echidna is as follows: Captain William Bligh, one of the survivors of the mutiny on H.M.S. *Bounty*, set out for Tahiti again in 1791 in H.M.S. *Providence*. On the way he called in at Adventure Bay* arriving February 9, 1792. During a 2-week stay one of Bligh's officers, George Tobin, recorded in his Journal (manuscript A562-3 held at Mitchell Library, Sydney) that ''The only animals seen were the kangaroo and a kind of sloth about the size of a roasting pig with a proboscis two or three inches in length. . . . On the back were short quills like those of the Porcupine. . . . This animal was roasted and found of a delicate flavour.'' Before the banquet Bligh made a drawing† of the animal accompanied by a description. He wrote yet another description in the log-book of the *Providence* prefacing his remarks with the statement that Lieutenant Guthrie ''killed an animal of very odd form'' (*fide* Lee, 1920). In both descriptions quills were mentioned; there was no mention, however, that a specimen was preserved. Even if one had been it could not possibly have been in Shaw's hands in time to be described and the description to be published in 1792 in London since the *Providence* sailed from Adventure Bay for the purpose of collecting breadfruit trees at Otahyty (Tahiti). There she arrived on April 10, 1792 and left there 3 months later sailing for Kingstown, St. Vincent, taking with her 2634 breadfruit trees; she arrived in Kingstown on January 23, 1793.

Some recent evidence suggests that the echidna described by Shaw came from Port Jackson and passed through the hands of three individuals before landing on Shaw's bench: Chief Surgeon John White, Governor Arthur Phillip, and Sir Joseph Banks in that order:

1. E. C. Slater discovered a painting in the Shaw Drawings‡ in the British Museum of Natural History, of an echidna, bearing various notes in Shaw's handwriting. One of these included an inscription in parentheses: '' 'This animal was found on a large red ant-hill; it seems to live on them therefore we called it the ant-eating Porcupine.' White.'' The latter is, presumably, John White who collected natural history specimens for his friends in England and for the Royal College of Surgeons. White could paint (Rienits and Rienits, 1963) and of his paintings Whitley (1975) remarks ''these evidently including the first painting of the Spiny Anteater.'' The painting found by Mr. Slater was reproduced in the Naturalist's Miscellany, Vol. 3 (1792) along with the statement in the text: ''In its mode of life this animal beyond doubt resembles the Myrmecophagidae having been

---

*This was Bligh's third visit to Adventure Bay; he was there in 1778 with James Cook in the *Resolution* and 10 years later as Captain of the *Bounty*. An apple tree he planted at that time was still there in 1792.
†This drawing passed into the hands of Sir Joseph Banks and was later reproduced by Home (1802b).
‡I am indebted to E. C. Slater of the Division of Wildlife Research, CSIRO, for permission to publish his discovery of this painting.

found in the midst of an ant-hill for which reason it was named by its first discoverers the ant-eating porcupine. It is a native of New Holland.''

2. Whitley (1975) unearthed in the Mitchell Library a letter written from Phillip to Banks in December 1791 which included a manifest referring to shipment on H. M. S. *Gorgon* of ''an eight gallon cag which contains a young Kangurroo, a porcupine, a rat kangurroo in which the false belly is placed different to what it is in the kangurroo and various other animals, and a skin of the Ma-ra-ong.''

3. The *Gorgon* sailed from Port Jackson December 18, 1791 (Cumpston, 1964) for England via Cape Town, arriving England July 7, 1792 (Fitzhardinge, 1961).

The above snippets of information suggest the following: White or his men found an echidna on an ant mound [probably that of *Iridomyrmex detectus*, an ant eaten by echidnas at certain times of the year (Griffiths and Simpson, 1966) and which is still common around Port Jackson]. White made a painting of the animal, gave a description of where it was found, and sent both to Shaw by the *Gorgon*. Phillip got the echidna from White, or obtained another, and sent it to Banks who gave it to Shaw in the latter half of 1792. This would be in time for him to publish a description of the specimen in that year. White's painting and information would also have arrived in time to be incorporated into the description.

External Features

Like that of the platypus the body is compressed dorsoventrally but lacks any suggestion of streamlining (Fig. 7); the dorsum is domed and the ventral surface is flat, even slightly concave (Fig. 8). The head appears to emerge from the body without any indication of being attached to a neck; at the posterior end of the body is a short stubby tail naked on its undersurface (Fig. 8). The snout is formed by prolongations of the premaxillae and mandibles and in an adult echidna about 45 cm long weighing 4–5 kg, the snout would measure about 7.5 cm in length. The external nares and mouth are at the distal end of the snout; there is no gape and as Captain Bligh remarked, the mouth is so small that ''it will not admit anything above the size of a pistol ball.'' The snout is strong enough to be used as a tool to break open hollow logs and to plough up forest litter to get at the prey—ants and termites (see CSIRO film, ''The Echidna,'' 1969). The small eyes (ca. 9 mm in diameter) are situated well-forward on the head, almost at the base of the snout, and look ahead more then they do sideways (Fig. 7). In spite of almost total decussation of optic fibers in the chiasma there is some evidence that echidnas could have a degree of binocular vision.

There is no scrotum, the testes are internal, and there is only one hole for the passage of feces, urine, and reproductive products (Fig. 8) as is the case in

Figure 7.   *Tachyglossus aculeatus multiaculeatus.* The Kangaroo Island echidna. *(From Griffiths, 1972.)*

Figure 8.   *Tachyglossus aculeatus acanthion*, a male specimen. Note absence of scrotum and the single cloacal aperture.

platypuses, marsupials and, curiously enough, in the beaver. The beaver and many marsupials are truly cloacate and all marsupials are monotreme in the sense that excretory and reproduction products pass through the one sphincter.

Both dorsal and ventral surfaces in specimens of the nominate race *Tachyglossus aculeatus aculeatus* are furnished with hair, but on the dorsal surface the pélage consists of a mixture of hairs and spines of which Hausmann (1920) says there are all gradational states from the finest of hair to the largest and most robust of spines (Fig. 23). As in all the mammals, the hairs occur in bundles of Nebenhaare and Haupthaare, the latter, however, are flattened along the whole length of the hair, not just distally as in *Ornithorhynchus* (Spencer and Sweet, 1899). The spines are circular in cross section and all authorities agree on morphological and developmental grounds that they are modified hairs (Maurer, 1892; de Meijere, 1894; Römer 1898; Spencer and Sweet, 1899): both hairs and spines form in long downwardly directed thickenings of the epidermis, each of which carries down the stratum germinativum before it. At the bottom of the extension a dermal papilla forms which projects upwards into the extension. The cells of the distal portion of the papilla form a germinal point from which hair or spine grows upwards through the tissues of the extension. The tip of the papilla keratinizes and finally appears above the general surface of the body. In cross section each hair or spine exhibits a keratinized cuticle, a cortex, and a central medulla as they do in all mammals. The keratin has the alpha configuration typical of the keratins of mammals (Gillespie and Inglis, 1965; see also p. 47). The Nebenhaare are formed as buds from the downgrowths of the principal hairs and they are identical at this stage to the anlagen of sebaceous glands which form at a later stage (Römer, 1898). Sebaceous glands are associated only with hairs and sweat glands are so rare that some researchers have failed to detect them. They are present, however, in the form of the Knäueldrüsen of Gegenbaur situated around the periphery of the areolae.

The spines on the sides and dorsal surface of the body point backward but those at the midline converge and cross one another alternately forming a very handsome pattern. At the rump the spines are arranged into two semicircular rosettes, one on each side. Spencer and Sweet (1899) claim to have observed that the muscles in the dermis, attached to the spines and hairs, were striated. Augee (1969), however, failed to confirm this and detected only plain muscle fibers attached to the hairs and spines.

The middle of the belly in both sexes has less hair and a thinner muscle layer than elsewhere on the ventral surface. In this region the pouch or incubatorium forms in the adult female during the breeding season. In this pouch she incubates the egg and carries the young (Fig. 9). The males do not develop a real pouch although sometimes when they are picked up they can contract the muscles of the belly and give a pouchlike impression. In both sexes two small dark hairy patches can be discerned at the lateral margins of the pouch area, known as the areolae or

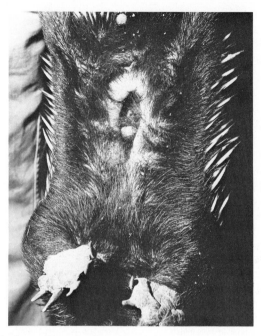

Figure 9.   *Tachyglossus.* Ventral surface of female incubating an egg in the pouch. *(From Griffiths et al., 1969; with permission of The Zoological Society of London.)*

milk patches. Some specimens of *Tachyglossus* exhibit an external pinna dorsolaterally on the head but in others it is hard to find (see Fig. 14).

The limbs are short and stout; the forefoot is pentadactyl, the digits being furnished with spatulate claws (Fig. 7). These are used to dig in forest litter, to burrow, tear open logs, and termite mounds, etc., in order to get at its insect prey. The animal is incredibly strong and when in danger can burrow down vertically and rapidly into very hard earth to which it clings like a limpet. When buried in this way it is hard, or impossible, for a savage dog to dislodge the echidna. This behavior certainly did not evolve as a defense against dogs (dingoes) since they arrived only ca. 8000 years ago with a late invasion of man, but it may have been adopted as a defense against the large carnivorous marsupials, *Thylacoleo carnifex, Thylacinus* species, and *Sarcophilus* species. This habit of rapid vertical burrowing makes it difficult to detect echidnas in the bush since they can disappear when they hear a noisy zoologist approaching. The difficulty is increased by the echidna's trick of kicking up dirt, bark, and leaves, with its hindlimbs over the back as it burrows down.

The hindlimb is pentadactyl; the claw on digit 1 is short, those on 2 and 3 are long (Fig. 26) and are used as grooming claws, and those on 4 and 5 are short. The femur is horizontally disposed, sticking out at right angles to the body. The tibia and fibula are twisted backwards, so much so that the feet and claws point backwards.

Jenkins (1970a, 1971) has developed a cineradiographical technique for study of posture and limb motion in *Tachyglossus* and has compared them with those in a selection of noncursorial mammals (opossum, tree shrew, hamster, rat, and ferret) and in two cursorial, cat and hyrax. As far as humeral and forearm excursion arcs are concerned in the noncursorial mammals the humerus is normally abducted from the parasagittal plane—the amount of abduction varying from 10° (in the shrew) to approximately 90° in the echidna. With the exception of the echidna humerus all the others usually function at angles of 10°–30° from the parasagittal plane; the cineradiographic record of the movements of the echidna humerus exhibited no evidence of a pattern of elevation and depression that could be related to the discrete phases of movement observable in the other noncursorial mammals. Jenkins says "apparently the humerus functions in variable positions in which the distal end may be elevated slightly above or depressed slightly below the shoulder joint." The echidna humerus functions in a transverse position largely by long axis rotation (Jenkins, 1970a). In the cursorial mammals studied, cat and hyrax, the humerus is adducted to within 10° of the parasagittal plane and the elbow joint usually functions below the level of the shoulder joint, quite unlike the conditions in the noncursorial mammals. From these and similar studies on hindlimb motion in the noncursorial and cursorial mammals, Jenkins (1971) concludes that "the observed variation in limb posture and excursion invalidates any concept of a single mode of posture or locomotion among terrestrial mammals."

As in *Ornithorhynchus*, the inside of the ankle in all males bears a hollow perforated spur only 0.5–1.0 cm long; juvenile females also can exhibit a small sharp spur which is lost later in life; thus if an echidna lacks a spur on the ankle it is certainly a female. The spur is connected to a duct that leads up the leg to a gland buried in the muscles just below the knee (Cabrera, 1919).

Under the skin the body of *Tachyglossus* is covered with an enormous muscle that is very thick over the dorsum but quite thin in the central belly. This is the panniculus carnosus; by the contraction of this muscle the echidna can achieve fantastic changes in shape, including rolling itself into a ball.

Shaw's "lingua teres extensilis" will be described in detail in Chapter 3.

Reproductive and Excretory Organs

The female reproductive organs are as in *Ornithorhynchus* with the difference that the ovaries are equally well developed, both are capable of producing ripe ova, and either of the uteri can carry eggs that eventually produce hatchlings*

---

*Twins have been found in the pouch but very rarely; whether or not they develop in one or in both uteri is not known.

(Semon, 1894a). The penis (Fig. 10) in a large adult is about 7 cm long, somewhat compressed, and about 1.25 cm wide. The glans is grooved and gives the impression of being bifid but not to the same degree as the penis of the platypus. The right and left portions of the glans each exhibit a pair of bulbous prominences, each bearing a circle of epidermal processes so that the whole distal portion of the penis gives the impression of a rod bedecked with four flowerlike rosettes.

The internal structure of the kidneys is known for *Tachyglossus*: the "kidney-shaped" cortex contains the Malpighian corpuscles consisting of a glomerular tuft of capillaries and Bowman's capsule which communicates with the proximal convolutions, the descending limb of Henle's loop, the thin segment of the descending limb, the ascending limb, and the distal convoluted tubule which finally leads to the straight collecting tubule. The proximal and distal convolutions are located within the cortex but the loops of Henle pass deep into the

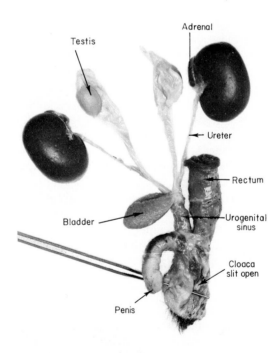

Figure 10.   *Tachyglossus.* Male reproductive organs. × 0.6. Probe indicates passage from cloaca to preputial sac. *(From Griffiths, 1968.)*

medulla and, along with the collecting ducts, form the medullary rays (Zarnik, 1910). The collecting ducts unite to form large ducts in the papilla which pass through the area cribrosa at the base of the papilla. There is only one papilla and it projects into a simple cavity, the pelvis. From here the urine is led to the urogenital sinus by the ureter. This description of Zarnick's would be just as applicable to the kidneys of many eutherians, but, as usual, the monotreme exhibits something unusual: the typical mammalian nephrons described above are occasionally accompanied by curious dwarf nephrons (Zwerg-kanälchen) in the cortex. These consist of a small Bowman's capsule and a short length of relatively uncoiled proximal tubule communicating with a collecting tubule. Zwerg-kanälchen are also found in the kidneys of many reptiles.

   Tachyglossus and *Ornithorhynchus* are ureotelic (Denton *et al.*, 1963); it would be remarkable if *Zaglossus* proved to be otherwise. Renal function of *Tachyglossus* is discussed in Chapter 4.

Mammary Glands

   The platypus never develops a pouch for incubation of her eggs but, as we have seen (Fig. 9), when the female *Tachyglossus* becomes pregnant she develops one on the ventral surface in which she incubates her egg and carries the young. In August, 1884 Haacke (1885) discovered the pouch, and that echidnas are oviparous, but we have yet to learn how the egg gets into the pouch. After hatching, the young one lives in the pouch and is nurtured on the milk secreted by two mammary glands, having ducts that lead to the exterior at two areolae each situated dorsolaterally in the pouch. These mammary glands appear as large crescent-shaped swellings under the skin of the pouch lips giving them a thick tumescent appearance (Figs. 9, 78). As the pouch young grows the glands enlarge and extend out laterally between skin and muscle of the abdomen. Although not as long and wide as the glands of *Ornithorhynchus* those of *Tachyglossus* are far thicker giving the gland a rounded appearance. From Fig. 11 it is apparent that the gland is made up of a lot of club-shaped lobules as in *Ornithorhynchus* but there is less connective tissue and the lobular structure is more obvious. The pouch is very sparsely haired and the areola exhibits only special hairs known as mammary hairs. The ducts lead to the exterior at the bases of the follicles of these hairs, as do sebaceous glands (Fig. 12). The expanded distal ends of the ducts or sinuses are even larger than those in the areola of *Ornithorhynchus* and the periphery of the areola is furnished with Knäueldrüsen.

   Tachyglossus males have mammary glands and areolae and it has been stated (Westling, 1889; Jacobson, 1961; Sharman, 1962) that they are as well developed as those of females. The facts are that at all seasons of the year the lobules are minute and hard to detect (Fig. 92A) but often a large fat pad is associated with them; to the naked eye this may be mistaken for mammary tissue.

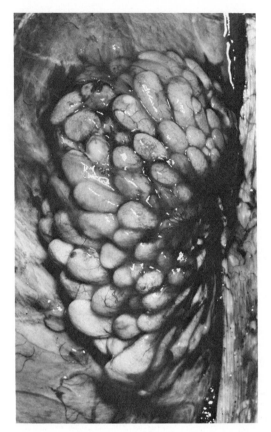

Figure 11.   Mammary gland of lactating *Tachyglossus*. × 1.7. *(From Griffiths, 1968.)*

Actually the mammary glands in male echidnas are small relative to those of females, as the mammary glands of men are relative to those of women.

Brain

The tachyglossid brain is even more like the brain of a mammal than that of *Ornithorhynchus* due to the presence of a very large lobulated cerebellum and the enormous gyrencephalic neocortex (Fig. 13).

The brain stem has been described in detail by Abbie (1934). The spinal cord swells insensibly into the medulla oblongata which bears outward evidence of the combined mass of the gracile and cuneate nuclei on its dorsal surface. Ascending fiber tracks from the dorsal funiculi terminate here and secondary fibers decussate and pass rostrally in the medial lemniscus. Other fibers from the gracile-

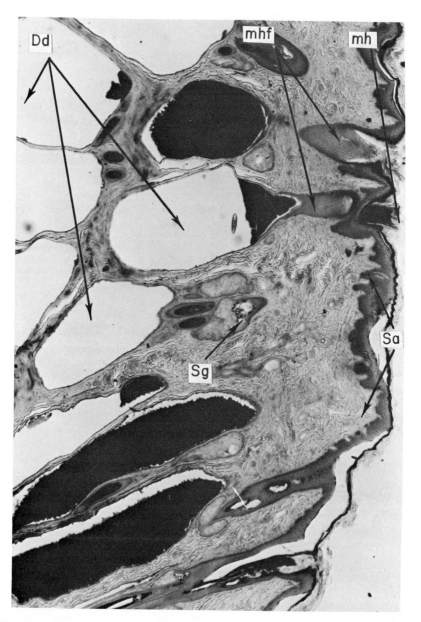

Figure 12. Longitudinal section of areola of echidna mammary gland. × 50. Dd, distended ducts; Mhf, mammary hair follicles; Mh, mammary hair; Sa, skin of areola; Sg, sebaceous gland. *(From Griffiths, 1968.)*

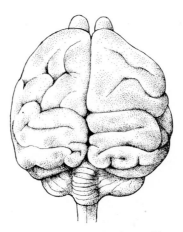

Figure 13.   Dorsal view of brain of *Zaglossus*. *(From Kolmer, 1925.)*

cuneate nuclei form synapses in the inferior olive or pass as external, dorsal external, and internal arcuate fibers to the corpus restiforme.

Cranial nerves V-XII emerge from the medulla and, as in *Ornithorhynchus*, the huge ribbonlike trigeminal nerve enters the brain anterior to the pons Varolii. It is a mixed sensory and motor nerve with four nuclei—chief sensory, dorsal root, mesencephalic in the superior colliculus, and the motor. As in *Ornithorhynchus* the chief sensory is enormous; fibers from it pass rostrally and caudally (descending root). The ascending fibers for the most part pass via a massive decussation to the opposite medial lemniscus which at this level also contains fibers from the gracile-cuneate nuclei; medial lemniscus fibers terminate in the nucleus ventrolateralis of the thalamus. The chief sensory nucleus also has connections with the pons and cerebellum. Impulses via the descending root pass to the pons, corpus trapezoideum, and the external arcuate system and from the latter to the cerebellum via the corpus restiforme. The descending root also conveys trigeminal impulses to the nucleus arcuatus trigemini, the superior olive, the inferior olive, and the lateral reticular nuclei.

The seventh nerve is of particular interest in the monotremes. The nucleus is very large and its size is correlated with the well-developed superficial facialis musculature (Hüber, 1930). In birds and reptiles there is a sphincter colli associated with the m. depressor mandibulae but this is restricted to the neck region; in Hüber's words the face is a "rigid mask devoid of expression." In the monotremes that musculature is not restricted to the neck region and is found in the face differentiated into groups around the ear, eye, and the snout (Fig. 14). All this musculature is under control of the seventh (the facial) nerve.

The lateral lemniscus with its complement of auditory fibers from the dorsal

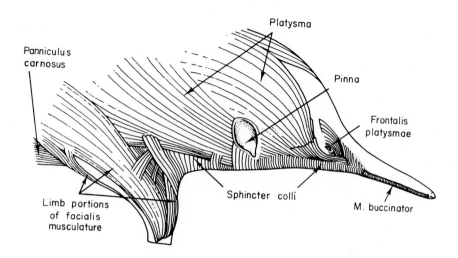

Figure 14. *Tachyglossus*. Dissection of fascialis musculature. *(From Hüber, 1930.)*

and ventral cochlear nuclei comes into contact with the medial lemniscus on the way to the midbrain as the two lemnisci do in *Ornithorhynchus*.

The cerebellum exhibits the lobus ventralis peculiar to monotremes, and is otherwise like that of *Ornithorhynchus* (Dillon, 1962); it is attached to the medulla by the three peduncles on each side—corpus restiforme, brachium pontis, and brachium conjunctivum.

The midbrain tectum exhibits, unequivocally, paired superior and inferior colliculi; the floor is made up of the paired cerebral peduncles and a large ganglion interpedunculare.

The thalamus is large due chiefly to the hypertrophy of the nucleus ventralis lateralis. Although not as extensive as that of *Ornithorhynchus* the trigeminal input via the medial lemniscus is considerable (Abbie, 1934) and accounts for much of the hypertrophy of the nucleus. A considerable amount of this input probably comes from the snout (see p. 195). The nucleus ventralis lateralis projects caudally beyond the lateral contours of the diencephalon as a "pulvinar" and, as in *Ornithorhynchus*, it pokes into the corpus striatum.

Medial to the caudal border of the pulvinar on its ventral surface a swelling is encountered—the lateral geniculate body that contains two nuclei receiving retinofugal optic fibers. Efferent fibers from these nuclei must pass to the neocortex since Bohringer and Rowe (1977) have demonstrated that there is visual representation at the occipital poles of the hemispheres.

Medial to the lateral geniculate body is the medial geniculate, the nucleus of which is putatively responsible for transmission of auditory impulses to the neocortex.

The mammalian cerebral cortex is made up of three entities: the pyriform or olfactory cortex located lateroventrally; the hippocampal located dorsally and medially; and the neocortex located in between the two. The latter is large and overshadows the pyriform and hippocampal cortices. Consequently, the pyriform cortex is displaced to a ventral position and where pyriform and neocortex meet, a longitudinal fissure, the rhinal sulcus, is formed. Similarly the hippocampal cortex is displaced, but medially and ventrally; where the two cortices meet another fissure is formed, the sulcus hippocampi.

The neocortex is well developed in all mammals but the gyrencephalic cerebrums of the Tachyglossidae have been and are a source of wonder to neurologists; thus Elliott Smith (1902) wrote, "The most obtrusive feature of this brain is the relatively enormous development of the cerebral hemispheres which are much larger, both actually and relatively, than those of the platypus. In addition the extent of the cortex is very considerably increased by numerous deep sulci. The meaning of this large neopallium is quite incomprehensible. The factors which the study of other mammalian brains has shown to be the determinants of the extent of the cortex fail completely to explain how it is that a small animal of the lowliest status in the mammalian series comes to possess this large cortical apparatus." Determinants of modern neurophysiology also fail to explain how echidnas come by this cortex.

Smith was at a loss to name the various gyri and sulci since they do not tally with those found in other brains, indeed quite often those on one side of the cortex do not match those on the other. This asymmetry has also been observed in the neocortex of the marsupial, *Vombatus ursinus* (John I. Johnson, personal communication). However, Smith distinguished seven sulci (on one side anyway) in the cortex of *Tachyglossus* and labeled them with letters of the Greek alphabet. Two of these sulci (Fig. 60) have proved to be sufficiently constant in position to be of some use as landmarks for cortical localization.

There is no corpus callosum but the two hemispheres are linked by an anterior and a hippocampal commissure as in the platypus and the marsupials (see Johnson, 1977, for review).

Skeleton

Chondrocranium

The development of the chondrocranium follows much the same course as it does in *Ornithorhynchus* (Gaupp, 1908), so much so that the illustration of the definitive chondrocranium of *Tachyglossus* (Fig. 15) would serve quite well for the 122-mm stage chondrocranium of *Ornithorhynchus* (p. 16). This is es-

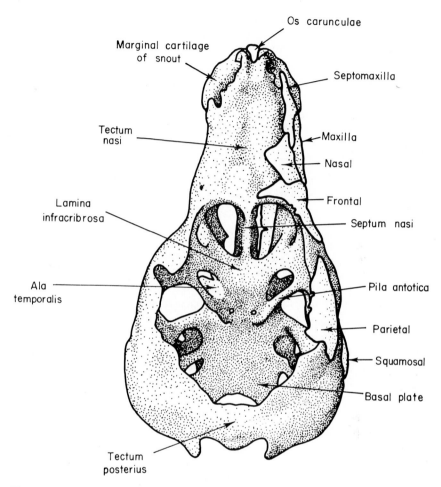

Os carunculae

Marginal cartilage
of snout

Septomaxilla

Tectum
nasi

Maxilla

Nasal

Frontal

Lamina
infracribrosa

Septum nasi

Ala
temporalis

Pila antotica

Parietal

Squamosal

Basal plate

Tectum
posterius

Figure 15. *Tachyglossus.* Gaupp-Ziegler reconstruction of the chondrocranium of a pouch young stage 48. *(From Gaupp, 1908.)*

pecially so if one imagined the marginal cartilages bearing on each side a crescent-shaped cartilage projecting laterally and caudally parallel to the snout; these would be the cartilages of the muzzle of the *Ornithorhynchus* chondrocranium. However, a few differences are detectable even in the earliest stages of development. The processus alaris of the *Tachyglossus* chondrocranium does not exhibit the little blob of cartilage at its distal end, nevertheless a lamina ascendens attached to the ala temporalis appears late in the development of the *Tachyglossus* skull; this plays an important part in filling in the lateral wall of the cavum epiptericum. The tympanic bone, lying medially to the posterior end of

Meckel's cartilage, as it does in *Ornithorhynchus*, has a triradiate shape, rather than a sickle shape as in *Ornithorhynchus*. A vestige of this triradiate shape is retained even in the tympanic of the adult, a circumstance that probably facilitates the strong union of tympanic to petrosal with consequent facilitation of transmission of sound waves through bone.

The development of the ectopterygoids is of interest. It has been known for some time that during development the pterygoid of the eutherian and metatherian skull exhibits dorsal and ventral moieties, the dorsal being a dermal element and the ventral an element of what appears to be a type of cartilage that eventually gives rise to dermal bone. These two elements fuse together in the metatherian and eutherian skull to form the pterygoid. In the monotremes, however, the two elements remain separate (Gaupp, 1908) giving rise to true dorsally located pterygoids and to the ventrally located ectopterygoids.

Reptilian and Mammalian Characters of the Monotreme Chondrocranium

Gaupp (1908) pointed out that the early chondrocranium of *Tachyglossus* is similar to those of reptiles since the trabeculae are separate, and, as we have seen, this is so in *Ornithorhynchus*. In other mammals the dual origin of the trabecular plate is obscured by simultaneous development of the central stem. The basal plate establishes contact with the trabeculae as it does in reptiles. In other mammals the sides of the basal plate are compressed by the cochlear capsules; it is to these that the trabeculae (or alicochlear commissures) are united.

The occurrence of the pila antotica in the monotreme chondrocranium is of outstanding interest; it appears in those of reptiles and birds and it ossifies to form the pleurosphenoids. In *Ornithorhynchus* and *Tachyglossus* they ossify and persist as vestiges along the sides of the sella turcica.

The posterior opening of the temporal fossa (not always present in the Tachyglossidae), is reminiscent of the same structure in reptilian skulls; it is never present in the skulls of Metatheria and Eutheria.

Another matter not yet discussed is that the eye cup in the monotreme chondrocranium is invested by a layer of cartilage (de Beer and Fell, 1936)—this is found in the eyes of reptiles and birds but is never encountered in those of the Metatheria and Eutheria; the scleral cartilage persists in the adult monotreme eye.

The reptilian character of the chondrocranium par excellence is, of course, the egg tooth. Its development is described in detail in Chapter 8.

In other respects the chondrocranium is much like that of other mammals: its general appearance conforms to that of the mammalian type; the condyles are located laterally on the exoccipitals, not close together at the ventrolateral corners of the foramen magnum as in cynodont therapsids; the nasal capsules in the definitive chondrocranium exhibit maxillo- and ethmoturbinals; the stylohyals are attached to true cristae paroticae; above all the cavum epiptericum is modified leading to a new side wall of the braincase, and, in the absence of major ossification of the pilae antoticae, to demotion of its medial wall to a mere membrane (dura mater).

Of course all mammalian chondrocraniums exhibit a lower jaw bearing the characteristics of a lower jaw in the reptile chondrocranium: the posterior part of Meckel's cartilage (future malleus) articulates with the quadrate cartilage (future incus). However, in the mammals the articular part of Meckel's cartilage is nipped off and thus becomes the anlage of the malleus. In this regard the monotreme chondrocranium is thoroughly mammalian; a point of interest, though, is that the newly-hatched monotreme and the newborn marsupial exhibit the reptilian articulation of the lower jaw which persists until late in pouch life (nest life for platypuses). Thus the very young monotreme and marsupial have no sense of hearing until later in pouch life when finally the malleus breaks away from Meckel's cartilage allowing the dentary-squamosal suspension of the lower jaw to be established, thus leaving malleus and incus free to become ear ossicles. The eutherian lower jaw, of course, goes through the articular-quadrate phase *in utero*. This fantastic tale of metamorphosis of jaw bones into ear ossicles is beautifully told by the therapsid fossil record, the ontogeny of mammals, and Hopson (1966) (see Appendix).

Tachyglossid Adult Skull

The skulls of *Tachyglossus* and *Zaglossus* are similar so a single description and illustration (Fig. 16) will suffice for the Tachyglossidae. Looking from the rear to the front the floor of the skull consists of a basioccipital and a fused basisphenoid-presphenoid complex (van Bemmelen, 1901) which includes the paired ridgelike ossifications of the pilae antotica along the sides of the sella turcica. Rostral to this the floor is formed from the unpaired median vomer which in *Zaglossus* must be relatively one of the longest vomers in the Mammalia. The anterior wall of the braincase is formed by the transverse mesethmoid, the floor of the nasal capsules, and a lamina infracribrosa which bears a cribriform plate extending horizontally forwards. This plate is pierced by many small pores for the passage of the olfactory fibers, from the olfactory epithelium, to the olfactory bulbs of the telencephalon. The olfactory epithelium is extensive (p. 182). The elongated snout is made up of extensions of the maxillae and premaxillae. The latter are incurved and meet at their distal ends but just posterior to their union is a space or lacuna through which the nasal passages open to the exterior. There are no teeth and anlagen of teeth do not appear during ontogeny. The roof of the mouth is formed in the same way as it is in *Ornithorhynchus*, by the union of mesially directed flat extensions from the maxillae and palatine bones to a downward median ridge of the vomer. Thus paired nasopharyngeal passages are formed along with an extraordinarily long false palate. There are no palatal vacuities and the internal choanae do not exhibit thickened rims. Abutting onto the rear ends of the palatines are the pterygoids, overlain by enormous ventral ectopterygoids. It has been suggested by Griffiths (1968) that the unusual size and robust nature of these bones are connected with the echidna's method of grinding up its food. This is achieved by the grinding action of keratinous spines, mounted on a "dental pad" on the dorsal surface of the posterior end of the tongue, against sets of transversely arranged keratinous spines on the palate. This pad is as wide as the concave region formed by the combined ectopterygoids and palatines and it fits the contours of the concavity very well. The live food (ants and termites in the case of *Tachyglossus*, earthworms in that of *Zaglossus*, Chapter 3) are crushed and homogenized between the keratinous spines on the pad and the keratinous spines on the epithelium which is applied to the ecto-pterygoids and false palate.

There are no jugals and the malar arch formed by zygomatic processes of the squamosal and maxilla is very thin. The dentary is reduced to a mere splinter but at the ventroposterior end of each ramus is an indication of an angle, and on the posterodorsal surface is a vestige of a coronoid process. The condyles are elevated relative to the coronoid processes, those of *Zaglossus* being elongated, robust, and standing higher above the long axis than in *Tachyglossus*; they articulate with shallow glenoid cavities on the ventral surfaces of the squamosals.

The posterior portion of the skull is invested with the paired exoccipitals and the unpaired supraoccipital surrounding the foramen magnum. The roof is formed by paired parietals, frontals, and nasals. The sides of the posterior end of the skull are formed by orbitosphenoids, squamosals, and the periotics or petrosals. In some specimens the passage, noted in *Ornithorhynchus*, between squamosal and periotic exhibits a post temporal opening, in others none is detectable; in the skull of the *Zaglossus* shown in Fig. 16, the post-temporal opening was present on one side but not on the other!

The periotic lies between the squamosal and basioccipital and exhibits the tympanic cavity or fossa on its posteroventral surface; in the roof, located posterolaterally, is the fenestra ovalis or f. vestibuli (Figs. 16 and 56). The osteology of this region is described in Chapter 7. The lateral wall of the cavum epipericum, to all appearances one large slab of membrane bone, is made up of three ossifications: a laterodorsal projection of the palatine, a lamina ascendens (alisphenoid) of the ala temporalis, and of multiple independent ossifications that have fused to one another and to the other two entities in the membrana sphenoobturatoria. The evidence for this interpretation of the nature of the bone forming the side wall of the brain case is discussed in Chapter 10. The foramen ovale for egress of the mandibular ramus (V 3) of the trigeminal nerve is sur-

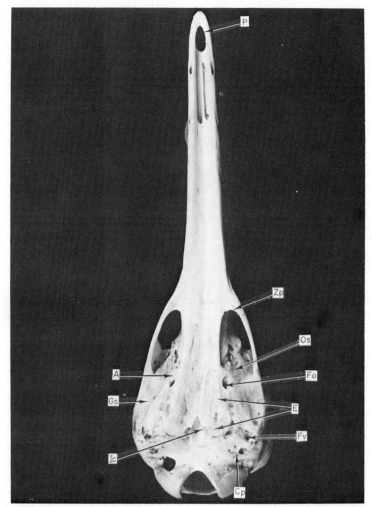

Figure 16. *Zaglossus.* Ventral view of skull. × 0.8. P, premaxilla; Zp, zygomatic process of maxilla; Os, ossification in sphenoparietal membrane; Fo, foramen ovale; E, ectopterygoid; Fv, fenestra vestibuli; Cp, crista parotica; Ic, internal choana; Gs, glenoid surface of squamosal; A, alisphenoid.

rounded by three bones and it is situated a long way rostral to the position it occupies in the skull of *Ornithorhynchus* (see p. 324). The three bones involved in formation of the foramen ovale are ectopterygoid, alisphenoid, and the membrane bone that has replaced the membrana-sphenoobturatoria (Fig. 119).

Comparison of the Postcranial Skeletons of Monotremes with Those of Cynodont Therapsids and Triconodonts

Both *Tachyglossus* and *Zaglossus* along with all other mammals including the platypus have seven cervical vertebrae*, the first two of which are modified to form atlas and axis. The atlas has the form of a ring whose anterior ventral surface is concave and articulates with the laterally located occipital condyles. The axis bears a rostrally directed dens or odontoid peg which morphologically is a derivative of the atlas (Jenkins, 1969) and which pokes forward into the center of the atlas ring. Jenkins (1970b) says of this arrangement, "The atlanto-occipital joint permits of extensive flexion-extension, while the atlanto-axial joint is primarily rotatory in function." In the monotremes the dens is of typical mammalian size and proportions but between it and the body of the axis is a bulbous ossification that bears facets for articulation with the atlas. Where this structure is united to the axis body is a distinct joint precisely as in the large axis body of tritylodontid reptiles (Jenkins, 1969). From this Jenkins surmises that when atlas-axis components of Triassic mammals are found they will prove to be similar to those of tritylodonts. However, other postcranial elements from the skeletons of the Triassic triconodonts, *Eozostrodon, Megazostrodon,* and *Erythrotherium*, are available and have been compared with those from therapsids and monotremes (Jenkins, 1973; Jenkins and Parrington, 1976). These authors stress that the early mammals are already specialized in many ways and suggest that on the basis of evidence presently available the tiny Triassic triconodonts (*Megazostrodon* probably weighed 20–30 g) and *Kuehneotherium* (see p. 313) have "adaptations for a diet and locomotor repertoire characteristic of many small insectivorous mammals today. They may be envisaged foraging for insects and grubs among litter and fallen plant debris, and clambering upward through low vegetation and even trees to seek out bark-dwelling and leaf-eating insects." These, then, are the kinds of mammals available for comparison with the monotremes—large, powerful, amphibious, and terrestrial burrowers. Nevertheless many structures and specializations are common to the triconodonts and the monotremes.

Comparison of the reconstructed shoulder girdles of eozostrodonts, cynodonts, and monotremes show that all have the same basic structural plan—dorsal scapulae, ventral coracoids, procoracoids, clavicles, and a T-shaped interclavicle. In the eozostrodonts and cynodonts the scapular blade is long and narrow, the infraspinous fossa is relatively deep, the glenoid is oriented posterolaterad and somewhat ventral, and there is no apparent articulation between the coracoids and interclavicle. The head of the humerus is hemispherical and

---

*According to Cabrera (1919) the vertebral formula for *Tachyglossus* is 7 cervicals (C), 16 dorsals (D), 3 lumbars (L), 3 sacrals (S), and 12 caudals (C); for *Zaglossus* it is 7 C, 17 D, 4 L, 3 S, and 12 C. Gregory (1947), however, has *Tachyglossus* with 7 C, 15 D, 3 L, and 2 S.

flanked by trochanters resembling those of metatherians and eutherians; the ulna articulates with a condyle (not a trochlea) which has a partial spiral configuration. In the monotremes, however, the scapular blade is expanded dorsally, the glenoid is laterally oriented, the infraspinous fossa is shallow, and the coracoids are articulated to the interclavicle (Fig. 6); all these modifications are undoubtedly advantageous to animals that have to dig burrows for shelter or to dig for food, since they add to the strength and stability of the girdle. Furthermore, the glenoid in the monotremes is deep and broad for the accommodation of the horizontally positioned wide head of the humerus; however, the ulna articulates with a condyle exhibiting a quasi-spiral configuration as it does in the eozostrodonts. This type of ulnar articulation is also found in the multituberculates and apparently the multituberculate scapula (McKenna, 1961) is similar to that of the Triassic triconodonts. McKenna also described an interclavicle tentatively diagnosed as multituberculate.

The pelvis of the Triassic triconodonts is essentially mammalian in that it has a narrow elongated iliac blade directed anterodorsad, an acetabular notch, an enlarged obturator foramen, and a relatively small pubis. The elongated iliac blade is found in the pelvises of monotremes, Metatheria, and Eutheria but not in those of cynodonts, but interestingly enough it does occur in the pelvis of the tritylodontid *Oligokyphus*.

Epipubic bones have been found to occur in another tritylodontid, *Tritylodontoides maximus* (Fourie, 1963); they have also been found in a multituberculate (Kielan-Jaworowska, 1969). Although epipubic bones have not been found in eozostrodonts, apparently due to the fact that the fossils have been partially damaged and are disarticulated, it would seem likely that they did occur, since in addition to the animals mentioned above, monotremes and marsupials have them (see Appendix).

ZAGLOSSUS

Discovery

Mÿnheer, A. A. Bruijn, a Dutch merchant and natural historian living at Ternate on the island of Halmahera in the 1870's, acquired a very strange skull said to have come from the Arfak Mountains of the Vogelkop of New Guinea; Bruijn, probably through the good offices of Luigi Maria d'Albertis, the legendary Italian zoologist who collected in New Guinea in the 1870's, presented the skull to the Museo Civico de Storia Naturale Giacomo Doria at Genoa. The Marchese Giacomo Doria himself and Professor W. H. Peters of Berlin, recognized that it came from an echidna and described it as a new species of *Tachyglossus*—*Tachyglossus bruijnii* (Peters and Doria, 1876). The outstanding feature of the skull was the extraordinarily long thin snout. Gill (1877) considered the difference from the skull of *Tachyglossus* so great that he created a new genus for it—*Zaglossus bruijnii*. In rapid succession five more names were proposed: *Proechidna*, *Acanthoglossus*, *Bruijnia*, *Anthoglossus*, and *Prozaglossus*, but *Zaglossus* had priority and has prevailed. The animal used to live in Australia but went extinct in the late Pleistocene (p. 74); it is now found only in the central cordillera of the island of New Guinea.

External Features

Paul Gervais (1877–1878) managed to acquire two specimens of *Zaglossus* which had been brought to Paris by a "voyageur," M. Leon Laglaize, who had procured them "avec le concours de M. Bruijn, a qui cette espèce de Monotremes est didiée, dans les Montagnes des Karons a une hauteur de 1450 metres." These animals were the subjects of the first published description, with illustrations, of the whole animal and skeleton. *Zaglossus* is a very large echidna, up to 1 m in length (Rothschild, 1913) and weighing up to 9.5 kg (Laurie, 1952). In shape it is like *Tachyglossus* but its extra long premaxillae and mandibles set it apart from that genus (Fig. 17); it also has relatively long legs and stands higher off the ground, features that give it the appearance of a small elephant. The hair is thick and woolly, the spines are short and, in fact, the pelage is very like that of the Tasmanian variety of echidna, *Tachyglossus aculeatus setosus* (Fig. 22). The eyes in *Zaglossus* are not obscured by encroaching hair as in *Tachyglossus* but are surrounded by bare, slightly wrinkled skin, and the beak or snout has an aquiline curve whereas it is straight or slightly retrousée in *Tachyglossus*. The

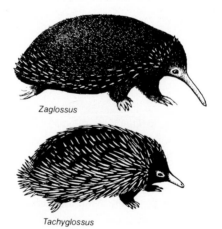

Zaglossus

Tachyglossus

Figure 17. *Zaglossus bruijnii* and *Tachyglossus aculeatus*. *(From Braxton and Knight, 1974; reproduced with permission of Australian Government Publishing Service.)*

combination of wrinkled bare skin around the eyes and of aquiline snout conveys an impression of benign condescension. *Zaglossus* has small but distinct external pinnae situated on the dorsoventral aspects of the head.

The tongue is long and vermiform as in *Tachyglossus*, with the difference that the anterior third bears a longitudinal groove, with closely apposed lips, on its dorsal surface; within the groove are three rows of backwardly directed keratinous teeth—hence *Zaglossus*. Another difference from the tongue of *Tachyglossus* is that it is, as far as I can determine, not protruded very far (see Chapter 3).

The limbs are pentadactyl but not all digits bear claws in all specimens; however, all specimens have the two long grooming claws (on digits 2 and 3) on the hind legs as in the long-haired subspecies of *Tachyglossus* (Fig. 26). The claws on the hindlimbs are directed backwards as in *Tachyglossus*.

There is no scrotum; the animal is cloacate and monotreme and in the middle of the ventral surface there is a relatively bare area exhibiting two hairy milk areolae, in both sexes, as in *Tachyglossus*. As far as I know, presence or absence of a pouch has not been documented for *Zaglossus* but Fuyughé villagers, living in the mountains of Papua, volunteered the information to me without leading questions, that the female carries its young in a pouch.

It would seem that all males have a spur on the ankle and perhaps juvenile females do also; not enough specimens have been examined to determine whether or not some adult females have a spur (p. 75).

Zaglossus has the enormously thick panniculus carnosus, contractions of which enable it to assume a variety of bodily contortions, roll itself into a ball, or tuck its head down and under the ventral surface. This maneuver is achieved by a quick upward stretch of the fore legs to enable the long snout to clear the ground. With the head tucked under and the snout pointing backwards along the thorax it looks like some huge hemipteran insect with a long proboscis. *Zaglossus* also can burrow vertically but its grip is not as tenacious as that of *Tachyglossus* and when it is pulled out of a hole it appears to be unafraid, walks about, and fossicks.

Reproductive and Excretory Organs

Apparently I am the only one to have had the opportunity to make observations on the reproductive systems of *Zaglossus* living naturally in New Guinea. The animals concerned were five adult males, one adult female coming into breeding condition, and one juvenile female; weights, sizes, and provenance of these animals are described in Chapter 8.

The female organs are as in *Tachyglossus*, both ovaries being equally well developed, and, from histological examination of the ovaries of the adult, both appeared to be capable of producing ripe ova (see p. 249). The ovaries are enclosed by the usual membranous infundibular funnels leading to convoluted Fallopian tubes, oviducts, and uteri which in turn communicate with the median unpaired urogenital sinus. The excretory system is as in the platypus and *Tachyglossus*. Likewise the male urogenital system is identical to that of *Tachyglossus* and Fig. 10 would serve as an illustration of *Zaglossus*, with the reservation that I made no observations on the morphology of the glans penis in *Zaglossus*.

Some aspects of oogenesis and spermatogenesis are described in Chapter 8.

The Mammary Glands

A normal lactating gland in *Zaglossus* has not yet been observed. However, the glands of the female coming into breeding condition, mentioned above, showed a structure identical to that of nonpregnant *Tachyglossus* at the start of the breeding season (p. 268). The gland is fan-shaped and exhibits elongated, flat lobules that show no sign of the expanded club-shape of the lactating *Tachyglossus* gland. There was no pouch but a pouch area could be seen since it exhibited less hair than the surrounding belly and two areolae with mammary hairs. Structurally the areolae were like those of *Tachyglossus*: the ducts of the glands opened at the base of a mammary hair follicle and at the periphery of the areola a ring of Knäueldrüsen of Gegenbaur was located. Kolmer (1925) failed to detect sebaceous glands in the areola and I could not find any either.

As in *Tachyglossus* the males have mammary glands; these were examined in two specimens and they proved to have the same structure as the glands of the above female, but were much smaller. The areolae also exhibited Knäueldrüsen but no sebaceous glands.

Brain

From Gervais' (1877–1878) beautiful illustrations it is apparent that the external morphology of the brain is identical to that of *Tachyglossus*, including the asymmetrical arrangement of the gyri on the very large neocortex (Fig. 13); however, sulci corresponding to alpha, beta, and zeta in *Tachyglossus* appear to be present. The internal anatomy and organization of the cortex of this brain are quite unknown.

2

The Different Kinds of Monotremes, Distribution, Movements, the Crural System, and Genetics

ORNITHORHYNCHUS

Kinds

The evidence for the view that there are three subspecies of *Ornithorhynchus anatinus* (Thomas, 1923) is so slight in terms of modern taxonomic standards that the subject cannot be debated; Thomas, for example, erected the subspecies *phoxinus* on the basis of one specimen collected in northern Queensland. When enough specimens, collected from representative regions of its range, are examined using biochemical and classical techniques the status of these subspecies will undoubtedly be determined. Fossils referred to the genus *Ornithorhynchus* have been reported, namely, *O. agilis* and *O. maximus*. These have been discussed recently by Mahoney and Ride (1975). These authors are satisfied that the holotype of *O. maximus*, a fragment of a right humerus, is tachyglossid and not ornithorhynchid. *O. agilis* syntypes consist of a right tibia and the posterior portion of a right mandibular ramus; of these J. A. Mahoney (personal communication) believes they are the same as those of living *O. anatinus*.

Distribution

The platypus lives only in the freshwater systems of eastern Australia and from observations made by numerous fauna and fisheries officers in Queensland, New South Wales, Victoria and Tasmania, by scientific colleagues, and from my own observations, they are abundant. In Queensland they extend at least as far north as the Annan River. Recently F. Carrick and T. Grant (personal communication)

43

trapped for platypuses in the following Cape York Peninsula rivers: Rocky River, Massey River, Peach Creek, Cockatoo Creek, and the Claudie River. They had eight observers with them but no platypuses were taken or seen during all-night watches. The temperatures of those rivers varied from 25–29.9°C, temperatures that platypuses can withstand with ease (p. 122). The rivers were otherwise productive, containing large fish which are said to support populations of crocodiles (*C. porosus* and *C. johnstoni*). That the methods for detecting platypuses were adequate is shown by the results of a control netting carried out in a backwater of Tinaroo Dam on the Barron River—nine platypuses were caught in 1 hr. Why platypuses are absent or scarce in Cape York's permanent rivers is not apparent; crocodiles may be the answer but the rivers support the relatively slow-swimming water rat, *Hydromys chrysogaster*, which would fall prey to crocodiles just as easily as a platypus would (see Appendix).

From the Annan River to the southernmost parts of Victoria, the platypus is found in most rivers flowing east of the Great Dividing Range including the freezing waters of creeks and rivers in the Australian Alps. Temple-Smith (1973) records that he has observed them in recent years in the following Tasmanian rivers and lakes: Arthur, Derwent, Forth, Gordon, Huon, Leven, Mersey, North and South Esk, Pipers, Ringarooma, Serpentine, Crater Lake, Great Lake, Platypus Tarn, and Lake St. Clair, which is just about all over Tasmania.

The north-westernmost distribution of the platypus seems to have shrunk since observations were made last century. Its presence has been reported in the Norman (Waite, 1896) but recently I enquired of people living by that river if they see platypuses but they had never heard of them. Some years earlier Armit (1878) watched a platypus "swimming about in a large waterhole situated about 150 miles west of Georgetown on the road to Normanton. I distinctly saw this animal's head and bill above water but was unable to capture it, as it dived on hearing the pack-horses trotting up to the hole to drink. My boys inform me that they saw this funny fellow in the upper Herbert; and it occurs on the Leichardt River." From my own observations and inquiries at the East Leichardt River and at the Leichardt itself it seems very unlikely that what were seen were platypuses since the rivers ran dry in the winter. Presently there is a large artificial water storage on the Leichardt, Lake Moondarra, but of this Hamer Midgley (a Churchill Fellow) who has an extensive knowledge of Queensland rivers and who is engaged in stocking this lake with native fish tells me he has never heard of, or encountered, platypuses in rivers flowing into the Gulf of Carpentaria.

The southwestern limits of its distribution also seem to have changed for the worse; the platypus has been reliably reported to occur in various rivers, apart from the Murray, in South Australia such as the North Para, Angaston Creek, and the Onkaparinga River. It is unknown at these places now, even to fishermen illegally, and consequently quietly, fishing for trout in the Mt. Bold reservoir on the Onkaparinga, and to the dam keepers there. Reports in the press of the occurrence of platypuses in the Murray in South Australia appear from time to time but it is certainly not abundant there.*

As far as the rivers flowing west of the Great Dividing Range are concerned the platypus is numerous in the tributaries of the Darling, Lachlan, Murrumbidgee and the Murray, but its distribution is discontinuous in the main channels of those rivers. It is common in the Murray down to Echuca but becomes less so downstream from there. It is frequently seen, however, in the Wakool and it has been taken recently at Mildura near the border of South Australia and Victoria (Pizzey, 1975). It was also observed near here but in the Darling River at Pomona about 1600 hr April 26, 1975 (Mrs. I. Doecke, Pomona, N.S.W., personal communication). This appears to be the first record of its occurrence in the lower Darling River.

Movements

Three studies of the movements of platypuses have been or are being carried out independently and more or less simultaneously: in January 1969 Griffiths and Elliott (unpublished) commenced marking platypuses primarily with the aim of following changes in the composition of the milk during the lactation period; Temple-Smith (1973) began in November 1969 a far more detailed research on movements, and Grant and McBlain (1976) did so at a later date. All three operations are continuing in the sense that marked platypuses released years ago may still be observed or caught.

Platypuses were caught in gill nets hanging freely in the water from long lines supported by floats. The method of marking used by Griffiths and Elliott (see CSIRO film "The Comparative Biology of Lactation," 1974) was to attach a sheep ear-tag, made of colored plastic and numbered, to the distal part of the tail using standard applicator pliers. These made a neat round hole in the tail through which the tag was clipped. The hole heals well and apparently the tag causes no great discomfort since females bearing tags for months have become pregnant, laid eggs, and lactated subsequent to marking. Temple-Smith retrapped marked platypuses in the pools where they were released and in other pools up to 1 mile away. However he found that only 5 out of 25 tagged platypuses were recaptured outside the pool in which they were originally caught. From this he concludes "site attachment appears to be strong in the platypus." The data of Griffiths and Elliott (Table 1) support this notion; out of 34 marked platypuses released at the places where they were originally captured, 12 were found at times ranging from 68 to 1330 days after release, in the same pool or area. Two of these places were at the entrances of the Cotter River into two larger water storages, Corin Dam

*Platypuses introduced by Fleay (1941) into the Rocky River, Kangaroo Island, South Australia survived and the descendants are thriving there.

TABLE 1

Apparent Home Range and Body Weight Changes of Platypuses Marked and Released at Place of Capture in the Murrumbidgee River and Its Tributaries

Place of capture and release	Sex	Initial body wt. (g)	Place of recapture	Time interval between release and recapture (days)	Body wt. at time of recapture (g)	Change in body wt. (g)
Retallack's Pool	♀	1250	Same pool	182	1250 (Lactating)	Nil
	♀	1150	Same pool	90	Not determined	—
	♀	1250	Same pool	190	1300	+50
	♂	1686	Same pool	330	1710	+24
	♂	1686	Same pool	491	1729	+43
	♂	1750	Same pool	(Recaptured second time) 97[a]	Not determined (Tag only taken in net)	—
Top end of Corin Dam	♀	650	Same area of dam	68	750	+100
	♀	675	Same area of dam	457	670 (Lactating)	−5
	♂	1500	Same area of dam (Sighting only)	501	—	—
Top end of Bendora Dam	♀	760 (Lactating)	Same area of dam	102	809 (Still lactating)	+49
	♂	1250	Same area of dam	445	1400	+150
	♂	1350	Same area of dam	1330	1450	+100
Pool at Duntroon	♀	898	Same pool	96	975	+77

[a] This animal apparently weathered severe floods or managed to get back to its pool after the floods that occurred between the time of marking and recapture.

and Bendora Dam, 7 km and 4 km long, respectively. Other points of interest that may be noted in Table 1 are that the very small female platypuses in Corin and Bendora Dams were adults, one weighing 675 g at initial capture and 1 year, 3 months later she was 5 g less in weight but found to be lactating. One of those females taken in November was lactating but when caught again 102 days later she was still lactating, indicating that the platypus suckles its young for months as does the echidna (p. 304). A large male apparently "rode out" severe flooding or managed to return to his home pool after the floods; the male recaptured 3 years, 8 months after tagging weighed only 100 g more; some platypuses showed little or no increases in weight over long periods of time. Temple-Smith found an example of this in the instance of an immature female tagged March 3, 1971 in a pool in the Shoalhaven River and taken again in the same pool 2 years, 4 months later by another set of platypus catchers, Carrick and Grant (Grant and McBlain, 1976); at release she weighed 855 g but at recapture only 800 g—body length was unchanged.

In a preliminary report Grant and McBlain described their observations on a single population of platypuses. They found that most animals moved less than 400 m from their point of capture during a series of 10-day observation periods and they recaptured several marked animals in the same pools over 3 years after they were released there. Other platypuses, they think, may move around much more since unmarked animals are still being caught in the area after the release there of 62 marked platypuses (40 females, 22 males). So far 18 females have been recaptured but only three males (see Appendix).

The Crural System and Its Biological Significance

At first sight inclusion of this subject in a chapter on movements and distribution may seem odd but there is evidence, to be discussed later, that it is concerned with defense of territories, and therefore of interference with movements.

Male platypuses exhibit a crural system consisting of a kidney-shaped alveolar gland, located on the dorsal aspect of the upper thigh muscles, connected by a thin-walled duct to a spur on the heel (Fig. 18). The keratinous* spur is hollow and its central canal is continuous with that of the duct. The gland secretes a venom that is passed to the spur; this venom can be injected into other animals and man by erection of the spur [see Temple-Smith (1973) for an account of the mechanism of erection] and its forcible insertion into the flesh of the victim by a

---

*Keratinous structures in birds and reptiles consist of two distinct molecular species of the protein, distinguished from each other by the mode of folding of the molecule; these are known as α and β keratin. In mammals, however, only one kind of keratin is synthesized—the α form. Interestingly enough the spur of the platypus consists of α keratin only [Drs. R. D. S. Fraser and T. P. McRae, Division of Protein Chemistry, CSIRO (personal communication). See also Fraser (1969).]

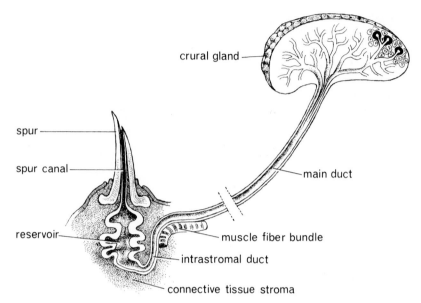

Figure 18.   *Ornithorhynchus.* Crural system. *(From Temple-Smith, 1973.)*

quick inwardly directed jab of the opposed hindlimbs. In man the venom causes agonizing pain, and a platypus can kill a dog with its venom apparatus. Temple-Smith has given a fascinating description of the effects of platypus attacks on man and animals in an appendix to his thesis. He has also made the first detailed study of this system in a large sample of animals. He found that in juveniles caught in late February and March, i.e., just emerging from the burrow, the spur is covered by an outer horny sheath and an inner fibrous layer. This condition persists for about 6 months but between 6 and 10 months of age the two outer layers commence to disintegrate (called exsheathment by Temple-Smith) and seem to expose the tip of the spur (Fig. 19). The process is complete at about 12 months of age.

The secretion formed in the crural gland passes via a series of primary and secondary ducts which unite to form a single main duct connected by an intra-stromal duct to a reservoir (Fig. 18). From the last named structure stored secretion passes to the exterior via the spur canal. The reservoir is not concerned with ejection of the venom since it is surrounded by the thick stroma of inelastic connective tissue and is lined by an almost inflexible keratinized investment. The main duct, however, is a flexible, distensible, membranous tube. Temple-Smith found that the duct, not the keratinized reservoir, is the main storage organ for the crural gland secretion and that there is an annual cycle of recrudescence and regression of the gland. From the data in Fig. 20 it is apparent

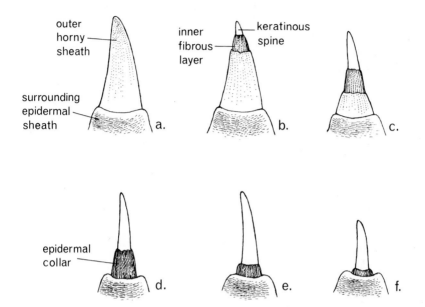

Figure 19. *Ornithorhynchus.* Morphology of the spur. (a) Spur of juvenile; (b) juvenile spur in early stage of exsheathment; (c) juvenile spur at a later stage of exsheathment; (d) recently exsheathed adult spur; (e) spur of adult; (f) spur of an old adult. *(From Temple-Smith, 1973.)*

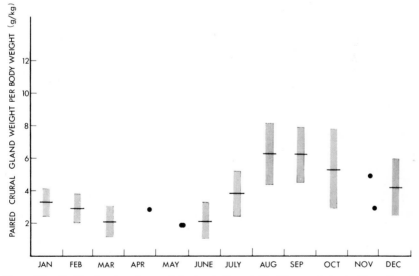

Figure 20. *Ornithorhynchus.* Seasonal changes in the weight of the crural glands. *(From Temple-Smith, 1973.)*

that the weight of the gland increases in late winter and early spring followed by a gradual decline in weight in late summer; maximum weights were encountered from mid-August to October.

The secretory epithelium of each alveolus of the crural gland consists of a single layer of cells attached to a basement membrane which is firmly appressed to a system of connective tissue septa located between the alveoli. The variations observed throughout the year in crural gland weights are accompanied by the variations in the heights of the secretory epithelium. Thus during the period of quiescence from April to early June the secretory epithelium consists of rather flattened cuboidal cells and the lumina of the tubules are small. This is followed by a phase of hypertrophy during which the epithelium becomes columnar and intracellular elaboration of secretory products take place; the nuclei of the cells are positioned close to the basement membrane. The lumina of the alveoli for the most part are empty at this stage.

From August–October, the time during which the maximum height of the secretory epithelium is encountered, the gland cells are densely packed with secretory granules consisting of membrane-bound vesicles filled with apparently protein since there is an enormous increase in the number of ribosomes linked with rough endoplasmic reticulum.

The secretory products were released from the cells by both eccrine and apocrine extrusion. The latter was observed early in the extrusion phase when the gland cells were packed with secretory vesicles. From such cells apical portions of the cytoplasm containing vesicles were pinched-off into the lumen of the alveolus. Later in the secretory phase the secretion was released into the lumen by rupture of the membrane of the vesicles at the distal surface of the cells giving rise to vesicles of apparent low electron density. A phase of extrusion-regression continues from November to March. In late stages of this phase leucocytes invade the connective tissue septa aground the alveoli and they penetrate the epithelium, passing into the lumina of the alveoli.

In fixed material under the light microscope the secretion appears as a coagulum of which histochemical tests show that the principal product of the secretory cells is protein. This secretion can be collected in amounts sufficient for analysis by withdrawal (under anesthesia) from the exposed duct into a hypodermic syringe; up to 4 ml can be obtained at a "milking" (Table 2).

The major protein components of the secretion consisted of prealbumins (MW = 70,000) and postalbumins (MW = 70,000 to 80,000) along with one to five larger proteins (MW = 90,000). Seasonal changes were observed in the protein complements of the secretion during regression and quiescence: a prominent prealbumin was present but postalbumin bands were either absent or present in low concentration. The concentration of postalbumins increase during May and June, i.e., when the hypertrophied gland cells are synthesizing secretory products.

TABLE 2

Volumes of Platypus Crural Gland Secretion Collected and Protein Contents of the Secretion[a]

Time of year	Number of animals	Volume[b] (μl)	Number of animals	Protein content[b] (mg/100 ml)
November–January	4	337.5 (100–700)	4	109.6 (101.5–121.5)
February–April	13	293.8 (0–800)	10	131.8 (82.5–155.5)
May–July	20	30.7 (0–200)	3	148.3 (115.0–155.0)
August–October	21	416.0 (0–4000)	10	151.2 (95.0–211.0)

[a] Data from Temple-Smith (1973).
[b] Range is given in parentheses.

As well as native secretions, fractions of the secretion eluted from Sephadex gel columns and solutions of lyophilized extracts of glands were tested for toxicity by injection of the various preparations into mice. It was found that toxicity was not related to the presence or absence of the prealbumins and postalbumins, and that no preparations nor the native secretion itself were particularly toxic when injected subcutaneously. However, expressed secretion, solutions of lyophilized extracts, and of fractions of the secretion were found to be very toxic when injected intravenously; the native secretion, however, was the most toxic of all.

The effects* of the intravenously injected preparations in the mouse were: hyperventilation within seconds of the injection, convulsions, dyspnoea, paralysis of the hindlimbs followed by uncontrolled swimming motions of the limbs, protrusion of the eyes, change of eye color from red to reddish purple (suggesting cyanosis and hypoxia) followed by jerky convulsions and death. Temple-Smith considers that the cause of death was respiratory failure since it was found in mice that had died from the effects of native crural secretion and that there were abnormal accumulations of red blood cells in the capillaries and larger blood vessels of the lungs; furthermore, the lung alveoli were constricted and the interalveolar septa were thick and edematous.

Attempts to associate the toxicity of the secretion with fractions eluted from Sephadex gel columns were equivocal; of 22 fractions obtained, toxicity was associated with fractions 1 and 2 only; both these contained cell debris and in the instance of fraction 2 high molecular weight proteins. When the cell debris was removed the fractions lost their toxicity suggesting that the toxic component of crural gland secretion is associated with the debris or an active component associated with it. The other fractions, as already mentioned, lacked toxicity but

*Those effects are identical to those shown by mice injected intravenously with the venom of shrews (Pucek, 1968).

contained the pre- and postalbumins, so presumably these are not connected with the toxicity of the native secretion, at least not in mice as Temple-Smith points out.

Temple-Smith's work has answered many questions concerned with the crural system but we lack a definitive answer to the question of what the toxin is. One suggestion advanced by Temple-Smith that it is a protein of high molecular weight or a mucoprotein could be tested by heating the secretion on a boiling water bath so as to coagulate the proteins, centrifuging off the coagulum and injecting the supernatant into mice. Apparently it is not a neurotoxin (Henderson, *fide* Temple-Smith) but release of histamine appears to be involved; this could be determined by estimation of blood histamine levels in animals injected with native secretion.

Temple-Smith noticed that the seasonal waxing and waning of the crural system closely paralleled that of the reproductive system (see Chapter 8) and that when freeing captured male platypuses from the nets there was a marked increase in aggressive use of the spurs during the breeding season. He also observed that aggressive encounters only occurred between males and that there is a greater incidence of spur marks or wounds in males than in females. From these observations he surmises that the crural system is used in intraspecific fighting between males during the breeding season, stating "In my view, the crural system is a potentially effective mechanism for ensuring spatial separation of male platypuses along the river habitat during the breeding season. Spur wounding and the associated subcutaneous and systemic reactions, including pain could act as a warning or deterrent to other males competing for or traversing the same area of river."

Juvenile female platypuses exhibit a spur sheath but compared with that of juvenile males it is rudimentary, and at no stage, according to Temple-Smith, is a spur detectable. This is unlike the situation in juvenile female echidnas (p. 69).

Genetics

Karyology

As far as direct studies of genetics in monotremes are concerned, nothing has been carried out simply because it has not proved possible to induce them to breed to order in captivity. However, research into genetics of an indirect nature is going on and has been published: determination of chromosome numbers and sex determination mechanisms, and determination of amino acid sequences in certain proteins, respectively.

The work on chromosome numbers is particularly lively and promises to be so for some time to come. It has been found that the chromosome numbers $2n = 53$ ♂, 54 ♀ published by Bick and Jackson (1967b) are not correct; the numbers are $2n = 52$ ♂, 52 ♀ (Bick and Sharman, 1975) as found in counts of mitotic

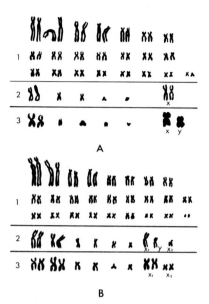

A

B

Figure 21. A, *Ornithorhynchus.* Mitotic karyotypes; 1 + 2 = female, 1 + 3 = male. Note the gradation, characteristic of sauropsidan karyotypes, from macro- to microchromosomes. B, *Tachyglossus.* Mitotic karyotypes; 1 + 2 = male, 1 + 3 = female. *(From Murtagh and Sharman, 1977; reproduced with permission of Australian Academy of Science.)*

chromosomes (Fig. 21A). Twenty-four of those chromosomes in both sexes have homologues of nearly equivalent size and shape; however, at the first meiotic division evidence was found that several of the pairs may not be exact homologues.

Two pairs of chromosomes in both sexes are of special interest, termed α and β by the authors, and each have one arm of equivalent length but the arms above the centromeres in the karyotypes appear to be incompletely homologous. The α and β chromosomes are always identifiable because the arms above the centromeres give a paler staining reaction than the homologous arms.

Examination of first division meiotic chromosomes showed the presence of a complex-chain multivalent and of an asymmetrical bivalent which exhibits the same staining reaction as the arms of the α and β chromosomes in mitotic metaphase. The total number of bivalents at first division of meiosis appears to be 21 (Murtagh and Sharman, 1977) so presumably the multivalent consists of 10 elements. The authors are of the opinion that the sex determining chromosomes appear to be incorporated in the large multivalent chain in the male.

Sex Ratio

Over the period May 19, 1967 to October 1, 1975 Griffiths and Elliott (hitherto unpublished), collected, in most months of the year, 47 male platypuses and 47 females by the gill-net technique described on p. 45; these were caught in the Murrumbidgee River and its tributaries in the Australian Capital Territory. It should be pointed out that up to February 9, 1972, 47 females had been caught and only 41 males; however, on the night of October 1, 1975 six males and no females were caught bringing the sex ratio to a neat 1:1! The area netted on this occasion was the top end of Bendora Dam where females are known to be at risk; perhaps their apparent absence was due to their staying in their burrows and not entering the water since this time of the year is the breeding season, so conceivably they were underground incubating their eggs or suckling their tiny young.

During the period November 1969 to November 1971 Temple-Smith (1973) netted 218 platypuses in the streams of southeastern New South Wales, 117 of these were males and 101 females so it looks as though the sex ratio of *Ornithorhynchus* is close to parity.

Proteins

Bennett and Baldwin (1972) and Baldwin (1973) have found in monotremes a subunit of lactic dehydrogenase (LDH) additional to the five electrophoretically distinguishable isoenzymes of that enzyme found in the somatic tissues of vertebrates. Starch gel electrophoresis of muscle and heart extracts of *Ornithorhynchus, Tachyglossus*, and of *Zaglossus* yielded the typical pattern of five LDH isoenzymes with the addition of subbands associated with LDH 2, 3, 4, and 5. The isoenzymes of the LDH 5 complex were separated, purified, and analyzed structurally and kinetically with the result that all three kinds of monotremes had an extra LDH subunit closely related to the subunit designated M in other vertebrates. Baldwin considers this subunit of LDH unique to the monotremes and that presumably it arose by duplication of the M subunit gene. Its functional significance will be investigated.

Baldwin (1973) and Baldwin and Temple-Smith (1973) have found that monotremes also differ from eutherians and metatherians as far as LDH of their testes is concerned. In many eutherians and birds a testes specific LDH known as LDHX occurs and it is presumed to play an important role in sperm metabolism. Baldwin and Temple-Smith found that the testes of the marsupials *Wallabia bicolor* and *Vombatus ursinus* had an LDH which could be classified kinetically as LDHX but the spermatogenetically active testes of platypuses and echidnas did not; they did, however, exhibit the LDH 5 complex found in somatic tissues, discussed above.

Cooper *et al.* (1973) state that the blood of the platypus has two kinds of hemoglobin; however, Whittaker and Thompson (1974) find that there are three, HbI, II, and III. HbI is the major component and Whittaker and Thompson have determined the amino acid sequences in the α-chain of the globin of HbI. In agreement with the findings for all other vertebrates above the level of teleost fishes, the number of amino acid residues totaled 141. Previous to the publication of this work on the platypus, Thompson *et al.* (1973) and Whittaker *et al.* (1973) determined the amino acid sequences in the α-chain globins of echidna hemoglobins. This information along with that for platypus HbI α-chain globin was used for a comparison of the differences ("changes") of monotreme α-chains from those of other mammals in order to estimate a date of divergence of the monotremes from "other mammalian groups." The method used to arrive at the estimate was that of Air *et al.* (1971) modified as follows:

1. The sequences used by Air *et al.* included those of man and of a monkey; in the present work Whittaker and Thompson substituted the dog α-chain sequence for that of the monkey to avoid undue emphasis on the primate linkage rate.
2. Allowance was made for hidden mutations by the use of the accepted point mutation units (PAM) of Dayhoff (1972).

From their calculations Whittaker and Thompson arrive at an estimate of the date of divergence of the monotremes from "other mammalian groups" of 180 ± 37 million years ago. However at the end of the discussion of their results a qualification is introduced: "Intraspecies variation in mutation rates also occur as exemplified by the echidna. Usually the mutation rate of the β-chain is faster than that of the α-chain (Air *et al.*, 1971), whereas in the echidna it is less. Calculations based on the echidna β-chain sequence would give a date of divergence estimate different from that given by α-chain and, in fact, more recent than the divergence point of the marsupials." In another paper Whittaker and Thompson (1975) recalculated their figures in the light of other information and came up with a divergence date of 211 ± 88 million years ago. However, in the same paper they give an amino acid sequence for the 144 residues of the β-chain of platypus HbI and using this data along with that derived from β-chain sequences of echidna hemoglobin (Whittaker *et al.*, 1972) and of the hemoglobins of many other mammals, they arrive at an estimate of 132 ± 33 million years ago. This discrepancy between estimates is stated to be due to the slower mutational rate of the platypus and echidna β-chains vis-à-vis the α-chains. In view of this Whittaker and Thompson suggest that assumption of a constant rate of mutation which is considered to be necessary for the estimation of divergence times, may not be valid. In another paper Fisher and Thompson (1976) came to the same conclusion after examining the amino acid sequences of platypus myoglobin.

OBDURODON INSIGNIS

This is the name of a new genus and species of extinct mammal which has been referred to the family Ornithorhyncidae (Woodburne and Tedford, 1975). The fossils are specimens of two teeth that appear to be different upper molar teeth. The holotype is a beautifully preserved specimen found in the Etadunna Formation on the west side of Lake Palankarinna, South Australia. The other tooth was found some 200 miles away at Lake Namba in the Northern Frome Embayment. The authors adduce evidence that *Obdurodon* was a member of the Ngapakaldi Fauna which, inter alia, consisted of various marsupials, crocodiles, and lungfish (Stirton *et al.*, 1968). The deposits in which the specimens of teeth and the fossils constituting the Ngapakaldi Fauna were found, are of Miocene age (see Appendix).

The holotype has an elongate almost rectangular shape seen in occlusal view and the crown is made up of two transversely oriented lophs separated by a wide deep transverse valley. In occlusal view the anterior loph consists of a single high lingual cusp and two lower labial cusps. The posterior loph is simpler consisting of a single labial and lingual cusp. A striking feature of the tooth is that it has six roots.

Woodburne and Tedford compared the holotype of *Obdurodon* with upper molars of *Ornithorhynchus* which have been described in detail by Simpson (1929) and by Green (1937). These teeth are basically similar in morphology to those of *Obdurodon* consisting of, in occlusal view, anterior and posterior lophs separated by a transverse valley; the base of the crown is equipped with multiple roots. However the molars of the living *Ornithorhynchus* are much smaller and more complex than in *Obdurodon*, exhibiting a proliferation of labial cusps and conules. Moreover the molars of *Ornithorhynchus* are elongate and lower crowned than those or *Obdurodon*.

It has already been stated that true teeth are found only in very young juveniles of *Ornithorhynchus* (p. 16), the teeth being replaced by large horny pads of keratin in the upper and lower jaws. These pads serve as very efficient grinders for the food. Woodburne and Tedford point out that since *Obdurodon* and *Ornithorhynchus* appear to share a basically similar root morphology "it is conceivable that the larger teeth of the fossil form were shed during life as in the Recent genus." They go on to say "nevertheless, *Obdurodon* does appear to have teeth of more regular less degenerate construction and it is tempting to maintain that the fossil form is dentally more representative of Tertiary or at least Neogene, monotremes than is the living genus."

A matter of interest that arises from this discovery of a genus of monotremes with teeth is what is the nature of the enamel? The structure of tooth enamel can be determined in thin sections by the use of a polarizing microscope. Using this method Moss (1969) has established that Cretaceous and living metatherians and eutherians have teeth with discontinuous or prismatic enamel. The teeth of sauropsid and therapsid reptiles have nonprismatic or continuous enamel (Poole, 1956). Moss also found that therapsid enamel is nonprismatic as are the enamels of two Triassic mammals (Moss and Kermack, 1967) named by those authors *Morganucodon* and "Welsh pantothere" but now known as *Eozostrodon* and

Kuehneotherium, respectively (see Parrington, 1971, 1974a). Moss also examined the enamel of what he calls "the rudimentary fetal teeth" of *Ornithorhynchus* and he gave it as his opinion that the enamel in this genus is nonprismatic on the basis of present available material. He stated that the tooth had unusual histology which he will report on at another time. However, in a footnote to their paper on egg teeth in monotremes Hill and de Beer (1949, p. 518) state "Dr. H.L.H. Green has examined the structure of the enamel and dentine of the molar teeth in the oldest mammary foetuses of Platypus available to him, and permits us to state that the enamel when seen in tangential section exhibits an unquestionable prismatic structure and that the dentine in section presents the characteristic radially striate appearance which is associated with the presence of dentinal tubules." It is not impossible that both accounts are right since both prismatic and nonprismatic enamel can occur in human teeth depending on the age of the individual (Moss, 1969). It would be of great interest to know the type of enamel in the teeth of *Obdurodon*.

Figure 22. *T.a. setosus*; dorsum, hairs, and spines. × 1.1.

TACHYGLOSSUS

Kinds

Six subspecies of *Tachyglossus aculeatus* have been described (see Griffiths, 1968): *aculeatus, setosus, acanthion, ineptus, multiaculeatus*, and *lawesii*. The criteria of differentiation have been, in the main, the degree of hairiness and the length of the claw on digit 2 on the hindlimb relative to that on digit 3 (Thomas, 1885). In spite of the suspicion that these criteria may be trivial and of the fact that some of the subspecies have been erected on the basis of examination of 3–4 specimens only, it must be admitted that when a member of a given subspecies is placed alongside that of another they look very different (Figs. 22–27). It was suggested by Griffiths (1968) that the reality or otherwise of these subspecies could be tested by determination of variation of the subspecific characters in echidnas collected from different areas in large enough numbers. Since that time I have collected and made observtions on live echidnas or on prepared skins of

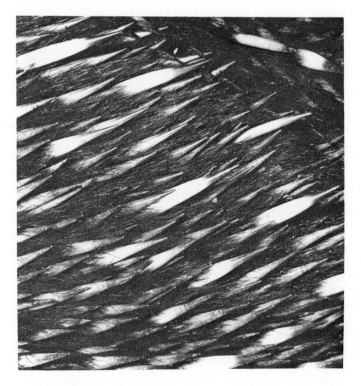

Figure 23.   *T. a. aculeatus*; dorsum, hairs, and spines. × 2.2.

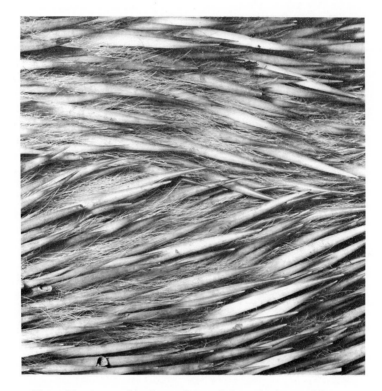

Figure 24. *T. a. multiaculeatus*; dorsum, hairs, and spines. × 2.2.

echidnas from Tasmania, the Canberra and Southern Highlands district and Australian Alps, eastern, southern, and western Victoria, Kangaroo Island and South Australia including the Fleurieu Peninsula adjacent to Kangaroo Island, the Northern Territory, Queensland, Western Australia, western and northern New South Wales, and New Guinea. In these specimens comparisons of pélage were made along with measurements of the length of claw on hind digits 2 and 3. The results are summarized in Table 3. From this it appears that in the southeastern part of Australia all echidnas have a long thick pélage of fine hair and stout spines (Figs. 22, 23), the longest and thickest hair being found in Tasmanian echidnas, while the ratios of the lengths of the claws on hind digits 2 and 3 are much the same throughout the region (1.20–1.24) with very little variation in the average absolute lengths of the claws (ca. 4.2 cm for claw 2 and 3.5 cm for claw 3). On the face of it there seems to be no difference between the echidnas of the southeastern part of the continent (*Tachyglossus aculeatus aculeatus*) and the echidnas of Tasmania (*T. a. setosus*) except hairiness. However, polymorphism of one of the hemoglobins of Tasmanian echidnas (see p. 70) serves to distinguish it from

Figure 25. *T. a. acanthion*; dorsum, hairs, and spines. × 0.82.

other echidnas. Moreover, the hair of *setosus* has a soft and shiny quality not found in that of mainland echidnas, in fact *setosus* was called the "échidné soyeux" by Péron and de Freycinet (1807–1816, Vol. 2, p. 13). In view of this and of its geographical isolation on Tasmania and nearby Bass Strait islands I should urge that *setosus* is a good subspecies.

Concerning the South Australian echidnas described in Table 3, Rothschild (1905) named them *multiaculeatus* after examination of a batch of 30 animals whose place of origin was stated to be the "southernmost part of South Australia." It is quite apparent, even from his scanty description, that he was talking about Kangaroo Island echidnas, the most distinctive of all echidnas (Fig. 7), with their very long fine pélage obscured by a plethora of characteristic extremely long thin spines. Furthermore there is nothing particularly "multi" or unusual about the spines of echidnas living on the mainland (Fleurieu Peninsula) a few kilometers away across the strait known as the Backstairs Passage (Table 3). In view of this I am certain that Rothschild received a shipment of echidnas collected on Kangaroo Island; since they are the most distinctive of all echidnas

Figure 26.   *T. a. aculeatus*; hindlimb. × 0.82.

and they are geographically isolated I feel they are a good subspecies and that the name *multiaculeatus* should not be applied to mainland South Australia echidnas, as Wood Jones (1923) has done, but restricted to Kangaroo Island echidnas. These and apparently all echidnas, other than subspecies *aculeatus* and *setosus* on the continent have short claws on the hindlimbs (Fig. 27) and the ratios of length of claw 2 to claw 3 are much larger (2.13–2.80) than in *aculeatus* and *setosus*. Echidnas found in the arid and semiarid parts of the continent, or in parts that are hot and dry for months on end, have a pélage of short bristles, sparse bristles, or no pélage at all (Fig. 25). These belong to the subspecies *acanthion*. Morphologically intermediate between these echidnas and the well-haired long-clawed echidnas of the southeast, a group of indeterminate subspecies status has been found in an area, roughly rectangular with Port Macquarie, Forbes, Tara, and Brisbane at the corners. In echidnas caught in this area one finds every gradation from short fine hair to short thick bristles (Table 3). Although the ratio of claw 2 to 3 is 2.20 identical to that in populations of echidnas found elsewhere, the range was very large (Table 3). It would seem that these indeterminate

Figure 27. *T. a. multiaculeatus*; hindlimb. × 0.82.

echidnas are the result of hydridization between *acanthion* and *aculeatus* and one could postulate that this is due to the accepted notion [see Braithwaite and Miller (1975) for a recent discussion] that the hybrids are a consequence of union of two closely related but usually allopatric subspecies, that because of their allopatry, have not evolved isolating mechanisms. This morphological evidence of a zone of hybridization is supported by evidence from a study of hemoglobin polymorphisms (see p. 70) in populations of echidnas.

Examination of a score of echidnas from Western Australia collected in the drier parts of the state north and east of Perth led to the conclusion that they could not be distinguished from echidnas in the arid parts of New South Wales, Queensland, etc. I should suggest, therefore, that these echidnas be referred to the subspecies *acanthion* and that Thomas' (1906) subspecies name *ineptus* for them be dropped. So far I have not had an opportunity to look at echidnas from the southwestern part of Western Australia; I should imagine they will prove to have a well-developed pélage of hair and that hybrids between them and echidnas from the drier parts of Western Australia will be found. If it proves that hairy echidnas do occur in Western Australia I have a feeling they would be hard to distinguish from *Tachyglossus* found in tropical northeastern Queensland and New Guinea (Table 3). The latter have been given the subspecific name of

TABLE 3

Length of Claws on Digits 2 and 3 on the Hindlimbs of *Tachyglossus* from Various Regions of Australia and New Guinea

Region	No. of animals	Length of claws				Ratio of length of claw 2 to claw 3	Types of pélage and spines
		Left leg	Right leg	Left leg	Right leg		
Eastern Tasmania	16	4.2 ± 0.46	4.2 ± 0.50	3.6 ± 0.38	3.5 ± 0.37	1.20	Long thick pélage of woollike hair, short stout spines (Fig. 22)
Eastern and Southern Victoria	10	4.2 ± 0.55	4.2 ± 0.60	3.4 ± 0.50	3.4 ± 0.50	1.24	Long thick pélage of fine hair, short stout spines (Fig. 23)
Australian Capital Territory, Southern Highlands, and alps of New South Wales	20	4.1 ± 0.13	4.1 ± 0.11	3.2 ± 0.11	3.4 ± 0.10	1.24	Long thick pélage of fine hair, short stout spines (Fig. 23)
Kangaroo Island	20	3.2 ± 0.10	3.2 ± 0.10	1.4 ± 0.07	1.4 ± 0.05	2.38	Very long thick pélage of fine hair, many long thin spines (Fig. 24)
Fleurieu Peninsula (mainland very close to Kangaroo Island)	5	3.3	3.3	1.6	1.5	2.13	Short bristly hair, long stout spines quite distinct from those of Kangaroo Islanders
Rectangular area of northeast New South Wales and southeast Queensland with Tara, Brisbane, Port Macquarie, and Forbes at the corners	28	3.5 (range 2.6–4.4)	3.6 (range 0.8–1.8)	1.6	1.7	2.20	Every conceivable gradation from long, short, thick bristles to short fine hair. Spines long and stout
Northeast Queensland, central and western highlands districts of New Guinea	8	3.0	3.0	1.3	1.3	2.30	Well-developed pélage of long hair. Spines for the most part long and thin but not as long or thin as in Kangaroo Island echidnas
Arid parts of Queensland, New South Wales, Northern Territory, and South Australia	16	3.1	3.1	1.1	1.1	2.80	Sparse bristles to none at all, spines long and for the main part stout, sometimes thin (Fig. 25)
Western Australia north and east of Perth	20	2.7 ± 0.09	2.8 ± 0.09	1.2 ± 0.05	1.3 ± 0.05	2.20	Very short hair to sparse bristles or none at all. Spines for the most part long and thin, sometimes stout

lawesii and are considered to be confined to the island of New Guinea (Ramsay, 1877, 1878). However, I could not tell them from echidnas caught in wet tropical country at Iron Range and Rockhampton, so, for the present, I propose to refer to these northeastern Queensland echidnas as *T. a. lawesii*. Perhaps *lawesii* is but a hairy version of *acanthion* adapted to a rainy climate; it is interesting to note that Collett (1885) in his original description of *acanthion* was impressed with its likeness to *lawesii*. Furthermore the one sample of blood examined from a specimen of *lawesii* had hemoglobin polymorphs the same as those of *acanthion* (p. 71). What is needed at the present time to determine these matters is more data on morphology and hemoglobin polymorphs of echidnas living in northern tropical Queensland and New Guinea.

Distribution

As far as present day distribution of subspecies of *Tachyglossus* is concerned the range of *T. a. aculeatus* is in a wide sickle-shaped area of country ranging from south of Port Macquarie to Horsham in Victoria; the outer boundary of the area is the southeastern coast of Australia between the points mentioned and the inner boundary is roughly the Great Dividing Range and its western slopes.

Although *setosus* is confined to Tasmania and the larger Bass Strait islands, it is plentiful, especially in Tasmania. *Multiaculeatus* is confined to Kangaroo Island and it too is plentiful. *Acanthion* occurs literally everywhere in the arid parts of Australia and it appears to be plentiful even in the Northern Territory. During the period 1970–1973 Mr. Laurie Corbett collected 20 live specimens and nine scats from places all over the Territory including the Simpson Desert.* One record (live specimen) is of particular interest since the animal was taken at Groote Eylandt in the Gulf of Carpentaria but according to local aboriginals it does not occur on Melville, another large island off the coast of the Territory (Andrew Griffiths, personal communication). This is unexpected since Corbett found it on the mainland close to that island and it has also been taken on nearby Coburg Peninsula and other places in Arnhem Land. The location of possible hybrids of *acanthion* and *aculeatus* has already been noted.

Hobart Van Deusen kindly allowed Griffiths (1968) to publish his researches on the distribution of *T. a. lawesii*, as a personal communication. Since then Van Deusen and George (1969) have published a detailed history of the discovery and of present day distribution of this echidna. It is found in a broad area of country stretching from Merauke, parallel to the Gulf of Papua, around to Rigo. The northern boundary of the area is roughly a line drawn through Merauke, Lake Daviumbu, Mount Bosavi (formerly Mt. Leonard Murray), vicinity of Bela in

*I am indebted to Mr. Laurie Corbett of the Division of Wildlife Research, CSIRO, for permission to publish his results on its distribution there.

the Mendi Valley, Baiyer River area, Jimmi River area, Goroka area, Krathke Mountains of the Eastern Highlands District, vicinity of Port Moresby, and the Rigo subdistrict. It is still reasonably common at the latter places and is sold at Koki Market, Port Moresby as food. It has not yet been reported as living in the administrative region of Papua known as the Gulf District, but no doubt it occurs there in the hills and mountains inland from the coast. I have already stated my opinion that *lawesii* occurs in northeastern Queensland.

Movements

Of 67 echidnas banded between October 1963 and May 1966 22* have been recaptured (Griffiths, 1968, 1972). The results are summarized in Table 4. It is shown there that the echidnas were released at three different places, two of which, Mt. Majura and Mt. Tidbinbilla, are known to harbor echidnas living naturally, and the third, Gungahlin, is cleared pastoral country unfavorable to echidnas. Only one of the echidnas was released at the place it was originally caught. This is a defect of the study but it was unavoidable since most of the echidnas that came to hand were picked up on roads or found in suburban Canberra gardens or in places obviously not their native habitats, so they were released at the three arbitrarily chosen places mentioned. The data at hand (last recapture was made December 19, 1970, so it looks as though no more data will emerge) substantiate the conclusions reached with less data by Griffiths (1968), i.e., if an echidna is released into a suitable habitat it will apparently stay for a long time in a small area around its point of release. However, one animal released at Mt. Majura moved 6 miles from point of release but it took over 5½ years to do so. This animal weighed over 5000 g when first captured and was probably at least 4 years old at that time. The evidence for this estimate of age is that one animal (Table 4) weighed 1910 g when first captured but after release it was recaptured 3.8 years later when it weighed 4600 g; hence the animal mentioned above had lived, in all probability, for 10 years in the bush.

Another defect of this type of study is one does not know what the echidnas have been doing during the time between capture and recapture. This has largely been corrected by Augee *et al*. (1975) who have followed the movements of echidnas within the habitats they were caught and have obtained information on individuals movements by radio-tracking. At capture the echidnas (18 in all) were marked and fitted with a transmitter weighing ca. 5–10% of the body weight. Of the 18, 13 were subsequently recaptured one or more times over a period of 25 months, the study area being visited four times a year, each visit lasting at least 2 weeks. The results are summarized in Table 5. From this it is seen that the maximum distance between sightings for an individual was 613 m

*Not 23 as previously published by Griffiths (1972).

TABLE 4

Movements and Body Weight Changes of Echidnas Banded and Released at Three Different Localities in the Australian Capital Territory

Place of release	Sex	Initial body wt. (g)	Distance between point of release and point of recapture (miles)	Time taken for recapture (days)	Body wt. at recapture (g)
Mt. Tidbinbilla	♂	—	0.5	350	5877
	♂	4200	Practically nil	346	4000
	♀	3875	Practically nil	40	Not read
	♂	4075	Practically nil	22	3500
	Same animal		Practically nil	28	3425
	♀	4250	0.5	818	4770
Mt. Majura	♀	4025[a]	0.4	509	5000
	♀	2850	Practically nil	240	3550
	♀	3450	0.3	475	3250
	♂	5270	6.0	2064	5364
	♀	4900	0.5	992	4020
	♀	3525	Practically nil	1055	3600
	♂	1950	0.2	271	2522
Gungahlin	?	1200	1.5	5	1159
	♂	2260	0.5	39	2780
	♂	1503	1.4	5	1423
	♀	1660	1.0	27	1817
	♂	1910	11.0	148	1590
	Same animal		4.5	1388	4600
	♂	1325	1.5	5	1325
	♂	1600	2.0	300	1425
	♂	1925	3.4	7	2025
	♀	2670	1.3	830	3750
Gungahlin or Mt. Tidbinbilla	—	—	9 or 10	1402 or 1440	—

[a] Originally caught at Mt. Majura.

and the greatest time interval between first and last sightings for two individuals was 25 months.

The authors found that seven of these animals remained in the area for at least 1 year. The number of sightings varied seasonally: 18 in February, 11 in May, 14 in August, and 27 in November.

The movements of individual echidnas were also studied for 10-day periods and some fascinating information surfaced. One animal tracked in late autumn–early winter, when ambient temperatures were mild, "overnighted" in shallow depressions in the earth under overhanging bushes. It was invariably found in

TABLE 5

Echidnas Recaptured at Least Once after Having Been Marked in the Study Area[a,b]

Sex	Date first captured	Diameter of range (m)	Maximum interval (months)
♂	November 1970	650	25
♂	November 1970	366	25
♀	February 1971	297	9
♂	February 1971	336	15
♂	February 1971	613	22
♂	February 1971	78	12
♀	August 1971	467	12
♂	August 1971	476	16
♀	November 1971	321	3
♂	February 1972	279	5
♂	February 1972	154	3
♂	May 1972	238	8
♀	May 1972	327	7

[a] Diameter refers to the greatest distance between locations; maximum interval to the time between first and last sightings.
[b] Data from Augee *et al.* (1975).

low sandhills and the overnight positions were always on the side of the hill exposed to the sun (north). Two others tracked during the winter months used more protected sites during the night: one of them sheltered under tussocks of the native "iris," *Patersonia glauca*, the thick foliage of the plant affording complete protection from rain (it rains a lot on Kangaroo Island in winter). The other animal sometimes used tunnel burrows for protection from inclement weather. Three burrows in all were used, each having been dug into the bases of the mound of a termite *Nasutitermes*. Other echidnas made use of hollow logs for shelter. The results of the short-term trackings are summarized in Table 6.

The maximum distance between sightings is taken as a measure of home range and the authors point out that the data in Table 5 indicate that the home range remains fixed for at least 2 years. The total number sighted in the study area (column 4 in Table 6) greatly exceeds the total number bearing radio transmitters, which indicates considerable overlap of home ranges. This point brings me to a minor criticism of the study. This large number of echidnas coming and going in a 20-ha area of bushland is an artifact due to continual stocking (unknown to the authors) of Flinders Chase with echidnas brought in from other parts of the island. This has been going on to my knowledge for 11 years and it probably has been going on for two decades. However, the study area is a natural habitat and the marked echidnas were released back into their own "backyards" so to speak. The results of the two studies described here lead to the conclusion

TABLE 6

Echidnas Tracked for 10 Days Each by Radio Telemetry (1972)[a,b]

Sex	Month of tracking	Diameter of range (m)	Known conspecifics in home range	
			Female	Male
♂	May	838	3	5
♀	July	750	2	10
♂	July	1067	3	4
♂	December	751	3	7
♂	December	636	3	10

[a] Diameter refers to the greatest distance between locations.
[b] Data from Augee *et al.* (1975).

that echidnas, even in dense populations, maintain home ranges when released into suitable habitats, and in general will stay put for years.

Genetics

Karyology

Bick *et al.* (1973) have confirmed the finding of Bick and Jackson (1967a,b) that the chromosome number is $2n = 63$ ♂, 64 ♀. In both sexes 56 of the chromosomes could be accounted for as 28 homologous pairs, but only 26 bivalents were detected at meiosis in the males. It was thought that a small multivalent observed was a trivalent but it has been found (see Bick and Sharman, 1975) that it is a bivalent made up of the α and β chromosomes as in *Ornithorhynchus*. These mitotic chromosomes vary greatly in size with every gradation from macro- to microchromosomes, the smallest pair being almost at the lower limit of the resolving power of the light microscope (Fig. 21B). Microchromosomes have not been detected in other mammalian cells but they are well known in the chromosome complements of Sauropsida. At mitosis two pairs of chromosomes can be seen in the karyotype of the female but only one of these pairs occurs in the male so they are deemed to be X chromosomes. The largest submetacentric pair is called X_1 and the smaller acrocentric pair is called X_2. The males have another submetacentric chromosome about two-thirds the size of the X_1 chromosome and since it is found only in males it is called the Y chromosome, thus the female sex determining chromosomes are $X_1X_1X_2X_2$ and the male sex chromosome constitution is X_1X_2Y (Murtagh and Sharman, 1977).

At metaphase of the first division of meiosis in male echidnas 27 bivalents and a multivalent chain of nine chromosomes can be detected. It is thought by Murtagh and Sharman that X_1X_2 and Y chromosomes are at the end of the chain and that the other six chromosomes (designated a, b, c, d, e, and f) are joined end-to-end by terminal chiasmata. Meiosis has not been studied in female echidnas but it is suggested that multivalent formation occurs accompanied by 29 bivalents, the chain consisting of six chromosomes not including the pairs of X_1 and X_2. Murtagh and Sharman surmise that regular segretation of the chain elements takes place so that males and females both produce two kinds of gametes: ♀ aceX_1X_2, bdfX_1X_2; ♂ bdfY, aceX_1X_2. Since only abcdef$X_1X_1X_2X_2$ and abcdefX_1X_2Y genomes were found in all 30 echidnas studied the authors advance, inter alia, the likely explanation that a gamete "recognition" mechanism might prevent penetration of ace eggs by ace sperm and of bdf eggs by bdf sperm.

Sex Ratio, Crural System

The sex ratio is a matter of debate that in part may be due to an inadequate criterion of which sex the echidna belongs to. The criterion is often only the presence or absence of spurs on the ankle, it being held that an echidna with spurs is a male. Over the years I have taken many juvenile echidnas that had small but sharp spurs on their ankles and that I had deemed to be males; some of these when dissected, however, proved to be females (Griffiths, 1968). In a remarkable statement Temple-Smith (1973) says that what I observed in the females was an empty sheath! I have also taken a female *Zaglossus* that had well-developed spurs (p. 75). Carmichael and Krause (1972) have described seasonal changes in the crural gland of *Tachyglossus*. They say it is "functional in the winter breeding months and involuted for the greater period of the year." At the height of activity the cells of the tubules are tall columnar and are packed with PAS-positive vesicles. Involution results in a reduction in the size of the gland from a diameter of 2.5 cm to 1 cm or less.

The apparent anomalous sex ratio of 2-3:1 in favor of males found by collectors in the past has been discussed by Bick *et al.* (1973). They feel that these ratios may be artifacts since the collections were made during breeding seasons only, "the animals being specifically sought (echidnas already fertilized and carrying intra-uterine or pouch eggs) may have occupied less accessible places and that hunters employed being unable to distinguish between the sexes collected in more accessible areas." Bick *et al.* themselves found that of 25 echidnas collected or observed in the bush since 1956 in various parts of Australia, 13 were females and 12 were males. I have had much the same experience; of 67 echidnas taken in a relatively small area, Canberra and the Southern Tablelands, and marked and released for a movements study (p. 65), 34 were diagnosed as males on the basis of their having a spur on the ankle, and 33 were females. Some of the former could have been females since as shown above some juveniles of that sex have a spur.

Recently in 1 week, October 26–November 3, 1973, i.e., after egg-laying and when the females would be expected to be carrying large pouch young or feeding large young in burrows (p. 299) and so to be out feeding often to keep up the supply of milk, 35 echidnas were taken at Kangaroo Island; 20 of these were females giving a sex ratio of 1.33 in favor of females. It may be predicted from the above evidence that if one collects large numbers at the right time of the year the sex ratios of echidna populations will prove to be close to unity.

Proteins

From electrophoretic analysis it has been found that the blood of *Tachyglossus* specimens collected from various parts of southeastern Australia, including Tasmania and Kangaroo Island, has two hemoglobins—HbI and HbII. Samples from New Guinea *Tachyglossus* and Western Australian echidna also exhibited HbI and II; polymorphs of both these kinds of hemoglobin have been detected (Cooper *et al.*, 1973). By analogy with other polymorphisms studied in animals from which genetic data can be obtained, it was assumed by Cooper *et al.* that the variation seen in HbI and II of echidnas is under genetic control. It was found that HbI and II have three polymorphs each—A, B, and C and that HbII variation occurs mostly in Tasmania where there are two polymorphic forms A and B with some *setosus* exhibiting HbIIAB. Variation of HbII, on the other hand, on that limited part of the mainland from which samples were collected, and on Kangaroo Island, is uncommon, almost all animals having HbIIA. All the Tasmanian animals tested had only type B for HbI. However HbI polymorphism did occur on the mainland and on Kangaroo Island. The results so far suggest that in the southeastern corner of the continent and adjacent islands there are four geographical groups of echidnas:

1. Tasmanian, characterized by being all HbIB and polymorphic for HbII.
2. Southeastern mainland group with a high frequency of HbIB. Apparently HbIA and HbIC are absent from this area but a larger sample would be necessary to determine the matter.
3. Two hybrid groups that contain all three HbI polymorphs. One of these groups is found in southern Victoria and could extend as far as Adelaide. The other hybrid group is found extending from north of Sydney at least as far as Narrabri and Port Macquarie. It will be recalled that the morphological evidence described above also suggests that there is a zone of hybridization which includes Port Macquarie and Narrabri.
4. A group from the rest of the continent, including Western Australia, with high frequencies of HbIA but no HbIB. The one echidna from New Guinea had HbIA. With the exception of one animal from Kangaroo Island and five out of a sample of 196 mainland echidnas all had HbIIA and so did the New Guinea animal.

These groupings based on hemoglobin types are reasonably consistent with the groupings derived from the morphological study, considering the vastness of the areas involved and the relatively small numbers of animals examined.

It would appear that the great majority of individuals of each subspecies have the following hemoglobins:

	HbI	HbII
T. a. setosus	B	A, AB, C
T. a. aculeatus	B	A
T. a. multiaculeatus	A, AC	A
T. a. acanthion	A	A
T. a. lawesii	A	A (one sample only)

Dodgson *et al.* (1974) have commenced a study of the amino acid sequences of the various polymorphs of hemoglobins of echidnas from different geographical areas. At present information is available on the sequence of the α and β chains of HbIIA (Thompson *et al.*, 1973) and HbIB in *T. a. aculeatus,* and on sequences in HbIA from an echidna caught in the Murchison district in Western Australia which I diagnosed as *T. a. acanthion.* The results may be summarized as follows: there are four differences in amino acid sequence between HbIA and B, while there are nine differences of HbIIA from HbIB and ten from HbIA. Concerning those differences Dodgson *et al.* stated that, at present, there is no evidence to suggest that they offer any functional advantage to the animals (see p. 111).

ZAGLOSSUS

Kinds

Max Weber (1888) was the first to observe that some *Zaglossus* specimens had five claws on the forefoot rather than three. This observation apparently sparked off a feverish search for more variation, so that by 1911, Kerbert (1911, 1913a,b) could list four subspecies of *bruijnii* and a new species *goodfellowi*. Later Thomas and Rothschild (1922) gave a key for the identification of two species and eight subspecies; two more species were added to the score by 1952. However, Van Deusen and George (1969) have substantiated Van Deusen's stand (cited in Griffiths, 1968) that only one species of long-beaked echidna, *Zaglossus bruijnii*, which is endemic to New Guinea should be recognized until such time that the range of variation of the characters used to distinguish the named (four) species be determined. The characters in question, color of hair, color, shape, size and distribution of spines, and the number of claws on the limbs, are almost the same as those used to determine subspecific status in *Tachyglossus*. One of the most obvious of the differences between specimens is the variable number of claws found. Allen (1912) found that in 25 specimens from the western end of the island that, although the animals are pentadactyl, digits one and five on fore- and hindlimbs lack claws. He used a shorthand notation to express this claw configuration:

$$
\begin{array}{ccc}
04320 & & 02340 \\
L & & R \\
04320 & & 02340
\end{array}
$$

with zero indicating a claw absent on digits 1 and 5. In addition to the 25 with that claw configuration six others were different:

1.		Claws on all five digits on all limbs	
2.		54321	12345
	L		R
		54320	02345
3.		54320	02345
		54320	02345
4.		04320	02340
		54320	02345
5.		04320	12340
		04320	02340
6.		54321	02340
		04320	02340

However, in a collection from the eastern end of the island 13 out of 14 examined (Thomas, 1907a,b; Thomas and Rothschild, 1922; Laurie, 1952; Van Deusen and George, 1969) had five claws on all limbs, but the 14th (one of Van Deusen and George) which came from near the border between Papua New Guinea and Irian Jaya, had the claw configuration:

	54321		12345	
L				R
	04321		12340	

Zaglossus specimens with three or four claws on their digits were deemed to be *Zaglossus bruijnii* by Thomas and Rothschild; a variety of this type of echidna, less hairy and with spines extending round to the ventral surface, was called *Acanthoglossus (=Zaglossus) goodfellowi* by Thomas (1907b). Echidnas with five claws were called *Zaglossus bartoni* by Thomas and Rothschild and a color variant of this kind was called *Zaglossus bubuensis* by Laurie (1952).

In 1972 and 1973 I had opportunities to examine 13 live *Zaglossus*, one pickled in alcohol, and one skin. The provenance of these animals was as follows:

Pangia (Southern Highlands District near Mendi)	3
Wau (Morobe District)	3
Ioma or nearby mountains (Northern District collected by Sir Hubert Murray the Lieutenant Governor)	1
Mt. Tafa near villages of Umboli and Mondo (Central District)	7
Mt. Suckling (Central District)	1

Three of these animals (one from Mt. Tafa, two from Pangia) had claw configurations:

$$
\begin{array}{cc}
54321 & 12345 \\
\text{L} \quad\quad & \text{R} \\
04321 & 12340
\end{array}
$$

the rest were five-clawed. Thus out of 29 animals from the eastern end of the
island four had anomalous claw configurations and one of those came from well
into eastern Papua (Mt. Tafa).

As far as color as a specific criterion is concerned all I can say is that one animal (claws on all digits) I took at Mt. Tafa was identical to a three-clawed *Zaglossus bruijnii* illustrated by Gervais (1877–1878); both had a dark bluish grey
pélage of fur on the back with fawn-colored patches on the ventral surface and
fawn-colored paws.

It is indisputable that the majority of individuals examined from the eastern
end of the island were five-clawed and at the western end they were three-clawed;
presumably this is due to genetic difference. In between these two geographical
extremes there appear to be hybrids. I should suggest, therefore, that we go back
to Rothschild (1913) and call the three-clawed echidnas *Zaglossus bruijnii
bruijnii* and *Z. b. goodfellowi*, the five-clawed, *Zaglossus bruijnii bartoni*, and the
rest hybrids of the subspecies. It is not unlikely that a study of hemoglobin polymorphisms in specimens of *Zaglossus* from western, central, and eastern New
Guinea will substantiate the notions of genetic differences and hybridization.
The trouble would be to persuade the authorities of the two countries involved
to let one collect the animals.

It will be seen from the above that I do agree with Allen, and Van Deusen
and George, to the effect that only one species of living long-beaked echidna,
Zaglossus bruijnii, be recognized.

Mahoney and Ride (1975) as already mentioned have made an inventory of
fossils referred to the Monotremata: *Echidna gigantea, Zaglossus hacketti, Z.
harrisoni, Echidna owenii, Echidna ramsayi, Echidna robusta,* and *Ornithorhynchus maximus.* The validity of these species is being determined by J. A.
Mahoney (personal communication) who believes that *O. maximus* and *E. robusta* are the same species and would call it *Zaglossus robustus*; *Zaglossus
hacketti* is a good species but Mahoney says of it "when and if, cranial material
of it is found it might demonstrate that it represents a new genus." *Echidna
owenii* and *E. gigantea* are the same species (Mahoney and Ride); the fossil is a
fragment of a right humerus lacking specific characters. Mahoney is of the opinion
that *Z. harrissoni* and *E. ramsayi* could be the one species and that that species
is probably *bruijnii.*

It may be mentioned here that a tooth referred tentatively to the Monotremata

(Stirton *et al.*, 1967) and named *Ektopodon serratus* is now known to be a tooth from a phalangeroid marsupial (Woodburne and Clemens, 1978).

Recently, well-preserved remains of *Zaglossus* have been found in South Australia (R. Wells, personal communication) and in Tasmania (Murray, 1976). The latter are the remains of a partially articulated skeleton of a *Zaglossus* found in a deposit of Late Pleistocene age. The specimen consists of a skull lacking zygomatic arches, axis and cervical vertebrae 3-4; thoracic vertebrae 5-14; lumbar vertebrae 1-3 or 2-4; one sacral vertebra; episternum and clavicles, fragment of the sternum and several broken ribs; right and left humeri; two right and one left scapulae (recovered from the floor of the deposit); right and left femurs; two fibulae; right and left tibias and right and left innominate bones. The author inclines to the view that these bones are similar to those of *Zaglossus bruijnii* but comparisons with that species, *hacketti*, and *robustus* are being carried out to determine whether or not the new find is a new species (see Appendix).

It appears from the fossil record that there were plenty of *Zaglossus* around in Pleistocene times, in fact Gill (1975) is of the opinion that *Zaglossus* was more common than *Tachyglossus* was in those times. However, *Zaglossus* went extinct in the late Pleistocene but *Tachyglossus* survived. My own feeling about the extinction of *Zaglossus* in Australia is that it was due to the animal's extreme dietary specialization: *Zaglossus* is an earthworm eater and not an anteater (see p. 96). I suggest, therefore, that the aridity of the continent in the late Pleistocene (Bowler *et al.*, 1976) led to a decline in the food supply of *Zaglossus* leading to starvation and extinction. Gill (1975), however, infers that the extinction was the result of an adaptation of *Zaglossus* to humid conditions and so presumably to inability to cope with an arid climate. Yet the Tasmanian *Tachyglossus* survived those conditions of low humidity. I suspect that it did so because it could catch ants and termites—items that the *Zaglossus,* owing to extreme anatomical specialization of its tongue for the purpose of catching earthworms (see p. 101), could not catch in numbers adequate for survival. *Zaglossus* survived in New Guinea, however, since it did not become as arid as the southern part of the land mass did (Bowler *et al.*, 1976) so, presumably, an adequate supply of earthworms was available.

Distribution

The geographical range of *Zaglossus* is from the Vogelkop Peninsula in Irian Jaya to Mt. Simpson in eastern Papua (Van Deusen, *fide* Griffiths, 1968; Van Deusen and George, 1969; R. Schodde, personal communication). Its altitudinal range according to Van Deusen and George is ca. 1200–2800 m. However, Temple (1962), J. H. Hope (1976), and G. S. Hope and Hope (1976) cite evidence of its presence in alpine meadows, at an elevation in excess of 4000 m at the base of Mt. Jaya (Mt. Carstensz) in Irian Jaya. This area has a cold wet climate where freezing commonly occurs above 3600 m at night but the moist misty air usually prevents very heavy ground frosts (Schodde *et al.*, 1975). These authors found that a hot bright sun may shine on the mountains in the morning but by 10 A.M. or 11 A.M. clouds form and rain falls. Precipitation in the form of snow may occur down to 4000 m. However in the rest of the island *Zaglossus* is an inhabitant of humid montane forests, warm during the day but can be freezing cold at night. The fact that *Zaglossus* can live in such a harsh climate as that found at the Carstensz Toppen argues that it will prove as good as or even better than *Tachyglossus* at thermoregulation. In my experience earthworms are abundant in the montane forests—it would be of great interest to compare the earthworm population of the New Guinea alpine meadows with those of the forests. Incidentally, G. S. Hope (1975) found no sign of *Zaglossus* on the high plateau of Mt. Albert Edward at about 3650 m and he suggests that this may be due to heavy hunting pressure.* However, at lower altitudes in the Wharton Range they seem to be reasonably abundant.

Van Deusen and George (1969) discuss two equivocal locality records of *Zaglossus* in Irian Jaya. One of these was Sorong which is a small island off the coast of the Vogelkop. The specimens were said to come from the hills on the nearby mainland but Van Deusen and George suggest they came from the Tamrau Mountains of the Vogelkop. The other record concerns the provenance of *Zaglossus bruijnii goodfellowi*. Thomas (1907b) believed this echidna came from the island of Salawati which is separated from the Vogelkop by a narrow strait of shallow water. Bergman (1961) was informed by the Rajah of Salawati that the animal did not occur on the island but lived on the Tamrau Mountains. Of this Van Deusen and George comment ''There continue to be many surprises in mammal distribution on New Guinea, however, and therefore it is dangerous to summarily dismiss records such as the Salawati one even though we know that there is much coastal trading by natives between mainland localities and the islands west of the Vogelkop Peninsula.''

Crural System

Of eight *Zaglossus* specimens I have examined and whose sex was determined by dissection, six were males and had spurs on the ankle. Of the females, 7.4 kg and 4.5 kg in weight, respectively, the smaller one, which had never bred (see p. 249) had spurs; this animal when first examined was deemed to be a male but dissection proved it to be a female. The other female, the larger one, had bred and she had no spurs. This suggests that juvenile female *Zaglossus* have spurs just as some juvenile *Tachyglossus* females do.

In the males the spur protrudes from a deep pit, the distal edges of which have a reddish inflamed appearance; at the base of the spur an exudate not unlike pus is

*J. H. Hope (1977) reports that apparently *Zaglossus* ''no longer survives on the eastern slopes of Mt. Wilhelm above the densely populated upper Chimbu valley though it was hunted there within living memory.''

encountered. In the adult female mentioned above a deep pit was present on each ankle and these were inflamed as in the males.

TABLE 7

Chromosome Numbers and Sex Determining Mechanisms of Monotremes[a]

	Chromosome number	Mechanism of sex determination	At meiosis	
			Length of chain	Number of bivalents
Tachyglossus	64♀/63♂	$X_1X_1X_2X_2♀/X_1X_2Y♂$	$6+X_1X_2Y$	27
Zaglossus	Presumed same	Presumed same	$6+X_1X_2Y$	27
Ornithorhynchus	52♀/52♂	XX♀/XY♂	$8+XY$	21

[a] Data from Murtagh and Sharman (1977). Reproduced with permission of Australian Academy of Science.

Temple-Smith (1973) studied the histology of one of the crural glands in one of the male *Zaglossus* mentioned above.* He found that the diameters of the aveoli and the amounts of interalveolar connective tissue were the same as in the platypus gland but the stored secretion was different in that it was basophilic. He also found that in this particular sample of *Zaglossus* gland there was a massive invasion of leucocytes throughout the connective tissue, gland cells, and lumina of the alveoli. It will be recalled similar infiltrations were found in platypus crural glands in the regression stage.

Genetics

Karyology

Work in progress indicates that *Zaglossus bruijnii* has a complex chain multivalent and 27 bivalents at first division of meiosis (Murtagh and Sharman, 1977) and the chromosome number is probably the same as in *Tachyglossus*, $2n = 63$ ♂, 64 ♀. Mitosis has not been observed yet but since there is a great similarity between meiosis in *Tachyglossus* and *Zaglossus* it is presumed that the diploid numbers are 64 ♀ and 63 ♂.

Thus all three living monotremes exhibit a complex multivalent which suggests that it was inherited from a common ancestor. It also suggests that similar methods of sex determination are present in all three genera. However, Professor Sharman tells me that no one will know what the sex determination mechanism really is until the results of an extensive program aimed at determination of homology and of X chromosome inactivation come to hand. The present views of Professor Sharman and his colleagues are summarized in Table 7.

*This animal had well-developed testes which exhibited full spermatogenesis.

3

Food and Feeding Habits: Digestive Organs and Digestion

ORNITHORHYNCHUS

Food and Feeding Habits

From time to time over the last 150 years observations on the food of the platypus have been published; a wide diversity of food items has been recorded but no accounts have been the same. This is doubtless due to the facts that the observations were opportune, were made at different times of the year, and that the freshwater ecosystems in which the platypuses were feeding ranged from the cold rivers of Tasmania to the warm rivers, creeks, and lagoons of northern New South Wales and eastern Queensland, the aquatic faunas of which are different. Thus Bennett (1835) found in the cheek pouches of Yass River (southern New South Wales) platypuses the comminuted remains of very small shellfish (presumably gastropods since the organisms were said to be adherent to water weeds) and insects along with mud and gravel. Semon (1894a) reported that the main food item of platypuses in the Burnett River in Queensland was the immature stages of a bivalve mollusc, *Corbicula nepeanensis*, along with some insect larvae and oligochaete worms. Allport (1878) found that caddis fly larvae were the main food item of platypuses in the Huon River (Tasmania) while Crowther (1879) reported that the food of Tasmanian platypuses is shrimps (*Anaspides*), water fleas, and "hard, black beetles." In passing it might be said that *Anaspides tasmaniae* will never be found in the cheek pouches of platypuses outside Tasmania since it lives only in a few streams of that island. Finally, Burrell (1927) summarized his observations of the food with the following list: immature molluscs, aquatic oligochaetes, the larvae of dragonflies, water bugs, and crustacea. He agrees with Bennett that mud and gravel ingested probably help to grind the food.

Quite recently systematic studies of the food of the platypus in the Shoalhaven River (southern New South Wales) have been commenced by F. N. Carrick, T. R. Grant, and R. Faragher, working as consultants for the Australian National Parks and Wildlife Service. The study involves regular sampling of the benthos of the river by the use of a dredge and of organisms adhering to the upper and lower surfaces of rocks and stones. The sampling will be accompanied by simultaneous identification of the parts of organisms found in the cheek pouches. This is a feasible way of getting a gross qualitative assessment of the food and of food preferences; a method such as examination of gut contents I have found to be useless since most of the identifiable parts of insects, molluscs, etc. never reach the gut as it is the practice of the platypus to eject them. It has already been mentioned that the platypus feeds underwater* and any prey snapped up is transferred to the cheek pouches along with adventitious sand, gravel, and organic detritus. When the pouches are full the platypus rests on the surface of the water; the contents of the pouches are transferred to the buccal cavity where they are comminuted by the grinding action of the horny pads on the maxillae and lower jaws, but the identifiable hard parts of the crustaceans, insects, and molluscs are ejected into the water through the series of horny serrations arranged along the margins of the lower jaw (see Appendix).

There is good evidence that overall food supply in seven different rivers in southeastern New South Wales varies seasonally: Temple-Smith (1973) found marked changes in body weight–body length regressions associated with the breeding and nonbreeding seasons. Differences in the regression lines found at those different times of the year are a direct result of variation in weight since it was found that significant variations in the length of platypuses taken at different times of the year were not detectable. The changes in body weight were found to be the result of fluctuations in the body fat reserves. The tail is the principal fat storage depot in platypuses and the changes in bodily condition are reflected in changes in the amount of fat in the tail. To express this quantitatively Temple-Smith calculated a tail-fat index—the ratio of the cross-sectional area of the midpoint of the tail to tail length. The highest mean tail-fat indices are found in the summer months and a gradual decline sets in in June culminating in low values from October–November (Fig. 28). It is suggested from these data that a decline of food supply occurs in the late winter months. This coupled to an increased demand for energy to maintain body temperature in cold water (see p. 123) brings about depletion of the fat reserves. It would seem unlikely that this seasonal waxing and waning of fat reserves would take place in the fat reserves of platypuses living in the warmer waters of Queensland.

---

*Occasionally platypuses eat items floating on the surface such as cicadas, *Melampsalta denisoni*, which often fall into the water out of trees lining the river banks.

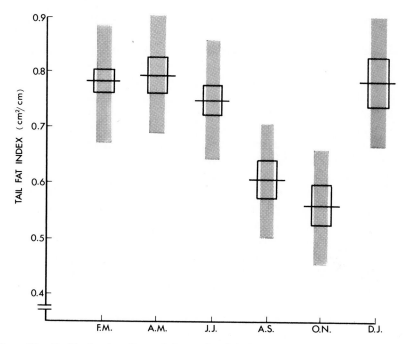

Figure 28. *Ornithorhynchus.* Seasonal changes in tail-fat index in adult females. *(From Temple-Smith, 1973.)*

Finally it should be noted that platypuses, as Temple-Smith points out, may gain some nutritional benefit from ingestion of organic detritus in the sediments of the rivers and lagoons. This is always a prominent item in the gut of the platypus and it is known that sediment contributes to the sustenance of the Bony-bream, *Fluvialosa richardsoni* (Lake, 1966).

The Alimentary Tract

The esophagus, lined by stratified squamous epithelium, leads into a very small stomach about the size of the end joint of one's little finger, also lined by stratified squamous epithelium (Oppel, 1896); there are no glands of any description so there is no peptic digestion in the platypus, nor in any monotreme for that matter. Schultz (1967) has made a study of the arterial and venous blood vessels to stomach and gut, which will prove to be of prime importance for anyone undertaking research on the absorption of nutrients from the alimentary canal.

Krause and Leeson (1974) find that the stratified squamous epithelium of the stomach exhibits three principal layers: stratum germinativum, stratum spinosum, and stratum corneum. The cells of the stratum germinativum contain free ribosomes, rough endoplasmic reticulum, a few mitochondria, scattered bundles of tonofibrils, and a few granules of varying electron density. These cells lie on a thin basal lamina and adjacent cells are connected by many hemidesmosomes. The cells of the stratum spinosum immediately above the germinal layer are elongated and the cell membranes are extensively interlocked. A few poorly developed intercellular bridges between the germinal layer and the stratum spinosum are apparent in this region. At the surface in the stratum corneum all cells are flattened and exhibit well-preserved nuclei; the number of tonofibrils, however, is much less than in the cells of the stratum germinativum. The cytoplasm of these cells is pale and exhibits large granules of varying electron density. Connective tissue papillae containing capillaries extend into the three layers almost to the surface of the stomach lining, throughout the organ.

There are no glands in the stomach but there are sets of Brunner's glands confined to the distal portion of the stomach (Krause, 1971a). Brunner's glands occur only in the Mammalia (both eutherians and metatherians) so their presence and their structure in monotremes is of phylogenetic interest. The glands are confined to the submucosa of the distal part of the stomach; each gland is made up of several elongated lobules which run parallel to the long axis of the stomach. The ducts draining the glands terminate at the junction of the stratified squamous epithelium and the epithelium of the small intestine. In eutherian mammals the glands are located in the submucosa of the duodenum but in *Didelphis virginiana* the ducts discharge into large stomata lined by either gastric or duodenal epithelium. The cells of the glands in the platypus fail to stain with Alcian Blue but they are periodic acid-Schiff (PAS) positive indicating that the secretion is a neutral mucopolysaccharide. The secretory parenchyma of Brunner's glands in the echidna, however, does stain with Alcian Blue as do those glands in rabbits and guinea pigs indicating the presence of acid mucopolysaccharides.

Electron microscope studies of the secretory parenchyma in the platypus show that the glands are serous rather than mucous. The serous nature of the glands in the platypus, however, is not unique since those of the mouse, rabbit, and opossum are likewise serous and not mucous. The characteristics of the serous glandular cells are: an abundance of secretory granules of varying electron density, granular endoplasmic reticulum, and many well-developed Golgi complexes.

The process of secretion of mucopolysaccharide by the cells of Brunner's glands has been largely worked out by Schmalbeck and Rohr (1967) in the mouse: the protein moiety is synthesized by the ergastoplasm and transported in small vesicles to the cisternae of the Golgi complexes where coupling of the protein to carbohydrate is carried out. Apparently the same processes take place in the platypus gland since numerous membrane-bound vesicles take their origin from areas of granular endoplasmic reticulum containing smooth membranes and they come into close association with nearby Golgi complexes.

Just what Brunner's glands, unique to mammals, achieve for the members of the Class is still debated. A concensus has it that they secrete an alkaline fluid containing mucin which acts to protect the proximal duodenal mucosa from the ulcerating effects of acid-pepsin secreted by the stomach. However, monotreme stomachs secrete no acid-pepsin; Krause suggests the possibility that the glands have other functions such as secretion of an intrinsic factor influencing the motility of the intestine or secretion of constituents that play a role in forming an aseptic lining for the intestinal mucosa.

There is no pylorus, the squamous epithelium of the stomach ending abruptly without transition at the junction with the small intestine which is very short—a little over 1.7 m in length. Schulz (1967) found that the mucosa is unique in that it is devoid of fingerlike villi and that it is thrown into numerous transverse folds. Krause (1975) confirmed that observation and has given an account of the histology of the folds: their surfaces consist of a lamina propria covered by a pseudostratified epithelium interspersed with goblet cells. The submucosa does not contribute to any appreciable extent to the structure of the folds. These features, apparently, are unique to the platypus. The crypts of Lieberkühn are lined by a simple columnar epithelium; apparently Paneth cells are not present in the crypts but they do occur in the intestinal glands of *Tachyglossus*. Groups of intestinal glands are drawn into common tubular ducts wich follow a tortuous course through the lamina propria and finally empty into the lumen of the intestine at the spaces between the intestinal folds.

Nothing is known of the digestive enzymes of the pancreas nor of the intestinal glands. The fact that the intestine is so short suggests a fast rate of passage of food which may explain the putative voracious appetite of the platypus. I say putative since no one has measured the daily dry matter intake of the platypus.

There is one lonely fact related to digestion in the platypus: the bile salt is sodium taurocholate along with a trace of taurodeoxycholate (Bridgewater *et al.*, 1962).

TACHYGLOSSUS

Food

Griffiths (1968) found that the echidna living in the southern tablelands of New South Wales was largely an anteater but termites at times contributed substantially to its sustenance. The ratio of ants to termites ingested was determined by examination of scats for the remains of the chitinous exoskeletons of the prey. These pass through the alimentary canal, apparently undigested and retain morphological characters that allow them to be identified. Since that study scats have been collected from other localities on the mainland of Australia, Tasmania, and Kangaroo Island; the contents of the alimentary tract from one specimen taken at Waigani near Port Moresby, Papua New Guinea, have also been examined. The technique for determination of ratio of ants to termites ingested is as follows: chitinous parts in the scats are freed from dirt and sand by floating them off in water; an aliquot of a well-stirred suspension of the parts is spread thinly in a glass dish and examined with a microscope. The jaws of ants are quite distinct from those of termites so that the numbers of ant jaws and termite jaws can be counted along straight line random transects in the aliquot. The ratio of the number of ant jaws to termite jaws indicates the ratio of ants to termites ingested. This gives only a rough idea of the amount of ant flesh to termite flesh ingested since the sizes of the different species of ants eaten varies considerably. The results are summarized in Table 8. From this it is seen that at three localities, the Cunnamulla district of Queensland, Central Australia in the Northern Territory, and the Mileura district of Western Australia*, the echidnas are, in general, termite eaters. The climates of these three localities are characterized by extremely hot summers (temperatures can range up to 50°C in the shade) and by low average annual rainfall of uncertain incidence—in some parts no rain may fall over a period of years. However, the five other habitats listed in Table 8 have cooler and wetter climates; here the echidnas are predominately anteaters and are almost wholly so in the cold wet northeastern coast of Tasmania. The reason why termites are preferred in arid areas is not known; it appears not to be a question of availability since there is plenty of ants, indeed one scat from the Simpson Desert in Central Australia contained ants only. The preference for termites in hot dry climates may be related to water content or salt content of the insects. Analysis for sodium content of pooled samples of various species of ants and of termites collected at Cunnamulla showed no difference between the two kinds of insects but there was a difference in water content: termites had 74.4% and ants 64.1% water by weight. Since echidnas deprived of water can live and grow in some circumstances on a diet of the termite *Nasutitermes exitiosus* (see p. 116) which contains 74% of water it is possible that termites are preferred in arid country for their relatively high water content. However, where the echidnas were living at Cunnamulla water of good quality is available year round from artesian bores. It may also be a matter of palatability or even comfort—termites do not bite like ants can. However, in northeastern Tasmania there are very few species of termite, none are mound builders, and only those living in rotten wood, forest litter, etc., would be available to an echidna. Furthermore, although at Eucumbene mound-building termites can be found, at those cold high altitudes they are rare. So it would appear that the habit of eating ants is enforced in these two habitats by a lack of termites.

---

*My best thanks go to A. E. Newsome and S. Davies for the gifts of the Northern Territory and Western Australia scats, respectively.

TABLE 8

Frequencies, Expressed as Percentages, of Ants and Termites Found in the Feces of *Tachyglossus* from Various Parts of Australia

Locality, height above sea level (m), terrain, annual rainfall (mm)	Number of scats	Percentage	
		Ants	Termites
Cunnamulla district, Queensland; 180; plains sparsely vegetated; 355	50	18	82
Central Australia, Northern Territory; 240–600; plains to mountains sparsely vegetated; 170–250	12	29	71
Mileura district, Western Australia; 300–600; plains to mountains sparsely vegetated; 170	18	30	70
Grenfell–Forbes district, New South Wales; 300–600; plains to mountains with thin sclerophyll forest; 530	19	60	40
Kangaroo Island, South Australia; sea level 240; wet sclerophyll forest to stunted mallee; 480–640	158	71	29
Canberra district; 600–1000, rolling woodland savannah to sclerophyll montane forest; 630	41	77	23
Eucumbene district, New South Wales; 1300; subalpine forest; 790	9	97	3
Northeast coast of Tasmania; sea level 600; dense sclerophyll forest and open heathlands; 740–1070	12	98	2

Nutritive Value of Termites and Ants

It has been reported (Griffiths, 1968) that soluble carbohydrate and possibly the amino acids cystine and methionine are limiting nutritional factors for echidnas fed the termite *Nasutitermes exitiosus*. The possibility that sulphur amino acids are limiting was based on the observations that in pronase digests of termites and ants very small amounts of sulphur amino acids could be detected and that in a single experiment addition of cysteine hydrochloride to a diet of *Nasutitermes exitiosus* increased nitrogen retention. Subsequently, however, it was found in a further series of tests that addition of cysteine to the termite diet of the echidnas did not improve nitrogen retention and that cysteine was found to be present to reasonable quantities in acid hydrolysates of *Nasutitermes*; evidently sulfur amino acids are not limiting factors when the diet consists of this termite. The observation that carbohydrate in the form of glucose, was limiting for growth and nitrogen retention (Griffiths, 1968) was confirmed in a further series of experiments: in a total of eight experiments on six echidnas to date, addition of glucose to a fixed daily ration of *Nasutitermes* induced very substantial increases in growth and decreases in nitrogen excretion. A typical example of the effect is given in Table 9 (see also p. 95).

The needs of echidnas for vitamins are unknown but recently Drs. D.E. Dykhuizen and B.J. Richardson of the Research School of Biological Sciences, Australian National University, commenced a study of the ability of various mammals to synthesize ascorbic acid. These authors have kindly allowed me to quote some of their findings: homogenates made from livers and kidneys of the herbivorous marsupials *Trichosurus vulpecula*, *Macropus giganteus* and *Macropus eugenii* were found lacking in the capacity to form ascorbic acid from added D-glucuronate as measured by the 2-6 dichlorophenolindophenol method. On the other hand, the insectivorous marsupial *Antechinus swainsonii* could form ascorbic acid from that substrate in the liver but not in the kidney. This is similar to the situation in the Wistar rat: homogenates of liver, but not of kidney, synthesize ascorbic acid. On the other hand homogenates of *Tachyglossus* kidney could synthesize ascorbate, but liver homogenates could not, a circumstance identical to that found in homogenates of kidney and liver in amphibians and reptiles (see Appendix).

It is possible that *Tachyglossus* ingests vitamin C contained in the ants and termites of its diet; it would be of interest to determine ascorbate levels in those insects.

Little is known of the nutritive value of ants for echidnas. Griffiths and Simpson (1966) found that echidnas living in the mountains near Canberra attack the mounds of the ant *Iridomyrmex detectus* during August, September, and October of each year, and ingest the inmates. These consist of workers, males, and females. During the three months mentioned the males and females rise from the depths of the mounds and come nearer to the surface where they are easily

TABLE 9

Effect of Addition of Glucose to a Fixed Daily Ration of Termites (200 or 400 g termites/day) on Growth and Nitrogen Retention (urea + NH$_3$ nitrogen) in an Echidna

Number of daily observations	Initial body weight (g)	Avg. daily increase in body weight (g)	Avg. daily intake of nitrogen and glucose (g)	Avg. daily urea + NH$_3$ nitrogen secreted (g)
10	2335	Nil	2.15	1.45
10	2380	20	2.15+	1.04
15	2340	25	10 g glucose 4.30	1.84

accessible to echidnas. The females are especially large and their bodies are loaded with fat. Late in October the males and females quit the mound for the nuptial flight and as soon as this happens attacks on the mounds cease although they still contain innumerable worker ants. From these circumstances Griffiths and Simpson surmise the echidnas attack the mounds to get the fat-laden females.

Structure of the Tongue and Ingestion of Ants and Termites

The prey of *Tachyglossus* is picked up by a very long vermiform tongue, which is absolutely vital to the animal; without its use the animal would starve for this organ both catches and masticates the food. In *Tachyglossus* the tongue, which can be extruded up to 18 cm from the end of the snout, is lubricated with a sticky secretion; any ants or termites coming into contact with the tongue stick to it and are drawn back into the buccal cavity on retraction of the tongue. Inside the buccal cavity the insects are comminuted by the grinding action of a set of keratinized spines, first described by Home (1802b), on the dorsal surface of the base of the tongue against sets of transversely arranged spines on the palate. Retraction of the tongue is brought about by contraction of two internal longitudinal muscles that stretch from near the tip back through the base to take their origin at the xiphoid process of the sternum. Fewkes (1877) found that each of these two muscles, the sternoglossus, is made up of two moieties, the m. sternoglossus superior and m. sternoglossus inferior both arising separately on the undersurface of the sternum. The m. sternoglossus superior passes directly forward into the tongue "forming with its fellow, the interior of the organ." The m. sternoglossus inferior according to Fewkes unites, soon after it arises from the sternum, with the m. sternoglossus superior and is inserted into the base of the posterior part of the tongue. A slip of the sternoglossus inferior "or a small one bound up with it and the Sterno-glossus Superior becomes a separate muscle, passing from the larynx to the tongue, and may be known as a Laryngo-glossus." Thus the fibers of the sternoglossi in the tongue proper are of dual origin, mostly sternoglossal superior and partly sternoglossal inferior in the form of the laryngoglossal fibers. Some 96 years later Doran (1973), independently without reference to Fewkes, confirmed the finding that some muscle fibers pass from the sternoglossus to the hyoid bone and from here into the tongue. Apparently he made no observations on the origins of the sternoglossi at the sternum. He proposed that the sternoglossus be considered as consisting of hyoglossal and sternoglossal parts. However Fewkes' terms m. sternoglossus superior and inferior, and laryngoglossal muscle are clear-cut and are established in the literature, so they should be retained. Doran found that the function of the m. laryngoglossus is to retract the hyoid bone during swallowing.

The keratinous spines at the base of the tongue, mentioned above, are located on a trapezoidal pad called the lingual pad by Doran. Fewkes described the musculature responsible for pressing this lingual pad against the palate, and for that giving it anteroposterior movement, an action leading to comminution of the prey. Doran also gives a description of the muscles responsible for pressing the lingual pad against the palate, which is substantially the same as that of Fewkes'; Doran says:

> the tongue is connected directly to the mandible by the genioglossus muscle and indirectly to the cranium and hyoid bone by the styloglossus and annulus inferior muscles. The genioglossus is strap-like and passes from the genial region of the mandible (above the origin of the geniohyoid) to the undersurface of the trapezoidal lingual pad. It differs markedly from the common fan-shaped genioglossus of most mammals. It appears to initiate tongue protrusion. The styloglossus arises on either side, from the area of the cranium where the styloid process would be if present, and unites in the midline as a posterior extension of the myloglossal raphe. Some of its fibres are attached to the ventral aspect of the hyoid bone and some pass into the tongue under the lingual pad. The annulus inferior arises from the lateral aspect of the pterygoid bones and unites with its fellow in a midline raphe ventral to the hyoid bone but above the hyoid attachment of the myloglossus. Together with the styloglossus it forms a sling, ventral to the tongue, which when contracted, raises the lingual pad against the palate.

Doran mentions no mechanism for the grinding action of the lingual pad; Fewkes, however, says the function of the myloglossus is to combine with the annulus inferior in pressing the posterior part of the tongue against the roof of the mouth and that of the styloglossus, in part, is also to press that part of the tongue against the palate. He continues of the styloglossus: ''it may also serve to draw the whole dental portion of the tongue backward, combining its function with that of the Sterno-glossus. It is then an opponent of the Genio-glossi postici, and the Genio-glossi.'' Later he says the ''Genio-glossi'' and the ''Sterno-glossi'' act in different directions pulling the tongue backwards and forwards. This grinding of ''dental portion'' against the spines of the palate is quite audible and the sound, a scratching noise, has been recorded along with photography of the tongue in action in a CSIRO film ''The Echidna'' (1969). The end product of the mastication is practically a homogenate of the ingested insects.

The spines on the lingual pad and those on the palate are backwardly directed making it possible for the one set to slide past the other when the lingual pad is pressed against the palate and is pulled backwards and forwards by the muscles described above. Doran and Baggett (1972), however, have published an illustration of a sagittal section of an echidna head showing backwardly directed spines on the palate and forwardly directed ones on the lingual pad. If this were really so the action of the musculature would simply lead to locking of the pad to the palate thus eliminating any grinding action.

The spines are lubricated by mucus secreted by glands located in the dermis of the pad, the ducts passing upwards through the very thick epidermis to open at the bases of the spines (Doran and Baggett, 1972). The spines themselves have the structure of hairs and interestingly enough Doran and Baggett point out that they are very like the horny teeth of the rasping organ of the cyclostome, *Myxine*.

In an earlier paper Doran and Baggett (1970) describe the discovery of a vascular stiffening mechanism in the tongue of *Tachyglossus*. The sternoglossi in the tongue proper are separated from each other in the midline by a longitudinally aligned neurovascular element. This consists of a centrally placed thick-walled artery, a large nerve ventral to the artery and dorsally and ventrally located thin-walled vascular spaces. These spaces are supplied with blood from the central artery via a series of bulbous extensions out from the artery communicating with smaller channels which in turn lead into the dorsal and ventral vascular spaces. In addition to the above each sternoglossal muscle within the tongue has its own intrinsic neurovascular bundle consisting of artery, nerve, and vascular spaces. The stiffening of the tongue is achieved by engorgement of the vascular plexus: blood flows through the central artery to the bulbs and communicating channels and fills the vascular spaces; arterial pressure prevents backflow and the tongue is thus engorged with blood and becomes stiff. Doran and Baggett found that the tongue in this condition was strong enough to break open channels in termite-ridden wood. The stiffening of the tongue apparently does not impede protrusion and retraction since these authors found it could be protruded about 100 times per minute. Protrusion is brought about by contraction of a well-developed system of loosely arranged circular muscles (made up of striated muscle) located around the sternoglossi for a major part of their length (Fig. 29);

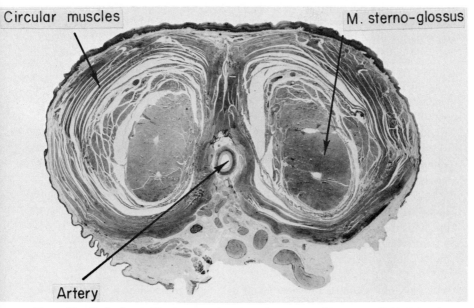

Figure 29. *Tachyglossus*. Transverse section of proximal region of the tongue. Heidenhain's iron hematoxylin. × 8.5. *(From Griffiths, 1968.)*

since the base of the tongue is fixed contraction of this circular musculature squirts the rostral part of the tongue forwards, all the more so when it is engorged with blood. The high rate of protrusion and retraction of the tongue together with its capacity for rapid movement, left, right, up, down, and to recurve on itself (CSIRO, 1969) helps to apprehend the prey scurrying in many directions at once. The muscles of the anterior end of the tongue bringing about these movements have been mentioned by Griffiths (1968); they were briefly described as radially arranged muscles invading the sternoglossi thus dividing these up ultimately into eight small longitudinal muscles. Since then it has been found the two sterno-glossi remain intact and that three pairs of longitudinal muscles arise *de novo* in the dermis of the anterior part of the tongue and pass forwards so that in transverse sections from near the tip one can discern eight longitudinal muscles in cross section (see Appendix).

Chemical Composition of the Sublingual Secretion

The muscular activities described above would be useless if the prey did not stick to the tongue. The sticky secretion already mentioned is elaborated by the paired sublingual glands; these are large flat lobulated structures of an elongated triangular shape, the narrow end pointing forwards. From this end emerges a duct that conveys the secretion to the buccal cavity. Paired discrete parotid and sub-maxillary glands are also present but the sticky secretion is formed only in the sublinguals. The salivary glands in eutherian anteaters are also very large but they are modified in different ways in different species: in *Myrmecophaga tridac-tyla*, the South American ant bear, the parotid and submaxillary glands of each side are fused into one gland of gigantic proportions stretching from the jaws all over the neck and chest. Of the closely related ant bear, *Tamandua tetradactyla*, Dalquest and Werner (1952) refer to its similarly huge salivary glands as parotids. They made the observation that the glands contained a thick transparent liquid of the consistency of Canada balsam. The product of the sublingual glands of the echidnas is of a similar consistency; if one touches a drop of the secretion with a finger and withdraws, the secretion will adhere and stretch into thin sticky strands up to 30 cm long.

Lew *et al.* (1975) have examined the chemical properties of the glycoproteins in the *Tachyglossus* sublingual gland product. The secretion was squeezed from the divided ducts into 95% ethanol and stored cold. With this treatment it became a hard white-colored solid. Pieces of the material were dried on filter paper, ground, and suspended on $0.5\,M$ sodium acetate of pH 5.6 and stirred for 4 days at $3°–6°C$. The insoluble matter (Preparation D) was centrifuged off and washed with buffer. Washings and supernatant were dialyzed against water and freeze-dried. Glycoprotein preparations from two separate batches taken from two groups of echidnas were made in this way. These glycoprotein preparations were purified by two methods: electrophoresis in polyacrylamide gel and by filtration on a column of Sephadex G-100. For the latter procedure the water-soluble glycoprotein material was dissolved in 0.1 M sodium acetate pH 5.6 and applied to a Sephadex G-100 column. The protein was located by sialic acid assay because absorption at 280 nm was insignificant owing to tbe low content of aromatic amino acids (Table 11). The elution pattern exhibited only one sharp peak of sialic acid. The sialic acid-containing fractions were combined dialyzed against water, and freeze-dried. This freeze-dried product was designated Preparation A. Purification of the glycoprotein in the freeze-dried product by electrophoresis was found to be possible only with one lot of the ethanol-treated solid.* The freeze-dried product from this lot was dissolved in 4 M urea and subjected to electrophoresis in polyacrylamide gel. This led to the separation of more than seven proteins as indicated by staining the gel with Coomassie Blue. Two of these bands corresponded with two bands stained with Schiff's reagent for carbohydrate. Gel sections of these two bands were cut and corresponding sections from gels combined. From these two preparations the glycoprotein was recovered in 0.1 M sodium acetate, dialyzed against water, and freeze-dried. These two purified preparations were called B and C, respectively. They gave the same elution pattern on the Sephadex G-100 column as did Preparation A. B and C were also tested for purity by electrophoresis in polyacrylamide gel followed by staining. Each was contaminated to about 5% or less with the other but there was no additional protein band. Furthermore all three glycoprotein preparations were apparently immunologically identical when tested against antisera, induced in rabbits by injection of soluble glycoprotein using the Ouchterlony double diffusion technique.

The chemical composition of these purified preparations are shown in Table 10. A and B have similar overall composition, while the values for Preparations C and D are somewhat lower. All contained N-acetyl-neuraminic acid (the sialic acid). Sulfate and phosphate were not present in appreciable quantity and neutral sugar was present in amounts less than 0.01%. Paper chromatography and cellulose powder thin-layer chromatography in two different solvent systems showed galactosamine as the only amino sugar in the preparations. Comparison of the sialic acid contents of the echidna preparations with submaxillary mucins of four eutherians (Table 11) shows that the echidna preparations have far less than those of ox, pig, and sheep but more than the mucin from dogs. Galac-

---

*The differences in electrophoretic behavior of batch 1 compared with that of batch 2 may be due to differences in storage time of the two batches in ethanol. When the soluble material from batch 2 was subjected to analytical polyacrylamide gel electrophoresis the glycoprotein had little or no mobility. Since both preparations contained about the same percentage of sialic acid and therefore probably had equally low isoelectric points this indicates that the molecular size of batch 2 was much larger than that of batch 1.

TABLE 10

Chemical Composition of Echidna Sublingual Glycoprotein[a]

Fraction	Amino sugar (%)	Sialic acid (%)	Nitrogen (%)
A	15	17	11
B	13	15	10
C	9	12	9
D	9	13	9

[a] Data from Lew *et al.* (1975).

tosamine contents for all echidna preparations are much the same as those of dog and sheep preparations but less than those of ox and pig.

Differences in amino acid composition between the echidna preparations were minor (Table 11) and except for a higher concentration of proline it is within the range of those of four eutherian submaxillary mucins. That the sublingual glycoproteins in echidnas are chemically quite similar to submaxillary glycoproteins in these eutherians may be at first sight surprising but as pointed out earlier the parotid gland in the eutherian anteater *Tamandua* secretes the sticky saliva for capture of ants; presumably the enormous fused parotid-submaxillary gland of *Myrmecophaga* elaborates a secretion of similar stickiness. It would seem quite likely that the glycoproteins of all mammalian salivas whether parotid, submaxillary, or sublingual will prove to have similar amino acid complements. It would be interesting to know if the glycoproteins of the sticky slimes of the ant bears also have a high proline concentration.

Structure of the Alimentary Tract

As in the platypus the stomach, except for Brunner's glands located in the submucosa of the pylorus, is nonglandular and exhibits a lining of stratified squamous epithelium (Oppel, 1896) consisting of stratum germinativum, stratum spinosum, and stratum corneum. The cells comprising the germinal layer and stratum spinosum are, ultrastructurally, much the same as in the platypus, but in the outer layers of the stratum corneum the elongated cells exhibit degenerate nuclei or none at all (Krause and Leeson, 1974).

Histologically and ultrastructurally the stratified squamous epithelium of the monotreme stomach resembles very closely that of the bovine ruminant [see Hydén and Sperber (1965) for an account of the ultrastructure of the rumen]. Since many substances are absorbed from the rumen, such as salts, water, volatile fatty acids, etc., into the bloodstream, it would appear quite likely that the echidna stomach will prove to have like absorptive properties especially since it

TABLE 11

Amino Acid, Sialic Acid, and N-Acetyl Galactosamine Contents of Echidna and Other Sublingual Glycoproteins[a]

	Preparation (g/100-g samples)			Bovine[b] submucin (major)	Canine[c] submucin	Ovine[d] submucin	Porcine[e] submucin
	A	B	C				
Lysine	0.31	0.37	0.42	0.29	0.5	1.43	0.63
Histidine	0.21	—	—	0.08	0.6	0.48	0.18
Arginine	3.90	3.00	3.19	2.63	3.1	3.25	0.19
Aspartic acid	0.84	0.53	1.03	0.99	1.1	3.02	1.05
Threonine	7.78	5.82	5.91	6.35	5.6	7.05	5.40
Serine	8.35	6.98	6.88	7.50	3.8	7.45	7.43
Glutamic acid	2.36	1.58	2.33	3.04	2.8	4.84	3.50
Proline	7.42	7.29	7.66	4.54	3.7	4.40	3.16
Glycine	4.50	4.02	4.47	4.71	5.8	5.20	5.22
Alanine	5.21	4.30	4.31	3.85	3.2	4.50	4.71
Half cystine	—	—	—	—	0.5	0.47	0.47
Valine	3.85	2.82	2.75	2.74	1.9	3.73	3.12
Methionine	—	—	—	0.05	0.1	0.25	0.11
Isoleucine	0.19	—	—	0.72	0.4	1.72	1.40
Leucine	0.28	0.14	0.40	1.70	1.7	3.03	0.56
Tyrosine	0.15	—	—	—	0.4	0.92	0.40
Phenylalanine	0.17	—	—	0.26	1.3	1.67	0.37
Sialic acid	17	15	12	35.9	8.6	26.0	19.8

(continued)

TABLE 11 (*continued*)

	Preparation (g/100-g samples)			Bovine[b] submucin (major)	Canine[c] submucin	Ovine[d] submucin	Porcine[e] submucin
	A	B	C				
N-acetyl Galactosamine	18	16	12	24.2	15.2	14.7	23.8
Neutral sugar	Trace	Trace	Trace	3.5	25.0	Trace	19.5

[a] Data from Lew et al. (1975). Reproduced with permission of *International Journal of Peptide and Protein Research.*
[b] Tettamanti and Pigman (1968).
[c] Lombart and Winzler (1972).
[d] Gottschalk et al. (1966).
[e] Hashimoto et al. (1964).

can hold with ease a 200-g meal of termites (Fig. 30) ingested in a matter of minutes (Griffiths, 1968). However evidence for an absorptive role can only come from experiments designed to detect the passage of nutrients into the venous drainage of the stomach; Schultz's (1967) account of the blood vessels of the alimentary tract will doubtless prove of great assistance in such a study.

The squamous stratified epithelium of the stomach continues without interruption by a pylorus into a relatively narrow tube corresponding in position to the duodenum in other mammals; I have termed this tube a pseudoduodenum (Grif-

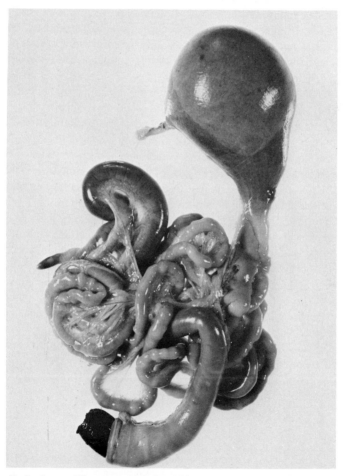

Figure 30. *Tachyglossus.* Alimentary canal of an adult killed after ingestion of a meal of termites. Note stomach distended with food, the appendix, great length of the gut, and characteristic elongated form of the feces. × 0.68.

fiths, 1965a). At this region is found a prominent set of Brunner's glands located between the squamous epithelium and the muscle layers. The cells of the glands contain mucopolysaccharide since they are PAS positive (Griffiths, 1965a) and they also stain with Alcian Blue (Krause, 1970). The ducts of the glands pass through the squamous epithelium and open into the pseudoduodenum.

With the aid of the electron microscope the cells of the parenchyma are seen to be intermediate between mucous and serous elements, not serous as in the platypus (Krause, 1970). These cells contain large secretory granules that tend to fuse into complexes. The endoplasmic reticulum is confined mainly to the perinuclear region of the cytoplasm but occasionally it extends into the apical cytoplasm. The many well-developed Golgi complexes occupy the supranuclear region of the cell; intercellular spaces occur between adjacent cells, but definite secretory canaliculi are absent. Krause points out all these features appear to be intermediate between the serous glands of mice, rabbits, and opossums and the mucous glands of man and guinea pig.

At the caudal end of the pseudoduodenum the squamous epithelium of the tube is replaced by numerous villi and glands of Lieberkühn (Oppel, 1897) clothed by a columnar epithelium; this is the anterior end of the small intestine. The distal portions of the villi are covered with a simple columnar epithelium, the central parts of the crypts with goblet cells containing prominent Indian-club-shaped mucigenous bodies, and the deeper parts of the crypts are lined with Paneth cells. Of these, Krause (1971b) finds that they occupy the basal region of the crypts throughout the small intestine and are often found scattered for a considerable distance into the colon. The cells contain secretory granules that stain intensely with eosin or with the red component of Masson's trichrome stain. Some cells contain less granules than others. With the use of the electron microscope it was found that the granules are of two kinds. Large dense granules found in the supranuclear and apical regions are most numerous, are rounded, and vary in size from 0.5–8 μm and are bounded by a limiting membrane. The other kind are elongated or irregular in shape and are confined to the basal and perinuclear regions of the cell. Unfortunately the functions of the Paneth cell granules are unknown. Krause holds the view that they may be responsible for elaboration of trehalase (see p. 95). One wonders why the granules have never been isolated by differential centrifugation. If they could be isolated in this way their enzymatic properties could be determined.

Digestion

Although the echidna stomach lacks digestive glands there is evidence that self-digestion of the live insect prey can take place in the stomach. Griffiths (1968) found that homogenates of the termite *Nasutitermes exitiosus* had strong amylase activity which presumably would contribute to digestion of glycogens.

As far as digestion of carbohydrates of shorter chain-length is concerned, Kerry (1969) has published further work on the occurrence and activities of disaccharidases in the intestines of *Tachyglossus* and of various marsupials. Adult *Tachyglossus* intestine exhibits only trehalase, isomaltase, and maltase activities (Table 12) whereas in four genera of carnivorous, insectivorous, and herbivorous marsupials the guts had very much higher levels of those enzymes as well as sucrase activity which is entirely missing from echidna guts. The moderate activity of trehalase found in echidna gut may be related to the facts that insects possess a high level of trehalose (Gilmour, 1961) and that trehalose is the principal disaccharide, or for that matter, sugar, of insects. The product of enzymatic hydrolysis of trehalose (2 molecules of glucose) is glucose but as we have seen there is not enough of this in *Nasutitermes*, at least, to promote optimal or near optimal growth and nitrogen retention. Thus to keep growing, echidnas must ingest more termite flesh than they otherwise would if the termites contained larger amounts of carbohydrate capable of being degraded to glucose (see Table 9).

Kerry also made the interesting observation that while the adult marsupials had little or no lactase in the gut, pouch young gut had considerable activity doubtless for hydrolysis of lactose in the milk; echidna milk has some lactose (see p. 293) so it is quite likely that the guts of monotreme young will prove to exhibit lactase activity.

Tachyglossus exhibits a combined pancreatic and bile duct that conveys lipases, amylases, proteinases, and bile to the anterior end of the small intestine; the bile salt as in *Ornithorhynchus* is sodium taurocholate (Bridgewater *et al.*, 1962). The bile promotes digestion of fat by forming emulsions of the triglycerides thereby increasing the surface exposed to the action of the lipases. Ligation of the combined pancreatic and bile duct in an echidna led to a marked decrease in the activity of amylases and proteinases in the succus entericus

TABLE 12

Intestinal Glycosidase Activities of *Tachyglossus* and Some Marsupials[a]

	μm Substrate hydrolyzed/min/g wet wt. mucosa				
	Maltase	Isomaltase	Sucrase	Lactase	Trehalase
Tachyglossus	5.50	4.40	Nil	0.01	2.65
Antechinus stuartii	42.60		10.20	0.06	8.10
Dasyurus maculatus	69.40	38.40	4.90	1.30	23.70
Perameles nasuta	18.90	12.00	5.00	0.68	11.10
Trichosurus vulpecula	41.20	22.90	6.80	0.47	7.20

[a] Data from Kerry (1969). Reproduced with permission of *Comparative Biochemistry and Physiology*.

(Griffiths, 1965a) but the deficiency of these enzymes and an absolute lack of bile had little effect on the well-being of the animal that was being fed termites. Self digestion of the termites and a slow rate of passage of the insects (Griffiths, 1968) undoubtedly were factors; feeds of 100–200 g wet weight of termites take upwards of 2 days to be cleared (Griffiths, 1965a), a circumstance that would allow ample opportunity for what enzymes were available to promote reasonable digestion. Doubtless the great length of the small intestine contributes to a slow rate of passage—it is over 3.5 m in length in adult echidnas (Fig. 30).

ZAGLOSSUS

Food

The food of the long-beaked echidna has been a matter of conjecture but during field trips to Papua New Guinea in July 1972 and September 1973 I found that, in all probability, the food is largely if not solely earthworms.* The first line of evidence was the testimony of Fuyughé villagers living in the valleys of the Wharton Ranges. In reply to the nonleading question "What does saangi (*Zaglossus*) eat?" they immediately and invariably said "embe" (embe, yimbi, or yombi means earthworms in the various dialects of Fuyughé). One man in response to the question "anything else?" replied "kuduf." Some days later I had occasion to collect scarab larvae and asked some women digging in a garden what they were—the answer was kuduf; so possibly *Zaglossus* may eat Coleoptera larvae as well as earthworms.

Further evidence came from examination of scats from seven different specimens caught between July 4 and July 14, 1972 in the mountains near the villages of Mondo (Mt. Tafa) and Umboli. When these feces were homogenized in water practically no arthropod remains floated to the surface as would have happened if the feces had been those of *Tachyglossus*. The sediment, however, mostly earth and sand, when it settled exhibited hundreds of the characteristic chaetae of oligochaete worms, but no arthropod remains.

Furthermore W. Ewers, then at the Department of Biology, University of Papua New Guinea, informed me that the stomach of a *Zaglossus* taken at Mt. Suckling in Papua also in July 1972, contained earthworms. It would seem that *Zaglossus* in the mountains of the eastern part of the island, eats earthworms in July at least.

---

*The food of the New Guinea *Tachyglossus* which was mentioned on p. 82 proved to be the same as that of Australian echidnas, namely ants and termites. Three genera of termites were detected: *Coptotermes elisae*, *Nasutitermes* sp., and *Amitermes* sp.

Structure of the Tongue and Ingestion of Earthworms

As in *Tachyglossus* the tongue is lubricated with sticky saliva from very large sublingual glands; it is also of paramount importance since it both apprehends and masticates the food but there are differences in its structure associated with ingestion of entirely different prey. As Gervais (1877–1878) found, the tongue has teeth or spines. These are housed in a deep groove that extends from the tip to about one third the way back along the tongue; the teeth are located in the anterior three quarters of the length of the groove. At the base of the tongue on its dorsal surface, as in *Tachyglossus*, is a lingual pad furnished with backwardly deflected keratinous spines or teeth. These structures were first illustrated by Gervais and an informative drawing is given by Van Deusen (1971) showing the facts, known to Gervais, that the spines or teeth in the groove point backwards and are arranged in three rows, a ventral and two lateral ones. These teeth are not located on a level with one another but are staggered, a ventral spine is followed by two lateral ones arising slightly posterior to it (Fig. 31 A and B). In this figure and in Van Deusen's the groove is shown open and the spines exposed; this is an artifact of fixation. In the anesthetized animal the groove is closed, and fairly tightly so, so that the tongue in this region is cylindrical and circular in cross section. If one runs one's finger down the tongue rostrally and opens the groove the very sharp teeth will penetrate the skin; it will also be apparent that the teeth are not rigidly fixed but can be pulled to a half-erect position. Perhaps the most remarkable thing about the tongue in the region of the groove is the great thickness of the three strata of the epidermis—stratum corneum, granulosum, and Malpighi. The latter exhibits inwardly directed extensions that penetrate deep into the dermis. This thick hide doubtless contributes to maintenance of the cylindrical shape of the tongue and keeps the groove closed. How the groove is opened for seizing a worm will be explained below.

Where the dermis penetrates into the stratum Malpighi or vice versa, long, sharp, barbed spines of keratin are located. These pass through the stratum corneum and project beyond the surface of the tongue (Fig. 32). They are found in the epidermis extending from the tip to the posterior part of the tongue (it is difficult to measure how far in fixed contracted specimens) and are located ventrally, laterally, and dorsolaterally but not on the medial dorsal surface nor on the inner surface of the groove. The barbed spines also occur in the epidermis of the skin in the *Tachyglossus* tongue. As far as function is concerned it is possible they are tactile sense organs. Another, more likely, possibility is they are concerned with picking up the sticky saliva so that it adheres to the tongue.

Figure 33 shows a transverse section of the tongue in the region just posterior to the groove. Superficially it resembles a transverse section of a *Tachyglossus* tongue, exhibiting two longitudinal sternoglossi. These were first described by Gervais and as in *Tachyglossus* they take their origin at the sternum. However,

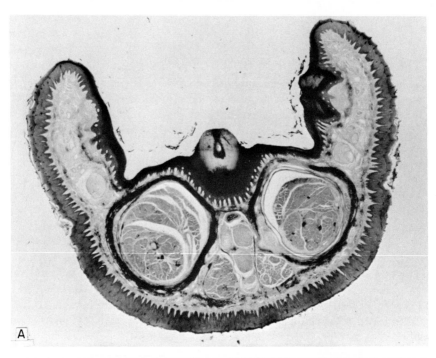

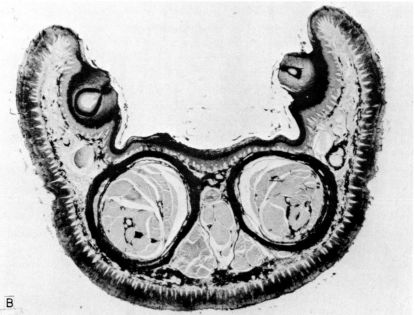

Figure 31. A, *Zaglossus*. Transverse section of tongue in the region of the groove with keratinous tooth (in section) in floor of the groove. The two large muscles in cross section are the sternoglossi and the two smaller situated ventromedially are the flexorglossi. B, Same as in A showing two keratinous teeth in section in walls of the groove. Heidenhain's iron hematoxylin. × 22.

Figure 32. *Zaglossus.* Section of portion of the skin of the tongue showing the keratinous barbed spines projecting beyond the surface of the skin. Heidenhain's iron hematoxylin. × 280.

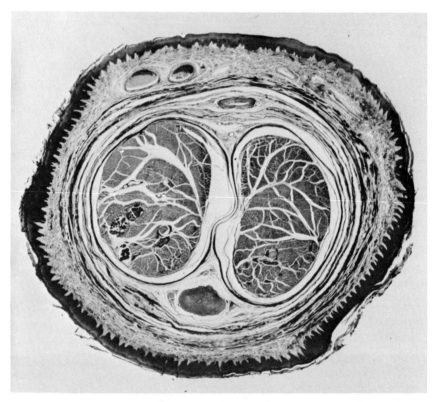

Figure 33. *Zaglossus*. Transverse section of tongue posterior to groove showing absence of striated circular muscles (cf. Fig. 29). Heidenhain's iron hematoxylin. × 21.

the tongue of *Zaglossus* lacks the sheaths of circularly arranged striated muscle found in the *Tachyglossus* tongue (Fig. 29). In their place are circularly arranged sheaths of connective tissue interspersed with smooth muscle fibers. These sheaths are found at all levels of the tongue posterior to the groove. Kolmer (1925) found that at the posterior end of the free portion of the tongue the sternoglossi were made up of helically arranged lamellae of muscles, separated by thin loosely arranged sheaths of connective tissues, around a more compact core of muscle. He found that if the muscle was cut transversely the helically arranged portion could be pulled out like a spiral spring, which when released, slid back to its former configuration. This then is the mechanism of protrusion; contraction of the helically wound part squeezes the central core and forces it forward. It is possible that the tongue of *Zaglossus* can be protruded for a considerable distance but I have never seen it extended more than about 2 cm from the end of the beak when the animal is feeding.

Near the posterior end of the groove another pair of longitudinal muscles take their origin in the connective tissue ventral and medial to the sternoglossi and pass forward to be inserted at the tip of the tongue; these I have named the flexorglossi muscles. As the tip is approached the sternoglossi progressively become smaller in cross section and the flexorglossi larger until at one level of cut all four muscles have the same diameter and are arranged at practically the same level in the tongue. The sternoglossi do not extend to the tip as the flexorglossi do. The function of the flexorglossi muscles is to bend the distal portion of the tongue downwards and so open the groove and expose the sharp teeth.

The sequence of events leading to ingestion of a worm have been described by Pocock (1912) who observed *Zaglossus* specimens kept in captivity at the Zoological Society of London: ''The mouth is a slit about half an inch deep at the end of the snout; and the food, which consists of raw meat mixed with milk, is taken in with the help of the tongue. They are very fond of earthworms but seem unable to imbibe them unless grasped either at the head or tail. In finding the extremity the echidnas are apparently guided entirely by the tactile sense of the lips; but when once it is found, the worm disappears as if drawn in by suction.'' I agree with that except for the suggestion of suction. My own observations are as follow: the echidna touches the worm with the distal end of the beak moving along until either the head or tail is reached. If the worm is too lively the echidna, with a very deliberate action, holds it down with a forepaw and probes with the beak even inserting it between contiguous claws one after the other until the end of the worm is located. The beak is then held vertically down over the end of the worm, the mouth opens and the tongue is extruded 1–2 cm. It has a curved profile, and the groove is open, displaying the three rows of sharp teeth, due to contraction of the flexorglossi muscles. The worm is then hooked on the exposed teeth by a slight forwards and upwards movement of the beak and it is hauled up into the beak, by retraction of the tongue, in a series of sharp jerks; doubtless this reflects rehooking of the worm at a higher level in the groove followed by retraction until the worm arrives at the posterior end of the buccal cavity. Here it is comminuted by the grinding action of the teeth of the lingual pad against three transverse rows* of backwardly directed keratinous teeth at the posterior end of the palate; the noise of the grinding is audible as it is in *Tachyglossus*. A well-defined central neurovascular element was not detectable in the *Zaglossus* tongue at any level of section but posterior to the groove one could see dorsally and ventrally located thin-walled vascular spaces similar to those described by Doran and Baggett in the *Tachyglossus* tongue. It may well be that a stiffening mechanism will be found to operate in the *Zaglossus* tongue also.

---

*Gervais (1877–1878) shows a palate with five sets of transversely arranged spines at its posterior end as well as 12 minor sets extending right to the anterior end. I have detected only seven minor sets on the palate of the eastern Papuan *Zaglossus*.

Histology of the Stomach and Length of the Gut

Although the food of *Zaglossus* is different from that of *Tachyglossus* the epithelium lining the stomach is stratified squamous epithelium consisting of stratum germinativum, stratum spinosum, and stratum corneum. In sections of the stomach stained with Heidenhain's iron hematoxylin it is apparent that the epithelium is of the echidna- and not the platypus-type in that the outer-layers of the stratum corneum exhibit degenerate nuclei or none at all.

As in *Tachyglossus* the alimentary canal is very long; in a 7.6-kg adult it was found to be 7 m in length, so presumably the rate of passage of the food would be slow allowing the digestive enzymes of the pancreas and succus entericus ample time for complete digestion.

4

Miscellaneous Physiology

ORNITHORHYNCHUS

Heart and Circulation

Structure of the Monotreme Heart

Birds and mammals have four-chambered hearts and two completely separate circulations, oxygenated blood from the left ventricle going to the body via the systemic aorta and all blood from the right ventricle going to the lungs via the pulmonary artery. The aortas, however, have different origins in the two classes. In birds during development the left fourth arterial arch is obliterated so that all blood from the left ventricle is directed into the right fourth arch which becomes the systemic aorta. In mammals the right fourth arch is obliterated and the left fourth arch becomes the systemic aorta. The monotremes have two completely separate circulations and the systemic aorta is derived from the left fourth arch as in other mammals (Hochstetter, 1896). This author and Hyrtl (1853) have described in great detail the arterial and venous systems in both genera.

There has been some confusion in the past about the nature of the atrioventricular valves but Dowd (1969) has recently studied the gross anatomy of the monotreme heart and he states that he is in complete agreement with the descriptions of Devez (1903) which he regards as definitive: in *Ornithorhynchus* and *Tachyglossus* the tricuspid left atrioventricular valves are membranous but their movements are limited not by chordae tendiniae but by extensions of papillary muscles from the ventricular wall. The right atrioventricular valve in *Ornithorhynchus* is partly muscular since the papillary muscles penetrate a long way into the flap of the valve; there is a septal cusp. In *Tachyglossus* the right valve is entirely membranous and the papillary muscles do not penetrate past their insertion onto the cusps; the right valve in *Tachyglossus* has a septal cusp.

Dowd also studied the coronary vessels and conducting system in the hearts of monotremes. In both genera there are two coronary arteries, right and left

103

"which have a typical mammalian course on the surface of the heart." A matter of interest is that the venous drainage of the heart walls consists of the great cardiac vein opening directly into the right atrium as in marsupials and birds, and of a coronary vein also opening into the right atrium—the coronary vein is not known in other mammals nor in birds.

A sinus node is present but only in *Tachyglossus* was it present as a discrete mass of specialized fibers—the nodal fibers being paler and having rounder nuclei than the nearly contractile fibers. In both genera, however, the atrioventricular nodes take the form of discrete masses of specialized nodal fibers. These were found to be continuous with the arterial myocardium and with the larger fibers of the atrioventricular bundle. The atrioventricular node in both genera is richly innervated but no specialized nerve endings were detected other than boutons terminaux; some axones appeared to spiral around the fibers.

Vascular Anatomy and Heat Regulation

Grant and Dawson (1978b) have confirmed and extended many of the observations of Hyrtl (1853), Owen (1868), and Hochstetter (1896) of the peripheral circulation in *Ornithorhynchus*. The recent study was carried out not as an attempt to give an anatomical description of the blood vessels of the platypus but to determine whether or not it exhibits specializations in its vascular system which might allow for conservation of heat: the forelimbs, manus, and webs, which account for 13.5% of the total surface area of the animal (p. 4) are supplied with blood from the subclavian artery. In the manus itself the branches of this artery are large in cross section but small in the web; many of the branches of the artery in the musculature were associated with one or more veins running in parallel—an arrangement known as venae comitantes. The venous drainage into the axillary region was found to be extensive, receiving blood from the veins of the arm, thorax, and head. Hochstetter (1896) commented on the size of the axillary venous drainage and suggested that it might serve as a reservoir of blood during diving.

The abdominal aorta branches into two iliac arteries supplying the rear limbs and tail. These arteries divide and subdivide into a series of vessels forming plexuses: (1) Arteries entering the muscles of the upper thigh and crural gland area. (2) Arteries to the main musculature of the leg including lower thigh, knee, and pes. Only tiny vessels are detectable in the latter. (3) Arteries doubling back along the abdominal wall and the region of the epipubic bones. (4) A complex of arteries running superficially over the muscles of the tail. (5) An artery passing along the length of the tail giving off small branches to individual blocks of muscle.

The venous drainage from all those beds of arterioles and capillaries is by a series of vessels that run, without exception, parallel and close to the arteries an arrangement classified by Grant and Dawson as a type 2 rete mirabile. In one animal a large arteriovenous anastomosis was found in the fat of the tail.

The arrangement of closely associated bundles of vessels taking blood in opposite directions is suggestive of a mechanism of counter current heat exchange. Some measurements of skin temperatures in feet and tail reported in Chapter 5 support the notion that such mechanisms operate and are quite efficient in heat conservation.

Lungs and Respiration

Lungs

As far as I know histological and ultrastructural studies of the lungs have not been carried out. Home (1802a) says the lungs are large and correspond to the size of the chest. There are two lobes on the right side and only one on the left; a diaphragm is present and described as very broad, muscular towards the periphery, but tendinous at the center immediately under the heart.

Respiratory Properties of the Blood

The erythrocytes are non-nucleated biconcave discs identical to those of other mammals (Briggs, 1936). Parer and Metcalfe (1967a) found that the mean hemoglobin level content of the blood was quite high, 18.3 g/100 ml with hematocrit at 52%. Similar values were found by Drs. R. W. Hosken and E. A. Magnusson (personal communication)—17.7 g/100 ml and 47.5%, respectively, and Johansen *et al*. (1966) found the hematocrit value to be practically the same, 50%. The latter authors and Parer and Metcalfe found that the O_2 dissociation curves of the blood at different CO_2 tensions exhibit the typical sigmoid shape found for other vertebrates (Fig. 34). That figure also shows that there is a pronounced Bohr effect (the alteration in equilibrium of the O_2 hemoglobin system by varying pCO_2 in the medium surrounding the erythrocyte). Parer and Metcalfe found the magnitude of the Bohr effect at the O_2 tension necessary for 50% saturation of the blood at pH 7.4 (Δ log p50/Δ ph) was 0.50–0.62, a figure in good agreement with that found by Hosken and Magnusson, 0.54 at pH 7.0–7.5. Johansen *et al*. (1966) consider that the pronounced Bohr effect is an adaptation to diving in that it allows complete utilization of the blood O_2 content while it keeps the blood to tissue gradient in O_2 tension relatively high. However, Clausen and Ersland (1968) disagree with that interpretation and point out that the Bohr effect also entails an increase in the tension necessary to saturate hemoglobin with O_2 owing to the fact that combined respiratory and metabolic acidosis develops during diving. Thus, they say, a larger Bohr effect implies that the arterial O_2 tension may fall below the critical tension for brain and heart, before lung O_2 has been effectively utilized. Clausen and Ersland also point out

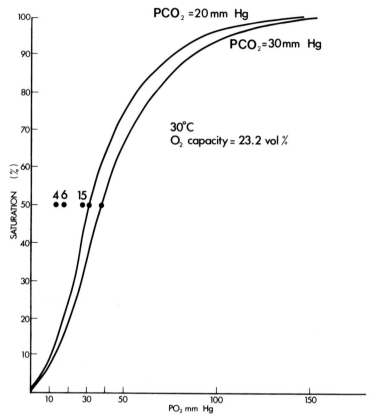

Figure 34. *Ornithorhynchus*. O_2-hemoglobin dissocation curves of blood. Additional values of p50 at 4, 6, and 15 mm Hg pCO_2. *(From Johansen et al., 1966.)*

that the figure for the Bohr effect in the platypus is within the normal range for eutherians including seals and porpoises, saying "Apparently these mammalian divers have not adapted themselves to diving through an increased Bohr effect." The parameters of unloading of O_2, however, are far from settled. Hosken and Magnusson find that at 31°C and pH 7.1 echidna hemoglobin has a p50 value of 15 mm Hg and that of the platypus about 22 mm, but when these hemoglobins are stripped of their 2–3 diphosphoglyceric acid the affinity for O_2 changes markedly so that the p50 is less than 3 mm Hg for both kinds of hemoglobin. What is urgently needed right now are determinations of 2–3 diphosphoglyceric acid in freshly drawn samples of platypus (and echidna) blood for comparison with those of other diving and nondiving mammals.

The effects of increased CO_2 tension on CO_2 content of oxygenated and reduced whole blood and on buffering capacity are shown in Figs. 35 and 36.

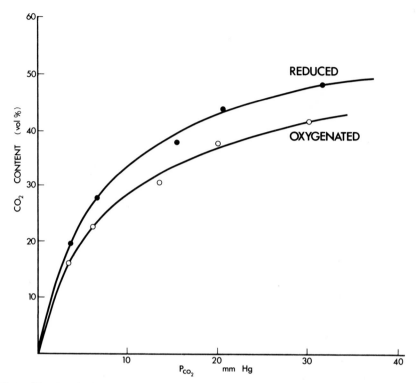

Figure 35. *Ornithorhynchus*. CO_2 dissociation curves of oxygenated and reduced whole blood. *(From Johansen et al., 1966.)*

These show that there is a considerable decrease in the capacity of oxygenated blood to carry CO_2 (Haldane effect).

The O_2 capacity of the blood as found by Parer and Metcalfe (22.7 vols %) and Johansen *et al.* (23.2 vol %) are in good agreement and are higher than those found in many other mammals but much lower than the 30–40 vol % found in cetaceans and pinnipeds by Green and Redfield (1933) and Irving *et al.* (1935). Clausen and Ersland found, however, that two diving mammals that they studied, water voles and beavers, had O_2 capacities of 19.4 and 16.1 vol %, respectively—figures slightly lower than those recorded for ordinary human beings*, 20.2 vol %. The hemoglobin contents of beaver and water vole blood were 11.9 and 14.4 g/100 ml giving O_2 capacities of 1.35 ml/g Hb in both species, and it can be calculated from their figures that the O_2 capacity of human blood is 1.36 ml/g Hb. That for seal's blood is 1.77 ml/g Hb from which they

---

*I have not been able to find O_2-capacity values for the blood of professional divers, such as the Ama of Japan or for our Torres Strait aborigines, values that would make for a meaningful comparison.

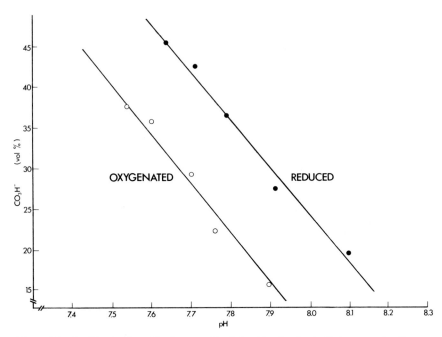

Figure 36. *Ornithorhynchus.* Bicarbonate–pH relationship showing buffering capacity of oxyge-nated and reduced whole blood. *(From Johansen et al., 1966.)*

conclude that the seal is exceptional and that its high O_2 capacity does not reflect a general adaptation to diving; the platypus' O_2 capacity, incidentally, works out at 1.25 ml/g Hb. Surely the point here is the high O_2 capacities shown by mammals like seals and whales are associated with the very long times that seals and whales stay down without breathing. The limit of endurance of the platypus under water was found by Johansen *et al.* to be about 3 min and they are sceptical of reports of voluntary submersions lasting 6 min (Burrell, 1927).

Physiological Responses to Submersion

Johansen *et al.* (1966) studied changes in blood gas concentrations and car-diovascular responses during involuntary submersion. Arterial O_2 and CO_2 con-tent showed amazingly rapid changes during submersion and equally rapid resto-ration of normal levels during the recovery phase. Figure 37 shows the time course of changes in arterial O_2 and CO_2 concentrations: in a little over 2.5 min the concentration of O_2 had dropped from the initial level of 20 vol % to 2 vol % while CO_2 increased from 40 to 54 vol % at the same time. Recovery from those effects of submersion was achieved in about 2 min. Along with rise in CO_2 content pH fell from ca. 7.47 to 7.2 but time to achieve predive blood pH was

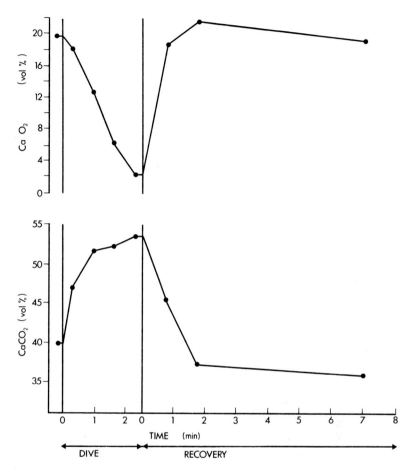

Figure 37. *Ornithorhynchus*. Changes in O_2 and CO_2 content of blood during submersion and recovery. *(From Johansen et al., 1966.)*

longer, about 9–10 min. Apparently the lower blood pH is tolerated since a free-living platypus would dive again before normal pH was reached.

As in all diving animals there was pronounced brachycardia; this is associated with reduction of flow of blood through muscles, skin, and sphlanchnic areas along with augmented flow to the central nervous system and myocardium. In the platypus brachycardia develops gradually during the submersion but when it is terminated there is a sudden release of cardiac inhibition.

It would seem, as pointed out by Johansen *et al.* that all the above physiological properties are suited to the feeding behavior of the platypus: short dives of 30–90 sec duration followed by quick recovery after each dive.

TACHYGLOSSUS

Lungs, Blood, and Respiration; Energetics of Locomotion

Lungs

The lungs of the mammals, including the monotremes, are different from those of Aves and Reptilia in many ways but principally in the great development of a branching treelike system of intrapulmonary bronchi (Goodrich, 1958). The bronchi divide and subdivide the form respiratory bronchioli leading to alveolar ducts, atria, alveolar saccules furnished with minute blindly-ending aveoli. In most mammals the bronchial tree is asymmetrical and the monotremes are no exception; the asymmetry is due to the presence of two lobes of lung on the right side and only one on the left side. Home (1802a,b) was the first to describe this in the platypus and the echidna and Narath (1896) has given a detailed description of the bronchial tree in *Tachyglossus*. The diaphragm is as in *Ornithorhynchus*.

Blood

The definitive study of the hematology of *Tachyglossus* is that of Bolliger and Backhouse (1960). Previous to those authors Davy (1840), Owen (1845), and Briggs (1936) had found that the erythrocytes are non-nucleated biconcave discs. Bolliger and Backhouse found that the discs are biconcave, not biconvex as stated by Griffiths (1968). Table 13 lists data on hemoglobin levels, erythrocytes, and leucocytes from 13 echidnas deemed to be in good health. The mean hemoglobin level was found to be 17.4 g/100 ml but the authors mention that echidnas kept in captivity for several months exhibit Hb levels 1–6 g % lower than the average. Apparently the two echidnas studied by Lewis *et al.* (1968) were relatively anemic since they had Hb levels of 14.2 and 13.2 g/100 ml, respectively. In three echidnas Parer and Metcalfe (1967b) found the mean level to be 17.6 g/100 ml.

Bolliger and Backhouse found that the granular leucocytes are polymorphonuclear as in mammals but the nuclei of the neutrophils have a larger number of lobes than those in man. From the data in Table 13 it can be seen that basophils are absent and that eosinophils are nearly so.

Roberts and Seal (1965) concluded from ultracentrifugal analysis that the serums of monotremes, marsupials, and placentals have a common pattern of protein components different from those of other vertebrates. Jordan and Morgan (1969) with the aid of cellulose acetate electrophoresis found that the serum of *Tachyglossus* gave seven clearly defined fractions with mobilities very similar to those found in human serum. On this basis they identified the fractions as α_1-, α_2-, α_3-, β_1-, β_2-, γ-globulins, and albumin.

TABLE 13

Hemoglobin Levels and Blood Counts in 13 Normal Echidnas[a,b]

Hb	RBC	WBC	N	E	B	L	Mon	MCH
16.5	6.9	8.7	12	0	0	87	1	23
16.7		6.6	40	0	0	57	3	
17.6	7.0	9.9	44	0	0	54	2	25
17.6	7.3		28	0	0	68	4	23
18.4			24	1	0	74	1	
16.6	6.9	5.0	15	0	0	84	1	24
18.6		7.6	32	0	0	67	1	
17.8								
17.5	7.5							22
19.4	9.2							21
16.6	6.8							24
17.5	7.0	7.8	34	0	0	65	1	24
17.5	7.0							25
Avg. 17.4	7.3	7.6	29	0.1	0	69	1.8	23

[a] Data from Bolliger and Backhouse (1960). Reproduced with permission of the Zoological Society of London.

[b] Abbreviations: Hb, hemoglobin (g/100 ml); RBC, red blood cell count (millions/mm³); MCH, mean corpuscular hemoglobin (micrograms); WBC, leucocyte count (thousands/mm³); N, neutrophils (%); E, eosinophils (%); B, basophils (%); L, lymphocytes (%); Mon, monocytes (%).

Respiratory Properties of the Blood

Parer and Metcalfe (1967b) and Tucker (1968) showed that the O_2-dissociation curve is of the usual sigmoid shape but the latter says the oxygen capacities of blood from three echidnas were 12.0, 19.1, and 21.6 vol % and that these values are comparable to the 21.6 vol % found by Parer and Metcalfe; the 12.0 figure must be a misprint for 21.0. The latter authors found the Bohr effect factor to be 0.49 which is close to the figure found by Tucker, 0.54. Parer and Metcalfe (1967c) found that concentration of CO_2 in arterial blood is 77 vol % whereas, they point out, it is only 44 vol % in rabbit arterial blood (Korner and Darian-Smith, 1954). However, I think this difference is not of any physiological significance since echidnas can tolerate very high concentrations of CO_2 in inspired air (see below) and rabbits can tolerate even higher concentrations (Hayward, 1966).

Drs. B. Boettcher and R. W. Hosken of the Department of Biological Sciences, University of Newcastle, New South Wales (personal communication) have separated hemoglobins IB and IIA, mentioned in Chapter 2, and are studying parameters of their O_2 binding: at 25°C and pH 7.1 HbIB has a log p50 value of 1.08 and a Hill number of 2.9 whereas HbIIA has a log p50 value of 0.99 and a Hill number of 2.3. Both hemoglobins have a ΔH of about -7.5 kcal/mole but HbIIA has a lower Bohr effect than HbIB. In view of this the authors consider that the two hemoglobins are functionally different. For some inscrutable reason Parsons *et al.* (1971) find by the use of electrophoresis, only one kind of hemoglobin in echidna blood. They find in a sample of blood from one adult echidna that 90% of the hemoglobin is alkali-resistant and that the blood of adult marsupials can exhibit 11 to 78% alkali-resistant hemoglobin depending on the species, yet they found in these only one kind of hemoglobin electrophoretically.

Respiration

Bentley *et al.* (1967) studied parameters of respiration in echidnas: end-tidal (alveolar) CO_2 concentration was 5.3% and that for O_2 was 14.5%, mean tidal volume 26.9 ml/breath, mean minute volume 137 ml/kg body wt., values considered by Bentley *et al.* to be within the wide limits of variation seen in eutherians. Parer and Hodson (1974), however, found minute volumes averaging 76 ml/kg and concluded that this figure is considerably below those of other mammals of the same weight citing in comparison, of all things, the rabbit whose minute volume is 557 ml/kg. Perhaps comparison with a slightly less active eutherian such as a three-toed sloth would lead to a different conclusion.

Bentley *et al.* found that minute volume could be altered in resting echidnas but the changes were related to changes in respiratory rate. In contrast to this it was found that respiratory minute volume could be increased by inspiration of high concentrations of CO_2 but the change was brought about by change in tidal volume not by increase in respiration rate. However, it took a lot of CO_2 to bring about any change: at concentrations as high as 5% CO_2 in the inspired air very little change in minute volume was detected and even at 7% CO_2 inspired, although the effect is appreciable, it was far less than that exhibited by dogs, seals, and men. Parer and Hodson (1974) confirmed that increase in minute volume follows increase in concentration of CO_2 in inspired air and that the increased ventilation was due to increase in tidal volume rather than in increased rate of respiration. Bentley *et al.* noted that their echidnas voluntarily tolerated concentrations of CO_2 as high as 6.9% in their inspired air when buried in the substratum (crushed corn cobs) of their box. Subsequently Augee *et al.* (1971) made a special study of the respiratory and cardiac responses of echidnas buried in earth and breathing asphyxial air created by their own burrowing habit (Fig. 38). Electrocardiograph leads were attached and the animals were allowed to dig down into 15 cm of loose earth in a bin; as they did so more earth was piled on them to bring the total soil depth to 45 cm. Subsequent measurement showed that each echidna had about 20 cm of earth above its dorsal surface. Air from near the snout was sampled and heart and respiration rates determined with the result shown in Fig. 38. Brachycardia was apparent but, after an initial fall, little

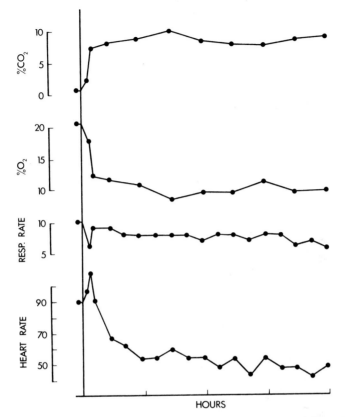

Figure 38. *Tachyglossus.* Changes in inspired air, respiration rate (resp./min), and heart rate (beats/min) in an adult buried under 20 cm soil. *(From Augee et al., 1971.)*

change in respiration rate. There was a rise in CO_2 and a drop in O_2 concentrations in the inspired air but in every case the trend was halted when CO_2 was in the range 10–12% and O_2 in the range 7–8%. When these levels were reached disturbance of the electrocardiogram indicated activity and the surface of the soil was seen to rise and crack open. This stopped after a few seconds and the "bellows" action was repeated at regular intervals leading to maintenance of constant levels of about 10% CO_2 and 10% O_2 in the inspired air for as long as the experiment was carried on.

The brachycardia was shown to be a direct result of changes in gas composition by passing various mixtures down to the earth around the snouts of the echidnas. At first room air was blown down for at least 20 minutes until heart rate was constant and inspired air contained no less than 18% O_2 and no more than 2% CO_2. An asphyxial mixture (13% CO_2, 12% O_2, 75% N_2) was then passed

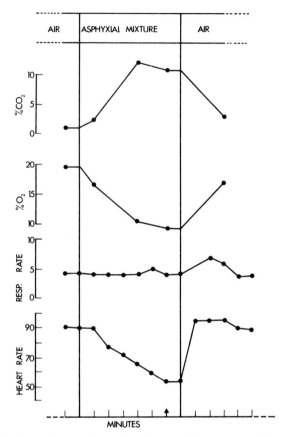

Figure 39. *Tachyglossus*. Changes in inspired air, respiration rate (resp./min), and heart rate (beats/min) in an echidna buried under 20 cm of soil through which air and an asphyxial mixture (10% CO_2, 12% O_2, 75% N_2) were alternately passed. Arrow indicates time at which the animal was observed to become active. *(From Augee et al., 1971.)*

with the result shown in Fig. 39. Brachycardia always occurred and the animals became active at CO_2 concentrations in the range 10–12%. These responses were abolished as soon as the asphyxial mixture was replaced by room air. The data in Fig. 40 show that the brachycardia followed inspiration of hypercapnic mixtures only and was not induced by hypoxic mixtures. Inspiration of pure nitrogen led to activity within 1 minute and the animal broached the surface of the soil. An interesting observation made by Augee *et al.* was that the echidnas made a small breathing cavity by packing the soil around the snout thus making an air pocket. Those authors conclude that the echidna is physiologically well-suited for burrowing. It would be of the greatest interest to determine the respiratory responses

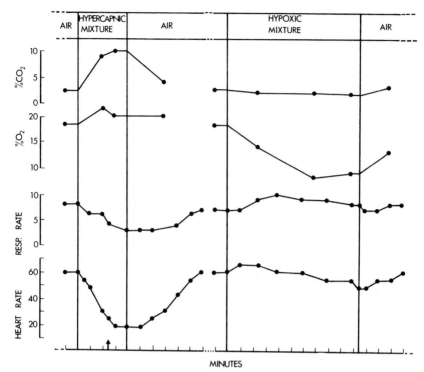

Figure 40. *Tachyglossus.* Changes in inspired air, respiration rate (resp./min), and heart rate (beats/min) in an adult buried under 20 cm of soil through which air, a hypercapnic mixture (13% CO_2, 22% O_2, 65% N_2), and a hypoxic mixture (9% O_2, 91% N_2) were alternately passed. Arrow indicates time at which animal was observed to become active. *(From Augee et al., 1971.)*

of platypuses and of rabbit kittens to hypercapnic and hypoxic mixtures for comparison with those of echidnas since all three animals are exposed to asphyxial conditions in their natural habitats.

Energetics of Locomotion

Edmeades and Baudinette (1975) measured O_2 consumption in two echidnas trained to walk at a variety of speeds on a treadmill. Steady state O_2 consumption increased linearly with running speed in both animals. Comparison of the slopes of the lines with those found for eutherians by other workers led to the conclusion that the rate of increase in steady state O_2 consumption with increasing walking speed in the echidna is similar to that in other mammals. However, the interesting observation was made that the total oxygen consumed/kg body weight while traveling a given distance (the "cost of transport") is less than in eutherians, due to the low resting O_2 consumption (see Chapter 5).

Kidney Function

The echidna kidney exhibits the tubular-loop anatomy (Zarnik, 1910) necessary for elaboration of a hypertonic urine by countercurrent multiplication, and the posterior lobe of the pituitary gland contains vasopressin (p. 151) which influences the permeability of the collecting ducts to water. In the absence of vasopressin the duct is impermeable to water and a dilute urine results. Presumably vasopressin has this action on the echidna collecting ducts since Bentley and Schmidt-Nielsen (1967) found that the kidney of *Tachyglossus* could form a urine of 2300 m Osm/kg water while plasma concentration was only 281 m Osm/kg water—a urine/plasma concentration ratio of close to 8.

From calculation of evaporative water loss, which incidentally does not increase with increase in ambient temperature (Schmidt-Nielsen *et al.*, 1966), and of the amount of water in the termite *Nasutitermes exitiosus*, Bentley and Schmidt-Nielsen were able to strike an approximate water balance for an echidna deprived of drinking water (Table 14). From these data they suggest that echidnas could maintain a positive water balance on a diet of termites without additional water, in an arid climate.

This estimate has proved to be remarkably close to a direct determination of the performance of echidnas eating termites with and without drinking water. From the data on No. 1 echidna in Table 15 it can be concluded that at 21°C and absolute humidity of 9 g/m³ an echidna on a fixed daily ration of termites will drink if water is available but it can get along quite well without it by concentrating its urine. However, if water is offered straight after a period of deprivation it drinks enormous quantities during the next few days. The performance of No. 2

TABLE 14

Approximate Water Balance of a 3-kg Echidna in Dry Air at 25°C on a Diet of 147 g Termites per Day Estimated from Various Sources[a]

Intake	
Free water: 77% of wet wt. of food	113
Metabolic water: fat	4
Protein catabolized (calculated from urinary urea)	5
	122
Output	
Evaporation: in dry air at 25°C	51
Feces	9
Urine	60
Total output (g H_2O)	120

[a] Bentley and Schmidt-Nielsen (1967). Reproduced with permission of *Comparative Biochemistry and Physiology*.

TABLE 15

Urinary Urea Concentration in Two Echidnas Fed Fixed Rations of Termites (*Nasutitermes exitiosus*) with and without Drinking Water[a]

Sex and number	No. of daily observations	Initial weight (g)	Avg. daily N intake (g)	Avg. daily free water intake (ml)	Body wt. change (g/day)	Urinary urea concentration (g/100 ml)
♂ 1	9	2170	2.11 (100 g termites)[b]	9	− 2.4	4.50
	8	2148	2.11	Nil	− 4.0	7.52
	9	2116	2.64 (125 g termites)	Nil	− 1.2	7.46
	4	2105	2.64	74	+67.0	5.80
♂ 2	9	2842	2.40 (114 g termites)	7	+ 6.0	6.20
	13	2896	2.36 (112 g termites)	Nil	+ 6.0	7.50
	7	2974	2.00 (100 g termites)	25[c]	+10.0	6.44

[a] TA ca. 21°C, absolute humidity 9 g/m³.
[b] Water content of termites 73–74 g/100 g.
[c] 52 ml taken in within 45 minutes of offering.

TABLE 16

Effect of Absolute Humidity on Body Weight, Urine Volume, Fecal Water, Urea, and Fecal N Excretion in an Echidna Fed *Nasutitermes exitiosus* with and without Drinking Water[a]

Number of daily observations	Humidity (g/m³)	Body weight (g)		Average daily ingestion of water			Average daily excretion of water			Avg. daily excretion of nitrogen	
		Initial	Final	In termites (g)	Drinking (ml)	Total	Urine (ml)	Feces (g)	Total	Urea N (g)	Fecal N (g)
15	15.6	5200	5200	147	Nil	147	91	25	116	2.27	0.71
7		5200	5500	148	40	188	83	43	126	2.09	0.81
8		5500	5050	148	Nil	148	116	26	142	2.63	0.77
9	3.3	5050	5500	148	80	228	80	33	113	1.70	0.92

[a] TA = 27–28°C.

echidna was even better although it was a larger animal eating less than No. 1. A further series of experiments was carried out on a third larger echidna at higher temperatures, moister and drier air, with the results shown in Table 16. At 27°–28°C in moist air (15.6 g H_2O/m^3) the echidna maintained its body weight on a ration of termites without extra water for over 2 weeks; when allowed the same ration of termites plus drinking water *ad lib* it drank large quantities and put on weight which was not all due to water since less nitrogen was excreted; urea concentration in the urine was the same whether water was allowed or not—2.5 g urea/100 ml.

In relatively dry air (3.3 g H_2O/m^3) in the absence of drinking water this animal lost weight rapidly, nitrogen excretion increased but there was no increase in the concentration of urea in the urine. When allowed water it drank more than double the quantity taken in at higher humidity and far less nitrogen was excreted. This bout of drinking appears to be for the purpose of rehydration since the total amount of water excreted in urine and feces is less than when the animal was deprived of drinking water.

These experiments are not altogether divorced from reality; echidnas cannot thermoregulate at ambient temperatures in excess of 35°C (Chapter 5) so those living in arid areas in Australia where summer temperatures of 45°–47°C are common have to retreat to a cave, breakaway, etc. to avoid the extremes of the climate. Temperatures and humidities in caves inhabited by echidnas in one such area, Mileura in Western Australia, have been measured by Davies (cited in Griffiths, 1968) and it was found that in summer (outside temperatures in the forties) temperature ranged from 25°–32°C and absolute humidities from 13.0–20.7 g/m³. As we have seen, echidnas in this part of Australia are largely termite eaters (Table 8) so it is not unlikely that they can survive for quite long periods in the cave environment without free water if plenty of termites are available even though they have to go out of the cave to catch the termites; in hot weather this is done in the cool of the evening. It is also not unlikely that the subspecies of echidna living in hot arid climates, *T. a. acanthion*, will prove to have kidneys capable of forming a more concentrated urine than those of *aculeatus* can, this being the subspecies used in the above experiments.

5

Temperature Regulation

ORNITHORHYNCHUS

A thorough study of thermoregulation in the platypus in air and in water has been carried out by Grant (1976). Many of his results are at variance with those of earlier workers, namely, Martin (1903), Robinson (1954), and Smyth (1973), but Grant's study was carried out on normal unstressed platypuses which were maintained in his laboratory for periods ranging from 3 months to just under 1 year. That he did succeed in keeping platypuses under controlled conditions is a *tour de force* in itself; admittedly platypuses have been kept in captivity in zoological gardens in the past but Grant showed that they can be kept alive and well while subjected to various manipulations and physiological insults in the laboratory, for months on end. It turns out that all that was needed was ingenuity, patience, and determination, especially the latter; Grant, at the outset, was enlightened by one senior expert that laboratory studies of the platypus would be impractical or meaningless.*

Body Temperature and Metabolic Rates in Air

Over a range of air temperatures, 5°–30°C, rectal temperatures were found to be a little more than 32°C (Table 17) as measured by radiotelemetry in the laboratory. Smyth (1973) found that the mean body temperature of recently caught platypuses was 31.2°C at an ambient temperature (TA) of about 23°C which is close to the figure found by Grant.

The capacity of the platypus to withstand TA in excess of 30°C has not been satisfactorily defined: Martin (1903) had one platypus that he exposed to a TA of 35°C but after 7 minutes of this treatment the animal fainted. Robinson (1954) likewise found that a platypus could not tolerate a TA in excess of 32°C; neither of these animals were normal in the sense that they were accustomed to labora-

---

*Grant's (1976) report is in the form of a thesis presented for the degree of Doctor of Philosophy; he has kindly given me permission to give his results here.

TABLE 17

**Body Temperatures (TB) of Platypuses Recorded during
Metabolism Experiments in Air of Various Temperatures**[a]

Number of animals	Number of determinations	TA (°C)	TB (°C)
3	12	5	32.4
3	20	10	32.1
5	39	15	32.0
5	39	20	32.1
3	30	25	32.1
3	21	30	32.8

[a] Data from Grant (1976) and Grant and Dawson (1978a). Repro-
duced with permission of *Physiological Zoology*, University of
Chicago Press. Copyright © 1978 by the University of Chicago.
All rights reserved. ISSN 0031-935X/78/5101-7719.

tory life. Augee (1976) states that he subjected two anesthetized platypuses to a
temperature of 35°C but he does not say for how long nor did he record body
temperature (TB) at that ambient temperature but he did observe secretion of
sweat (see below). Smyth (1973) tested three of his recently caught animals at
TA 30°C and found that TB rose to 33.5°C at which stage they lay on their backs,
their forepaws were vasodilated and they lost consciousness. When TA was
lowered to 25°C the animals behavior and TB returned to normal. This feeble
exhibition of thermoregulation at TA 30°C is very different from the performance
of Grant's platypuses at the same TA (Table 17), a temperature at which they
were held for some hours.

It has been known for some time that platypuses have an abundance of sweat
glands (Poulton, 1894; Montagna and Ellis, 1960) but as mentioned above Augee
(1976) found that these glands, unlike those of the pig (Ingram, 1965), actually
secreted sweat and they were found at all locations tested: ventral surface, dorsal
surface of the tail, base of the limbs, and the beak; they are particularly numerous
at the margins of the frontal shield. Sweat could be elicited at TA 22°C but it was
not detectable at 5°C. However, injection of adrenalin induced sweating at this
low TA as well as at room temperature.

As the TA is lowered metabolic rate increases except in the thermoneutral zone
(TNZ). Smyth found that it lay between 15° and 25°C but this determination was
the result of one experiment on one animal. Grant found that the TNZ lay
between 25° and 30°C as determined on his normal platypuses. The minimum
resting metabolism measured between those two temperatures was found to be
45.4 kcal/kg$^{3/4}$/day which is higher than that of echidnas and of many marsupials
and eutherians (Table 23).

Body Temperature and Metabolic Rates in Water

Grant's enclosure for his platypuses consisted of a tank with a shelf above water level at one end leading to a tunnel which communicated with a nest box. The platypuses were allowed to enter or leave the water at will. Water temperature was adjusted to 5°, 10°, 15°, and 25°C as required. With this setup it was found that TB in water was a little higher than TB in air, temperatures of between 33° and 34°C being maintained during activity in water of temperatures down to 5°C while their TB in the tunnel or nest box at TA ranging from 16°–25°C varied from 31.9°–32.9°C. The time spent in the water was variable, most excursions lasting only 5 minutes, some lasting as long as 1 hour. The average time per 24-hour period measured over several 24-hour periods is shown in Table 18.

That the figures for TB in water and for times spent in the water are not artifacts of captivity is shown by a comparison of these data with field data collected by Grant. The study area was a pool in the Shoalhaven River in New South Wales containing a number of marked animals; a large number of sightings of these marked platypuses was made; many of the sightings were brief and variable in duration as in the laboratory study, and again as in the laboratory study, platypuses were observed in the water for periods in excess of 1 hour both in winter and summer. The maximum time any one individual stayed in the water

TABLE 18

The Total Number and Average Duration of Excursions Made into Water of Various Temperatures by Two Platypuses over Several 24-Hour Periods of Observations[a]

Total number of excursions	Avg. number hours/day in water	TW (°C)
	Male	
133	3.9	5
149	4.9	10
111	4.5	15
92	6.9	25
	Female	
100	8.1	5
108	6.9	10
100	7.1	15
53	9.7	25

[a] Data from Grant (1976) and Grant and Dawson (1978a). Reproduced with permission of *Physiological Zoology*, University of Chicago Press.

TABLE 19

Body Temperatures of Platypuses Taken from Nets in Various Areas of New South Wales[a]

	TB (°C)	
	Females	Males
Summer, TW 20–24°C	32.1 (n = 18)	32.2 (n = 12)
Winter, TW 6–9°C	30.4 (n = 12)	30.4 (n = 2)

[a] Data from Grant (1976).

was 190 minutes, the temperature of the water being 21°C. Likewise, the TB of platypuses trapped in nets at night during the summer was found to be almost the same as the TB of platypuses in water in the laboratory (Table 19). In winter TB of platypuses taken from the nets was about 2°C lower than in summer. This difference could be ascribed to reduced insulation of the fur since the animals struggle and the nylon strands of the nets can penetrate to the skin allowing contact with the cold water. The struggling could also enhance heat losses by increase in convection currents.

Following the observation that a platypus in the wild will stay in the water for as long as 190 minutes Grant determined TB in platypuses confined to water for 3 hours at three different water temperatures: 5°, 10°, and 15°C. There were four animals, two males and two females, and all four turned on the amazing performance shown in Figure 41. At the end of a 3-hour sojourn at 5°C all four had a TB higher than TB in air; this performance permanently dispels the widely-held notion that platypuses are poor thermoregulators in water.*

As in air metabolic rate in water increases with decrease in temperature (Fig. 42) but at 5°C the metabolic rate was only 1.4 times that of the level at water temperature (TW) 20°–30°C but it amounted to 3.2 times the rate at the TNZ— 145.0 kcal/kg$^{3/4}$/day; in all experiments the increase in metabolic rate in water was about twice the resting metabolic rate in air of the same temperature.

Control of Heat Loss in Thermoregulation

The fact that the rise in metabolic rate when a platypus is swimming in water at 5°C is only 3.2 times the resting level at the TNZ suggests that it has marked

---

*Platypuses living in the wild are subjected to much lower temperatures than 5°C; Temple-Smith (1973) found that the mean water temperature of one of his study pools near Canberra for the month of July was 1.67°C.

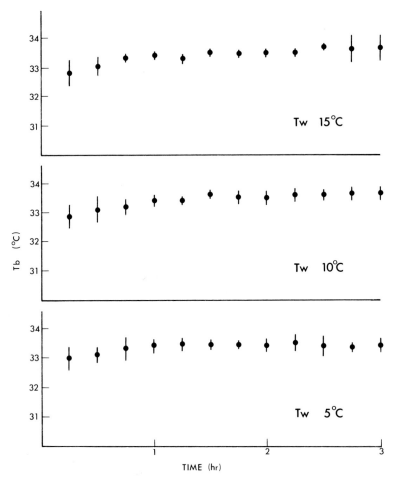

Figure 41. *Ornithorhynchus.* Average TB of four platypuses during exposure to various water temperatures. *(From Grant and Dawson, 1978a; by permission of the University of Chicago Press. Copyright © 1978 by the University of Chicago. All rights reserved. ISSN 0031-935X/78/5101-7719.)*

capacity for restriction of loss of heat to the environment. Accordingly Grant examined various parameters involved in increase and restriction of heat loss: evaporative heat loss (EHL), conductance of tissues and fur, and insulation in air and water.

EHL. From the whole body EHL remained constant up to TA 20°C, after which it increased steadily, the maximum measured being at 30°C and up to 60% of the heat production at that temperature. The actual figures range from 0.54 at

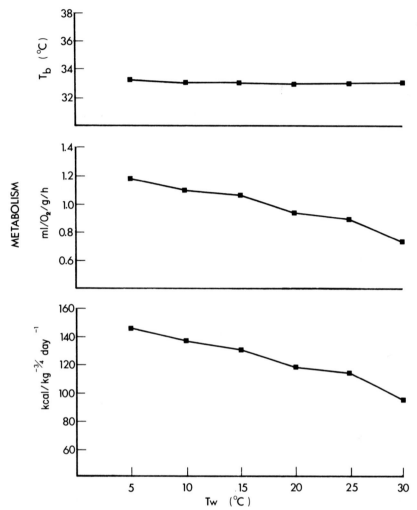

Figure 42. *Ornithorhynchus*. Effect of different water temperatures on TB and metabolism. *(From Grant and Dawson, 1978b; by permission of the University of Chicago Press. Copyright © 1978 by the University of Chicago. All rights reserved. ISSN 0031-935X/78/5101-7719.)*

5°C to 1.2 cal/g/hr at 30°C, which are quite low compared with the values for large mammals that sweat. However, Grant points out that any level of evaporative cooling at the surface will be important since the surface area from which the water can be evaporated is large in relation to volume in the small mammal like a platypus. The low EHL in the platypus can also be considered to be due to the relatively low heat production.

CONDUCTANCE.   This is a measure of heat loss from an animal to its environment as a function of the surface area of the body. Grant determined surface area of the platypuses empirically and from the data calculated that the Meeh factor had an average value of 9.0. Conductance (heat loss by conductance, convection, and radiation through the tissues and integument) was measured as

$$\text{Conductance} = \frac{\text{heat production--total EHL}}{(\text{TB-TA}) \times \text{surface area}}$$

Since in water EHL would be nil or almost so it was neglected when calculating conductance as the measure of heat loss through tissues and integument in water. It was found that the conductance was relatively constant at temperatures below 15°C in air, the mean value being 0.116 cal/cm²/°C/hr. Conductance increased markedly at temperatures of 25°C and above. The conductance value mentioned is much lower than those found for bandicoots (Hulbert and Dawson, 1974) and for sheep and cattle (Blaxter and Wainman, 1961). Such low conductance indicates that the platypus has good insulation and the ability to control heat loss by restriction of blood flow through the peripheral circulation.

In water conductance was 2.4–3.0 times that of conductance in air over the temperature range 5°C–20°C. Nevertheless absolute insulation in water was found to be far better than that of other amphibious mammals (p. 128).

INSULATION.   The furred integument of the platypus plays a major role in the regulation of heat loss. The thermal insulation or thermal resistance of the pelt was calculated from the equation

$$I = \frac{\text{TS-TA}}{H}$$

where TS and TA are skin and ambient temperatures, respectively; and H is heat loss through the pelt so that I is expressed as kcal/m²·hr·°C. H was measured in an apparatus consisting of three heat flow discs, the output of which was calibrated against the heat flow of expanded polystyrene foam of known thermal conductance.

It was found that the ventral fur in air had the greatest insulative properties (Table 20), dorsal fur was next, and fur on the dorsal surface of the tail exhibited the lowest value; only the dorsal surface of the tail is furred and relatively sparsely at that. It is also apparent that the heavily molted pelt was a less efficient insulator than the nonmolted.

The difference between the mean values for dorsal and ventral pelts was significant as were the differences between the two body pelts and the tail pelt. These were due to differences in depth of the ventral and dorsal fur and to the sparseness of the pélage on the dorsum of the tail.

TABLE 20

Ornithorhynchus **Insulation Values for Complete Pelts in Air and for a Heavily Molted Pelt**[a]

Kind of pelt	Lightly or nonmolted $(kcal^{-1} \cdot m^2 \cdot hr \cdot {}°C)$	Heavily molted $(kcal^{-1} \cdot m^2 \cdot hr \cdot {}°C)$
Dorsal	0.346	0.308
Ventral	0.411	0.346
Tail	0.269	0.170

[a] Data from Grant (1976) and Grant and Dawson (1978b). Reproduced with permission of *Physiological Zoology*, University of Chicago Press. Copyright © 1978 by the University of Chicago. All rights reserved. ISSN 0031-935X/78/5101-7719.

Insulation values for body pelts in water were 30–40% of the values obtained in air and the insulation of the tail was only 8% of that recorded in air. Grant found that the absolute value of the insulation of the pelt and the layer of air above it was constant but that it made up a decreasing proportion of the total insulation when air temperature fell below 20°C. Below that temperature total insulation remained constant but the body tissues made up about 60% of the total. At temperatures higher than 20°C the percentage of the insulation due to the tissue changed. This prompted Grant to look for anatomical structures in the blood vascular system which might allow for counter-current exchange of heat between arteries and veins. Some evidence that the structures observed (p. 104) might play a part in control of heat loss came from determination of skin temperatures of all the unfurred surfaces of the body. Grant determined empirically that these accounted for 32% of the total surface area of the body and that the temperature of these surfaces was allowed to fall to within a few degrees of ambient water temperature. Thus heat conservation at these surfaces is efficient. It is interesting to note, however, that Grant found the skin temperatures on the forefeet were always higher than those on other furless surfaces. This is probably due to the fact that the forelimbs are the main organs of propulsion in the water so blood supply to the forelimbs must be kept up to ensure that adequate oxygen and substrate levels are delivered to the muscles involved in swimming.

Grant also determined the temperature of the skin on the under-surface of the tail in sleeping platypuses. When asleep they curl up [a fact first noted and illustrated by Bennett (1835)] so that the naked ventral surface of the tail is applied to the ventral surface of the body, the most thickly furred. In this position it was found that after exposure to TA 5°C for 3 hours TS was 21.0°–24.0°C; at TA 10°C, 24.6°–25°C; at TA 15°C, 24.2°–29.8°C; and at TA 20°C, it was

29.8°–31.8°C. Thus the curled posture adopted by the sleeping platypus is effective in conservation of body heat. It also is probably of the greatest importance for incubation of eggs and maintenance of body temperature of the newly hatched (p. 218).

Comparison of Conductance, Insulation, and Thermoregulation of the Platypus with Those of Other Amphibious Mammals

The insulation of the fur of the platypus pelt in water falls to 60–70% of the value found in air. This compares favorably with the decrease of 90–96% in insulation of polar bear fur and with the same percentage decrease in insulation found for beaver pelt [Scholander *et al.* (1950) and Frisch *et al.* (1974), respectively]. This is in spite of the fact that in air beaver pelt has an insulation of 0.86 kcal/m² · hr · °C, whereas it is 0.38 in the platypus, but in water the absolute value for platypus fur is 0.127 kcal/m² · hr · °C and for the beaver fur it is only 0.035–0.086 kcal/m² · hr · °C. The absolute value for polar bear insulation in water is likewise much less than that of the platypus—0.044–0.112 kcal/m² · hr · °C.

Another amphibious mammal with which the platypus compares favorably as far as maintenance of body temperature goes is the muskrat, *Ondatra zibethica*. This mammal, like the platypus, can maintain its normal body temperature at 37.5°C in air as cold as 5°C, but in water the muskrat cannot maintain its body temperature below a TW of 20°C (Hart, 1962). At TW 5°C, for example, TB is 32°C, less than that of *Ornithorhynchus* at 5°C which, of course, rises from 32 to 33°C when placed in water of that temperature. The metabolic rates for the two different mammals at that TW and TB are 225 for the muskrat and 145 kcal/kg$^{3/4}$/day for the platypus. One reason for the poor performance of the muskrat in water is that conductance is 1.4 times that of the platypus.

Australia has an amphibious eutherian, the water rat, *Hydromys chrysogaster*, which lives in burrows cheek by jowl with those of platypuses in the river banks; it enters the water to catch its food just as the platypus does. *Hydromys* has a wider distribution on the continent than the platypus has so it can be found in the rivers, creeks, and lagoons wherever the platypus occurs. Thus it is called on to feed in the same very cold water that the platypus does. Preliminary results (Fanning and Dawson, 1977) show that *Hydromys* has even poorer thermoregulation in water than the muskrat has. Thus the lowly "primitive" platypus exhibits sophisticated mechanisms of thermoregulation far more effective than those of eutherians living in similar habitats and with less expenditure of energy. As Grant points out "a low level of TB in monotremes may allow the group to possess a relatively low TC and an increased activity range without the necessity for the expenditure of large amounts of energy just to maintain a eutherian level of body temperature."

TACHYGLOSSUS

Thermoregulation in Nonfasting Echidnas

Body Temperatures and Metabolic Rates at High Ambient Temperature

Augee and Grant (1974) have confirmed earlier findings of Martin (1903) and Robinson (1954) that when TA exceeds 30°C, TB in *T. a. aculeatus* rises and at TA 40°C there is an explosive rise to TB 38°C followed by death. Augee (1976) also confirmed Martin's finding that metabolic rate increases with increase in TB. In lightly anesthetized echidnas at high TA there was no sign of panting (i.e., increase in respiratory rate) but in three out of four echidnas there was an increase in respiratory minute volume—in the fourth echidna it fell.

A specimen of *T. a. acanthion* from "tropical Queensland" was tested at high TA and at over 40°C its TB rose to 40°C. It was returned to room temperature but was found dead the next day. Nevertheless this is a rather better performance than that exhibited by *T. a. aculeatus* at high TA; it may have been due to the fact that *acanthion* exhibits much higher conductance values than hairy *aculeatus* and *setosus*.

It has been known for some time that *Tachyglossus* has very few sweat glands—Leydig (1859) failed to detect any in the skin of the back, belly, and the soles of the feet. Pinkus (1906) found one and only one in samples of the skin of three pouch young of different sizes. He found none in the skin of adults except in that of the "Mammartasche." This term was applied in the last century to an artifact brought about by fixation—shrinkage of the skin around the mammary areolae forming a small invagination that some anatomists thought was real (see p. 279). Peripheral to the areola is a ring of large sweat glands called Knäueldrüsen by Gegenbaur (1886). J. Hales and R. Gemmell of the Division of Animal Physiology, CSIRO (personal communication), stated that after an intensive search of the skin on back and belly of a series of echidnas, they detected a few on the belly skin, but so few that they consider them to be of no consequence in heat loss.

Augee and Grant (1974) in agreement with these observations found that no water was secreted by echidna skin unlike that of the platypus even at TA as high as 40°C, nor in response to application of adrenaline.* Thus, when exposed to

---

*Grant (1976) regards it as somewhat of an anomaly that platypuses have sweat glands whereas *Tachyglossus* living in hot arid climates have none. It is possible, however, that sweat glands in platypuses have little or no role in cooling but are concerned with increasing skin temperature or warming the layer of air trapped in the fur just above the skin. Bligh (1961) found that in sheep, sweat secreted brought about an exothermic reaction with the keratin of the wool which leads to a rise in

high TA echidnas can do very little about it, speaking in a physiological sense—they can't pant or sweat, heat production increases at high TB, and any increase in conductance is, apparently, not effective in preventing further rise in TB (see p. 132). However, they can and do avoid the extremes of their environment by retreating to the relatively equable conditions of caves and other refuges (Davies, cited in Griffiths, 1968).

Body Temperature and Metabolic Rates at Low Ambient Temperature

In contrast to its performance at high TA, *Tachyglossus a. aculeatus* is capable of very good thermoregulation at low TA. Parameters of thermoregulation under these conditions have been established by Schmidt-Nielsen *et al.* (1966), Augee and Ealey (1968), and Augee (1969). Although Schmidt-Nielsen *et al.* published their work before that of Augee, both groups were working simultaneously and independently; Augee's work appeared later because he was working on a far larger number of animals (15 echnidas), whereas the other group had access to three only; furthermore Augee examined aspects of thermoregulation (see p. 132) not studied by Schmidt-Nielsen *et al.* The findings of the two groups are in good agreement.

The TNZ lies between 20°–30°C ambient (Schmidt-Nielsen *et al.*, 1966; Augee, 1969) and at thermoneutrality body temperature is 32.2°C. From the O_2 consumption data of Schmidt-Nielsen *et al.* it can be calculated, assuming an average body weight of 3 kg (weights ranged from 2.5–3.5 kg), that the standard metabolic rate (SMR) is 34.0 kcal/kg$^{3/4}$/day. Augee's group of echidnas had an average body weight of 4.4 kg (range 1.9–7.25) and it included two of the largest *Tachyglossus* ever reported—7.1 and 7.25 kg body weight, respectively. He found O_2 consumption was lowest at 25°C and the mean figure at that temperature was 0.22 ± 0.1 ml O_2/g/hr. From these data it can be calculated the SMR was 32.3 kcal/kg$^{3/4}$/day.

Schmidt-Nielsen *et al.* found that the range in body temperature at TA 24° over a 24-hour period of continuous recording was 1.9°C in their most stable animal and 4.1° in the most unstable. Augee *et al.* (1970) found a diurnal rhythm in TB of like dimensions in echidnas living outdoors. These performances in thermoregulation are much better than that of *Tupaia glis* (Bradley and Hudson, 1974) or of *Lasiorhinus latifrons* which has a diurnal variation in TB of up to 3.3°C at ambient 25°C (Wells, 1973).

---

skin temperature. If this happened in platypuses entering cold water the warmth engendered might establish a thermal gradient within the fur capable of reducing heat loss. The idea that entering cold water induces sweating may sound odd, but it is conceivable that sudden entry into freezing cold water is a specific stimulus leading to secretion of adrenalin and hence of sweat. It will be recalled that Augee (1976) found that platypuses at TA 5°C (air) failed to sweat but at that temperature injection of adrenalin induced sweating.

When the ambient temperature is lowered to 0–5°C, TB remains at the usual level of 29–32°C even if the exposure-time at 5°C is 12 hours (Schmidt-Nielsen *et al.*, 1966). Augee's echidnas put up a far more impressive performance: the average morning and afternoon body temperatures over a period of 21 days at 5°C were, respectively, 26.6°±1.0 (147 observations) and 28.6°±1.5 (147 observations). O_2 consumption at TA 5°C was increased 3.3 fold over consumption at TA 25° (from 0.15 to 0.50 ml O_2/g/hr) and heart rate increased from 70 beats/min to 116.

Augee (1969) carried out a further series of experiments on the effect of lowering TA on three different subspecies of echidnas exhibiting three very different degrees of hairiness—woolly *T. a. setosus* from Tasmania, hairy *T. a. aculeatus* from Victoria, and practically hairless *T. a. acanthion* from Queensland (see Chapter 2 for descriptions of pélages of different subspecies).

The régime of exposure to different temperatures in temporal sequence was as follows:

TA (°C)	Days
20	16
15	7
10	14
5	21
10	7
15	7
25	7
30	7

As was expected, at TA 5°C TB was highest in the Tasmanian echidnas (28.2°C), 27.2°C in the Victorians, and the lowest temperatures were found in the Queenslanders (23.5°C) which also had the highest O_2 consumption at TA 5°C. Peripheral vasoconstriction theoretically could be considered to be a factor in preventing heat loss in the Tasmanian and Victorian echidnas but Martin (1903) concluded that his echidnas had little powers of vasoconstriction. J. Hales and R. Gemmell (personal communication) have restudied the problem using the sophisticated radioactive microsphere technique. They have found no consistent evidence to date of changes in vascular flow to the skin, the body, or the extremities, even at TB as high as 33°–34°C.* This supports Augee's arguing on

---

*Baird *et al.* (1974) have found that *Tachyglossus* has the *capacity* for peripheral vasoconstriction since injection of various drugs (5-hydroxytryptamine, noradrenaline, acetylcholine, and two kinds of prostaglandin) either caused deep body temperature to fall or else had no effect. Where body temperature fell it was found to be due to peripheral vasodilation.

the basis of the putatively poorly-developed powers of vasoconstriction in the skin of echidnas—his experiment indicates that the insulating effect of hair in preventing heat loss is an important factor in temperature regulation. That heat loss was different in the three subspecies is shown by the differences in conductance exhibited during the first periods of exposure to temperatures of 20°, 15°, 10°, and 5°C: *T. a. acanthion*, 0.336 cal/cm²/hr/°C; *T. a. aculeatus*, 0.254 cal/cm²/hr/°C; and *T. a. setosus*, 0.195 cal/cm²/hr/°C.

However, on reexposure to 10°, 15°, 20°, 25°, and 30°C ambient temperatures after the 21 days at 5°, it was found that conductance in the three subspecies was now much the same—0.17 cal/cm²/hr/°C in spite of the fact that there were no discernable changes in the pélages. Thus, there was a basic difference in the conductance at the beginning of the experiment, and this was largely removed by acclimation, as Augee says "bringing the two lesser efficient (in terms of heat conservation) subspecies to the level of the more efficient *T. a. setosus.*" Augee offers an ingenious explanation (see below) of how this may come about.

Thermoregulation in Fasting Echidnas

Adults

If *Tachyglossus* is deprived of food, and if ambient temperature is lowered to 5°C, it eventually becomes hypothermic and torpid* with TB falling to ca. 6°C (Augee and Ealey, 1968; Augee, 1969). Small echidnas become torpid within 3–9 days of exposure to TA 5°C; larger ones take longer, upwards of 32 days; in fact large echidnas over 4 kg in weight do not enter torpor unless their body weight falls below 4 kg during the fasting period (Augee *et al.*, 1970). Data from three echidnas that entered torpor within 3–9 days are given in Table 21.

It is apparent from this that TB was only a fraction of a degree above ambient and O_2 consumption rate can fall to as low as 0.03 ml/g/hr. Allison and Van Twyver (1972) also induced torpor in an echidna which had electrodes, chronically implanted, in the olfactory bulbs, pyriform cortex, hippocampus, and thalamus, to record electrical activity in those organs; respiratory and heart rates were also recorded. They confirmed Augee's observation that heart rate falls from about 70 beats/min to seven at a deep body temperature (brain) of 13.1°C. At this brain temperature the electroencephalographic activity (EEG) at all loci was greatly reduced as happens in other hibernators. At arousal from torpor, brought about by increase in TA, brain temperature increased as did EEG; for example pyriform cortex rhythmic activity increased from ca. 4 waves/sec at brain temperature 13.1° to ca. 20 waves/sec at brain temperature 30°C.

---

*From time to time it is stated in the literature (lately, Hudson, 1973) that platypuses do not become torpid; however, Fleay (1950) reported that occasionally one of his platypuses went cold and became torpid, exhibiting body temperatures close to ambient.

TABLE 21

Minimum Body Temperatures, Heart Rates, and Oxygen Consumption of Torpid Echidnas at TA 5°C[a]

Body weight (g)	Body temperature (°C)	Heart rate[b] (beats/minute)	O₂ consumption (ml/g body wt./hr)
2800	5.5	7	0.03
3300	5.7	13	0.14
2200	5.8	7	0.03

[a] Data from Augee and Ealey (1968); reproduced from *J. Mammal.*
[b] Normal heart rate at TA 25°C and TB 32.2° is 70 beats/minute.

Augee and Ealey observed that torpid echidnas kept at TA 5°C can exhibit rise of TB from 6° to the normal value when slightly disturbed. Figure 43 illustrates the time course of O_2 consumption during one such arousal. At the 20th hour after initiation of arousal the animal had a TB of 28°C, TA being maintained at 5°–6°C throughout. For some inscrutable reason Hudson (1973) regards this type of arousal as not spontaneous; his views will be dealt with later.

The ability to arouse from torpor is lost after repeated periods of TA 5°C, but arousal can always be elicited by raising TA (Augee and Ealey, 1968); if TA is suddenly raised from 5° to 25°C it takes the echidna about 11 hours to achieve a TB of about 28°C, O_2 consumption rate being about 0.3 ml O_2/g/hr at that TB.*

Augee noted that during entry of *T. a. aculeatus* into torpor the conductance (coordinates: O_2 consumption ml/g/hr versus temperature differential TB-TA) plot is similar to that for *T. a. aculeatus* mentioned above. However, during arousal there is a marked deviation from this basal conductance line; the deviation increases and reaches a maximum at ca. TB 18°C. From then on as TB increases the conductance approaches the basal conductance line. Since echidnas apparently have very limited powers of peripheral vasoconstriction Augee considers it unlikely that the changes in conductance could be solely due to changes in the circulation in the skin and that a likely explanation of the phenomenon is that a shift in the site of thermogenesis from a peripheral to a central location takes place. Some evidence that this is so came from measurements of temperature, during arousal, in the deep musculature and in the rectum. At central deep body temperature below 17.5°C the central temperatures were lower than the rectal and at TB 19.5 the central temperature was always higher. Augee suggests that, in the absence of brown fat in echidnas, the large peripheral muscles, the panniculus carnosus and the platysma part of the facialis musculature, could be

*The values for O_2 consumption given by Augee and Ealey (1968) in their Figs. 3 and 4 are too large by a factor of 100; the correct figures will be found in their text and in Augee's (1969) thesis.

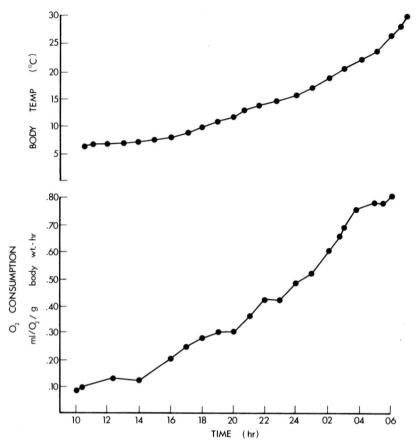

Figure 43. *Tachyglossus.* TB and O_2 consumption during arousal from torpor of an echidna 3500 g in weight. *(From Augee and Ealey, 1968; reproduced from J. Mammal.)*

the site of the initial inefficient thermogenesis. Beneath this is a layer of fat and central to this is the deep musculature. Augee then suggests that the decrease in conductance or acclimation exhibited by the three different subspecies of echidna when re-exposed to temperatures higher than 8°C could be due to passive improvement of insulation brought about by a shift from peripheral to central thermogenesis. This is an attractive notion; however, Augee and colleagues are following another line of research in connection with change in conductance associated with changes in the composition of lipids in the membranes of mitochondria (see p. 141).

In a later paper Augee *et al*. (1970) studied entry into and arousal from torpor in five animals kept in outdoor concrete enclosures furnished with a pit 1 m square and 30 cm deep filled with soft earth. A biotelemetric system, designed by the authors, permitted continuous recording of TB without disturbing the animals, over periods of up to 5 months. The echidnas, except one large one weighing over 5.5 kg, became torpid from time to time during the winter months of June, July, and August when air temperatures frequently fell to as low as 3°C. Before becoming immobile the echidnas entered the pits and dug in the earth until they were partially covered—the temperature of the soil was ca. 8°–10°C. The longest continuous period of torpor lasted 9½ days and intervals between periods of torpor ranged from 30 hours to 11 days. As Augee *et al*. pointed out this pattern of periodic arousal has been found in all hibernators so far studied. Arousal, as in the laboratory, was slow: at soil temperature of 10°C and TB 12°C it took one echidna 9 hours to reach a TB of 32°C.

Augee *et al*. describe an instance of torpor in an echidna living in the bush. The animal, 2 kg in weight, was discovered in a depression at the base of a stump during the winter month of June; air temperature was 13.5°C (10°C the previous night), ground temperature 10 cm down was 11°C, and the echidna's rectal temperature was 12.5°C.

From all their data Augee *et al*. conclude that torpor is natural to the echidna ant that it can arouse without an external source of heat just as all other hibernators can. They also mention that induction of torpor by lowering TA to 5°C is probably more rigorous than anything of this nature that an echidna would probably encounter in the bush (see p. 136). In spite of this qualification Hudson (1973) remarks ''It is common practice in hibernation studies to ignore the relationship between environmental demands and the physiology of hibernation. For example, though Augee and Ealey (1968) noted that the burrow temperature of the echidna, *Tachyglossus aculeatus,* seldom fell below 10°C (the lowest recorded ambient temperature was 8°C), most of their laboratory measurements were made at 5°C. It should have been expected, a priori, that they would not obtain arousals at 5°C (TB < 6°C), just as observed. As expected, their body temperatures telemetered from field animals indicated a spontaneous arousal at a ground temperature of 9°C and a TB < 10°C. In addition the arousal of the animals in the field was much faster, requiring less time than the 20 hours an animal in the laboratory needed to increase its body temperature from a little over 5° to 28°C. It is likely that Augee and Ealey were measuring hypothermia in the laboratory and hibernation in the field.'' However, some of those statements are not in accordance with the facts; torpor was not measured in the field but in concrete enclosures furnished with pits filled with earth; burrows were not dug, the animals were only partially covered by earth; arousal in the laboratory with TB 6°C and TA 5°C was spontaneous which simply goes to show echidnas know nothing of a priori considerations; the lowest ambient temperature recorded was not 8°C but ca. 3°C—the lowest soil temperature was 8°C. In any case why not study the echidna's performance at 5°C? Hudson himself quotes the instance of *Sicista betulina* which can allow its body temperature to fall to 5°C and then exhibit spontaneous arousal. One of the most interesting features of echidna torpor was demonstrated by the procedure of exposure to 5°C; this was the long time taken to attain normal body temperature— 20 hours, whereas the marmot, a hibernator of the same size as *Tachyglossus* takes only 2–3 hours under similar conditions. Benedict and Lee (1938) found arousal in this animal to be accompanied by a great burst of heat production, but the echidna (Fig. 43) exhibited a slow and steady rate of heat production. Finally it should be mentioned that echidnas live in the Australian Alps up to the 1700-m level where it gets far colder than Augee and Ealey's cold room. For example, echidnas have been observed by members of the Division of Wildlife Research, CSIRO, at a study area known as Snowy Plains where records of temperature have been kept since 1967; air temperatures in the Stevenson screen as low as −20°C have been noted. Presumably the ground is as cold, or even colder, than in the screen; however, echidnas living there would have to cope with extremely cold weather, presumably by going torpid. There would be plenty of scope for spontaneous arousal from this condition at Snowy Plains since the sound of the winter winds there, roaring through the trees, reminds one of the close proximity of an express train.

The ability to enter torpor is connected with activity of the adrenal glands; this will be discussed in Chapter 6.

Body Temperature in a Suckling Echidna in the Bush

On October 28, 1972 a very small echidna was dug out of what appeared to be a rabbit breeding stop on the side of a hill near Canberra. I acquired the animal and returned it to the stop which unfortunately had been damaged by the digging. On October 30 it weighed 474 g, was haired, had very short spines, and was incapable of moving about and feeding itself. From observation of similar-sized young in the laboratory (p. 301) it was deemed to be a suckling recently dropped from its mother's pouch. A day later it had dug itself a small burrow laterally and downwards from the stop; the earth in which it had dug was quite hard and the final depth achieved was about 30 cm. The little animal lay rolled-up at the bottom of the borrow and so tight was the fit that it was hard to extract it. On this day it weighed 474 g and cloacal temperature was 28.5°C. On the fifth day it was found that the animal had gone torpid; the cloacal temperature was 12.8°C and the temperature of the soil of the burrow was 7.2°C (Table 22). On the seventh day there had apparently been a spontaneous arousal, cloacal temperature was 27.2°, and soil temperature of the burrow was 12.8°C. Thereafter it went torpid again exhibiting temperatures of 13°–14.4°C (Table 22). It is apparent from these data that the suckling had not been fed during the 16-day sojourn in the burrow since body weight showed no upward turn. There was some evidence that the burrow had been visited by an animal on one occasion

TABLE 22

Body Temperatures and Body Weights of a Suckling Echidna Living in a Burrow

1972 date	Body weight (g)	Cloacal temperature (°C)	Burrow temperature (°C)
October 30	474	—	—
October 31	474	28.5	—
November 1	460	—	—
November 2	455	—	—
November 3	445	12.8	7.2
November 4	445	13.3	—
November 5	435	27.2	12.8
November 6	435	14.4	—
November 7	430	14.4	13.3
November 8	425	14.4	12.8
November 13	410	13.0	13.0
November 15	410	14.0	14.0

during the first week—possibly by the mother who may have been put-off by the smell of humans on the suckling.

When cloacal temperature was low the animal was immobile and remained tightly rolled-up; it required considerable force to open it sufficiently to get the thermometer into tbe cloaca. At the time (November 5, 1972) the animal had a temperature of 27.2°C, it was lively, and dug in energetically when returned to the burrow.

On November 15 the suckling was removed from the burrow, brought back to the laboratory, warmed up, and fed cow's milk which it drank from a spoon. From this it would appear that an echidna when dropped from its mother's pouch is well prepared to cope with infrequent feeding by going torpid to save energy. Whether or not suckling echidnas in burrows do save energy by going torpid as compared with the state of normothermia can only be resolved by measurements of metabolic rates [see Hudson (1973) for a discussion of the savings of energy during torpor and expenditure during arousal in adult eutherian hibernators].

Comparison of Thermoregulation in Prototheria, Metatheria, and Eutheria

In recent years one gains the impression from the literature that members of a given taxon of the Class Mammalia have standard metabolic rates different from those of members of another taxon. Thus it is frequently stated that the SMR's of protherians are less than those of metatherians and the SMR's of metatherians are lower than those of eutherians. This stems from the fact that MacMillen and Nelson (1969), Dawson and Hulbert (1970), and Arnold and Shield (1970) determined the SMR's of a wide selection of marsupials of different sizes and compared the result with that found by Brody (1945) and Kleiber (1961) for a narrow selection of atypical eutherians (sheep, pigs, cats, dogs, mice, etc.) which happened to have high SMR's. So the dictum that the SMR's of prototherians and metatherians are lower than those of eutherians is rapidly being entrenched in the literature. However, if one makes another selection of the data available it can be seen (Table 23) that the SMR's of prototherians are as high as, or even higher, than those of many marsupials and eutherians.

A considerable amount of evidence collated by Kayser (1961) and Dawson (1973) attests to the fact that many eutherians, in contrast to platypuses and echidnas, are not good thermoregulators at low TA. Within the space of a few days some exhibit fluctuations in TB as large as 17°C in response to changes in TA; the latest recruit to the ranks of the poor thermoregulators is the tree shrew, *Tupaia glis*, which at TA 20°C has a mean nocturnal TB of 33.8°C and a mean diurnal TB of 38.9°C, a day-night difference of 5.1°C (Bradley and Hudson, 1974). It has also become apparent (see Dawson, 1973) that in spite of lower SMR and body temperature that some marsupials are excellent thermoregulators employing means of thermoregulation identical to those of the eutherians that thermoregulate well. Yet the marsupial *Lasiorhinus latifrons*, a cryptic and burrowing wombat that lives in the same habitats as some echidnas, exhibits in the face of mild heat stress a performance as poor as that of *Tachyglossus* (Wells, 1973). At TA 35°C *Lasiorhinus* lies on its back, breathes through its mouth and, as TB rises to 38°C (normal TB is 35°C), salivates, loses muscle tone, and becomes limp. Like echidnas they cannot sweat, as shown by a negative result for water when cobalt thiocyanate papers were applied to the body including the axillary regions, at high TA. However, like echidnas they are good thermoregulators when challenged with cold and they can withstand bodily hypothermia.

From the foregoing it is apparent that a mammal's capacity for thermoregulation has nothing to do with whether or not it belongs to the Prototheria, the Metatheria, or the Eutheria.

In this context of comparison of thermoregulation in the Mammalia some observations on the effect of temperature on metabolism of echidna and platypus tissues *in vitro* are of interest. Aleksiuk and Baldwin (1973) studied the effect of exposure to temperatures, over the range 5°–40°C, on the metabolic rate (measured as O_2 consumption) of homogenates of skeletal muscle and of liver from platypuses and echidnas. The data summarized in Fig. 44 show that there is an exponential relationship between temperature and metabolism in each case, and that there is no evidence of a homeostatic region in any of the curves. This is

TABLE 23

Body Weights, Temperatures, and Standard Metabolic Rates of Monotremes, Some Marsupials, and Some Eutherians

Species	Body wt. (g)	TB at thermoneutrality (°C)	SMR[a] (kcal/kg$^{3/4}$/day)
Monotremata			
Ornithorhynchus anatinus	—	32.1	45.4[1]
Tachyglossus aculeatus	3000	32.5	34.0[2]
Tachyglossus aculeatus	4400	32.2	32.3[3]
Metatheria			
Antechinomys spenceri	24.2		44.5
Antechinus maculatus	8.5		44.1
Dasycercus cristicaudata	89	37.7	32.7
Dasyurops maculatus	1780		39.9
Lagorchestes conspiculatus	2700		47.2
Lasiorhinus latifrons	25200	35.0	25.7[4]
Macrotus lagotis	940		40.5
Perameles nasuta	690	36.1	48.9
Pseudantechinus macdonnellensis	43	34.2	33.1
Sarcophilus harrisii	500	36.0	48.4
Trichosurus vulpecula	1980	36.2	43.0
Eutheria			
Bradypus griseus	3400	33.1	24.7
Bradypus griseus	2800		33.5
Bradypus tridactylus	3500	32.2	25.0[5]
Choloepus hoffmanni	4500	35.0	36.8
Dasypus novemcinctus	3000	34.0	27.3[6]
Tamandua tetradactyla	3500	33.5	39.5
Tenrec ecaudatus	790	33.0	35.7

[a] Numbers in parentheses are the following references: (1) Grant (1976); (2) Calculated from the data of Schmidt-Nielsen *et al.* (1966); (3) Calculated from the data of Augee (1969) and personal communication; (4) Calculated from the data of Wells (1973); (5) Calculated from the data of de Almeida and de Fialho (1924); (6) Calculated from the data of Scholander *et al.* (1943). All other data taken from Dawson (1973).

quite unlike the curves for metabolic rates of tissues from poikilothermic animals but the form of the curves is very similar to those for metabolic rates of liver homogenates from rats. These authors in another paper (Baldwin and Aleksiuk, 1973) studied the effect of temperature on lactate and malate dehydrogenase activities in preparations of skeletal muscle, heart muscle, and liver. Both these activities in platypus and echidna preparations are characterized by thermally induced changes in enzyme–substrate affinity which tend to stabilize reaction

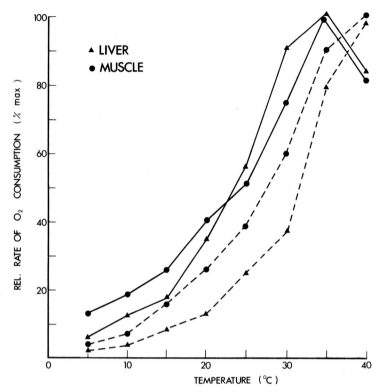

Figure 44.   Effect of temperature on relative rate of O_2 consumption by liver and skeletal muscle homogenates from platypuses and echidnas. ▲, Liver; ●, muscle; ———, platypus; ----, echidna. *(From Aleksiuk and Baldwin, 1973; reproduced by permission of the National Research Council of Canada.)*

rates at K_m substrate levels between 25°C and 35°C. At temperatures below 20°C reaction rates decrease (Table 24).

Baldwin and Aleksiuk state "it is tempting to speculate that this phenomenon may be of adaptive advantage in slowing down metabolism, thereby aiding the conservation of energy stores during hibernation." These authors also found that Arrhenius plots for both enzymes are linear over a temperature range of 10°–40°C revealing activation energies of 19 and 17 kcal/mole for platypus and echidna LDH, respectively, and of 16.5 kcal/mole for MDH in both genera. They say the linearity of the plots supports the notion that rate stabilization at low substrate concentrations is brought about by alterations in enzyme–substrate affinity. If the observed independence of temperature of reaction rates between 25° and 35°C were the result of a thermally induced change in the rate limiting step for the overall reaction the Arrhenius plots would be nonlinear. However, McMurchie

TABLE 24

Effect of Assay Temperature upon Reaction Rates at Saturating and at Minimal K_m Levels of Substrate for LDH and MDH Enzymes from Platypus and Echidna[a]

	Temperature range (°C)	$Q_{10} V_{max}$	$Q_{10} K_m [S]$
Platypus LDH	10–20	3.2	1.5
	25–35	2.7	0.9
Echidna LDH	10–20	3.2	1.4
	25–35	2.7	1.0
Platypus MDH	10–20	1.7	1.6
	25–35	1.6	1.1
Echidna MDH	10–20	2.2	1.6
	25–35	1.9	1.0

[a] Data from Baldwin and Aleksiuk (1973). Reproduced with permission of *Comparative Biochemistry and Physiology*.

and Raison (1975), without reference to the above, found that activation energies of the succinoxidase system change with change in temperature and that this is associated with structural changes in mitochondria. In rats these structural changes occur in the membranes of liver mitochondria at 23°C and 8°C and the activation energy of enzyme systems associated with the mitochondrial membranes increases below those temperatures (Raison and McMurchie, 1974). The structural changes were detected by electron spin resonance spectroscopy and are referred to as phase changes; it has been suggested that they signify the initiation and termination of a thermal phase separation between lipids in the fluid, liquid-crystalline phase above 23°C and a rigid gel phase below 8°C. In the one eutherian hibernator (*Citellus lateralis*) that has been studied the change in membrane structure occurs at about 23°C when the animal is in the active state as in nonhibernating homeotherms, but in the hibernating state (TB 4°C) no change is detectable between 2° and 35°C. In active echidnas a phase change in membrane lipids was detectable ca. 17°C in liver mitochondria (more oleic acid present at the expense of stearic), but in hibernating echidnas with TB 10°C a phase change occurred at slightly below that temperature. McMurchie and Raison (1975) also determined the activation energy (Ea) of the succinic oxidase system which is associated with the membranes of the mitochondria, in active and in torpid echidnas: for an active echidna Ea was 8 kcal/mole between ca. 35°C and 17°C, below that it was 18 kcal/mole; in mitochondrial preparations from two torpid echidnas it was 12 and 19 kcal/mole determined at temperatures between 35° and 10°, and 35 and 5.8°C, respectively. Below 10° and 5.8° the slopes of the

Arrhenius plots changed but the authors did not work out of the Ea's below those temperatures. For a nonhibernating homeotherm the Ea of the succinoxidase system above the temperature of the phase change is 4 kcal/mole; it would be of great interest to know the Ea of the succinoxidase system determined below the temperature of the phase change in a hibernating eutherian for comparison with that of echidna mitochondria.

Although Baldwin and Aleksiuk (1973) found no change in activation energies of LDH and MDH it would seem very probable they would have found changes in the slopes of Arrhenius plots if they had made these calculations on O_2 consumption data in their first paper. This however would not explain why the Ea of LDH and MDH did not change; it is possible that only the activation energies of enzymes with a metal prosthetic group, capable of reducing molecular oxygen directly, i.e., an oxidase not a dehydrogenase, are changed by changes in temperature. Unfortunately we do not know what enzyme had its activation energy changed by temperature in the succinoxidase system used by McMurchie and Raison since the system consists of a chain of electron donors and receptors terminating in cytochrome oxidase, the ultimate reducer of molecular oxygen. It is possible that all metal-catalyzed oxidation systems involving reduction of molecular oxygen undergo change in activation energy as temperature is lowered; this notion is supported by the observation that the Arrhenius plot of the oxidation of uric acid by molecular oxygen catalyzed by cuprous chloride is not linear at temperatures above 17°C (Griffiths, 1952).

ZAGLOSSUS

Nothing is known of temperature regulation in *Zaglossus* but some measurements of cloacal temperatures have been made. Van Rijnberk (1913) found in two specimens, living in a zoo in Amsterdam, that over a 24-hour period cloacal temperature in one specimen varied between 25.5° and 30.5°C, and in the other between 27.0° and 32.0°C; ambient temperature was not measured.

I have made a few measurements of cloacal temperatures but in recently caught animals. The animals are those described in Chapter 8. The average temperature of the cloaca (nine readings on seven animals of both sexes) was 29.1°C (range 27.0°–32.0°C) at an average TA of 19.2°C (range 18.0°–21.0°C). The highest cloacal temperature recorded was 32.0°C and this occurred at the highest TA—31°C. Conversely the lowest cloacal temperature occurred at the lowest TA, 18.0°C, but two other animals at this ambient temperature exhibited cloacal temperatures of 29.5°C and 29.0°C, respectively. From these observations it is apparent that deep body temperature of *Zaglossus* will prove to be close to that of *Tachyglossus*, i.e., 32°C at TA 22°–25°C (see Appendix).

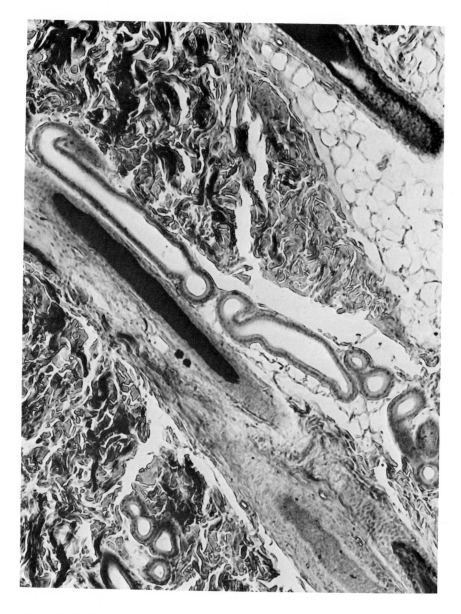

Figure 45. *Zaglossus.* Apocrine sweat glands in skin of the dorsum. The elongated dark-staining bodies are portions of hairs in section. Heidenhain's iron hematoxylin. × 120.

Some years ago Kolmer (1929) reported that the skin of a specimen of *Zaglossus* exhibited sweat glands. An examination of samples of belly and back skins from the seven adults mentioned above showed that there is a general distribution over the body of a large number of apocrine sweat glands (Fig. 45). It is therefore quite possible that *Zaglossus* will offer a better performance in the face of heat stress than *T. a. acanthion* which has no sweat glands (Augee, 1976), but for comparison we need present data on the performance of both genera, acclimated to TA higher than "room temperature," at a test temperature of 40°C. One factor that predisposes one to think that *acanthion* will turn in the better performance is conductance: *Zaglossus bruijnii*, as already mentioned, is a very hairy echidna with few spines, rather like *T. a. setosus* (see also Toldt, 1906), so it is quite likely it will exhibit conductance values considerably lower than those of *acanthion* which has little or no hair.

What we also need at present is a determination of the capacity of unanesthetized healthy platypuses, maintained in the laboratory, to withstand heat stress.

6

Endocrine Glands and the Glands of the Immune System

ORNITHORHYNCHUS

Endocrine Glands

The Pituitary Gland

The hypophysis is a large pear-shaped body lying appressed to the tuber cinereum of the hypothalamus. It consists of two separate endocrine glands both of ectodermal origin but from different tissues: the pars nervosa, which is the swollen ventral extension of the infundibulum whose cavity is continuous with the third ventricle of the brain, and the pars anterior or distalis which is a derivative of the buccal cavity ectoderm known as Rathke's pouch. From the description of a single gland by Hanström and Wingstrand (1951) it is apparent that anatomically the pituitary resembles very much that of a reptile: the pars distalis is elongated and lies parallel to tbe median eminence of the tuber cinereum (Fig. 46) and the pars nervosa exhibits an outgrowth of the infundibular cavity lined by ependymal cells as do the partes nervosae of birds and reptiles (de Beer, 1926; Saint Girons, 1970). There is a well-defined hypophyseal cavity (a rest of the cavity of Rathke's pouch) separating the pars distalis from the pars intermedia which invests the pars nervosa. The pars tuberalis is extensive covering the ventral and lateral regions of the median eminence.

As far as the cytology of the pars distalis is concerned Hanström and Wingstrand found that the rostral end exhibits acidophil cells staining with carmine-red or with orange G; the caudal region consists mostly of basophils and a very few acidophils. As far as I am aware no studies of the ultrastructure or of the hormones of the platypus pituitary have been carried out.

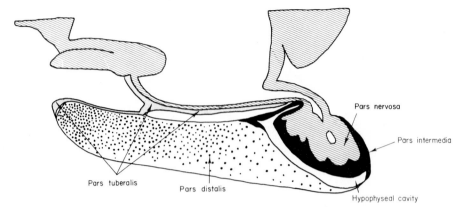

Figure 46. *Ornithorhynchus.* Sagittal section of pituitary. *(From Hanström and Wingstrand, 1951.)*

The Adrenal Gland

ANATOMY. The anatomy of the gland in platypuses and echidnas will be described here since it is very much the same in all three genera (Kohno, 1925; Basir, 1932; Wright *et al.*, 1957; Griffiths, 1968). Externally it is pear-shaped and in the adult it lies in close contact with the anteromedial surface of the kidney just anterior to the exit of the ureter (Fig. 10). In *Ornithorhynchus* it is about 18 mm long by 8 mm thick, 9 × 4 mm in *Tachyglossus*, and 13–18.5 × 4–6.5 mm in *Zaglossus*. Internally, instead of the central medulla of chromaffin tissue surrounded by the various concentric zones of cortical tissue of the eutherian and metatherian adrenal, the chromaffin tissue is located in the caudal half of the gland and the cortical in the anterior half (Fig. 47). There is no well-defined boundary between the two tissues and they interdigitate especially in *Zaglossus*. Frequently in the latter genus, but not in the other two, the distal extensions of the chromaffin tissue are surrounded by large open sinuses. Cross sections of this area of interdigitation look very like those of the adrenals of reptiles (Gabe, 1970) and of birds (Sturkie, 1965) in which the chromaffin tissue is scattered throughout the cortical tissue in the form of separate clusters or islets of cells. However, the similarity is only apparent since the extensions of chromaffin tissue in the monotreme adrenal are never cut off from the main caudal body to form islets. The cells of the chromaffin tissue are arranged to form anastomosing solid cords and are of two kinds: those at the periphery of the cords are elongated, radially arranged, and contain many granules staining strongly with Heidenhain's iron hematoxylin; those at the core of the cords are polygonal, not oriented in any particular way, and exhibit far fewer granules (Griffiths, 1968).

Wright, Jones, and Phillips can distinguish three kinds of cells, occurring in groups, in the cortical tissue in the platypus adrenal but in no way is there a

Figure 47. *Zaglossus*. Longitudinal section of right adrenal. Chromaffin tissue is aggregated at the caudal end (lower part of photograph) of the gland. Note open sinuses around portions of the chromaffin tissue. Heidenhain's iron hematoxylin. × 11.

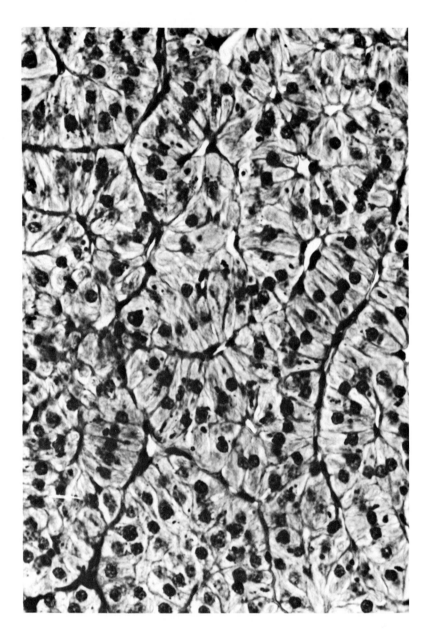

Figure 48. *Zaglossus*. Cells of the cortical tissue showing strong resemblance to those in the interrenal glands of diploglossan lizards. Heidenhain's iron hematoxylin. × 580.

regular segregation into concentric zones as in the eutherian and metatherian adrenal. The tachyglossid cortical tissue is quite unlike that in the platypus and it bears a startling resemblance to cortical tissue in reptilian interrenal glands. This is especially so in the adrenal of *Zaglossus* where the cortical tissue consists of very tall columnar cells, with strongly defined outlines, closely packed to form convoluted and anastomosing cords one or two cells in thickness (Fig. 48). This is precisely the arrangement and the type of cell found in the interrenals of diploglossan reptiles; Gabe's (1970) illustrations of these would serve for the *Zaglossus* adrenal. Wright *et al.* and Griffiths found practically no store of lipid granules in the cortical cells but Augee (1969) found in his echidnas, after they had been exposed to cold, that the cortical cells exhibit ample stores of lipid granules and that there is a characteristic distribution of lipid within the gland: it occurs in cells at the interfaces of cortical and chromaffin tissues and at the periphery of the cortical tissue.

BIOSYNTHESIS OF ADRENOCORTICAL STEROIDS. Weiss (1973) identified and measured the levels of four steroids in the peripheral blood of the platypus: cortisone 8.2 $\mu g/100$ ml plasma, cortisol 5.4 $\mu g/100$ ml, and combined cortico-sterone and 11-dehydrocorticosterone at 1.8 $\mu g/100$ ml.

In homogenized platypus adrenal incubated without extra substrate for 30 min, cortisone, cortisol, and 11-dehydrocorticosterone were detected in the concen-trations of 5.1 μg, 1.3 μg, and 1.5 $\mu g/100$ mg adrenal tissue, respectively. However when [^{14}C] pregnenolone was added to the homogenates of platypus

TABLE 25

Percentage Yields of Conversion Products from [^{14}C] Pregnenolone by Platypus Adrenal Homogenates[a]

	Platypus no.		
Steroid isolated	P1	P2	P3
Cortisol	25.8	45.4	69.0
Cortisone	45.0	24.3	4.2
Corticosterone	2.1	1.0	14.6
11-Dehydrocorticosterone	11.6	10.2	2.1
Aldosterone	0.02	0.05	0.02
11-Deoxycortisol	—	0.7	1.0
17-Hydroxyprogesterone	0.5	1.1	0.6
11-Hydroxyandrostenedione	2.0	3.3	1.3
Progesterone	—	0.4	1.2
Pregnenolone	4.2	1.0	0.6

[a] Data from Weiss (1973).

adrenal nine conversion products were formed. The identities and percentage yields of these compounds are given in Table 25. The facts that cortisol, cortisone, corticosterone and 11-dehydrocorticosterone could be isolated from both plasma and adrenal tissue, and that these steroids plus the others mentioned in Table 25 could be synthesized in adrenal tissue with pregnenolone as substrate indicate that synthesis of adrenal steroids is similar to that in eutherians.

TACHYGLOSSUS

Endocrine Glands

The Pituitary Gland

ANATOMY. Hanström and Wingstrand (1951) have described the glands from two specimens of *Tachyglossus a. setosus* and one of *T. a. aculeatus*. As in *Ornithorhynchus* it is a pear-shaped body dependent from the ventral surface of the infundibulum. The arrangement of the various parts is like that shown in Fig. 46 for the platypus. Minor differences from the platypus pituitary are: a downward projection from the pars tuberalis makes contact with the pars distalis, and the pars nervosa exhibits a sacculate appearance due to extensions of the infundibular cavity and its extensions are lined by ependyma, the processes of which intermingle with those of the pituicytes—the characteristic glial cells of the pars nervosa (Griffiths, 1968).

Hanström and Wingstrand found that the pars distalis was differentiated as in *Ornithorhynchus* but in a different way; the rostral portion consisting mainly of basophils and carmine-red acidophils and the caudal part mainly of Orange-G acidophils. Fink *et al.* (1975) confirmed that the pars distalis is differented into rostral and caudal zones by the staining properties and the ultrastructure of the secretory cells. The ultrastructural identification was made on the basis of size of secretory granules and three types of cells could be distinguished: those with granules 100–200 nm in size (type I), 200–300 nm (type II), and 300–400 nm in size (type III). Types II and III were found in both parts of the gland but type II predominated in the rostral portion and type III in the caudal part. Type I cells were few in number and were scattered through the gland.

Fink *et al.* also examined levels of luteinizing and adrenocortical hormones in the rostral and caudal regions of the pars distalis but found, no doubt due to the small number of glands available, that there was no significant difference. The concentration of ACTH in the whole pars distalis was found to be very low—0.1 mg/g tissue.

THE NEUROHYPOPHYSEAL HORMONES. The pars nervosa of the pituitary gland of all vertebrates contains an extraordinary number of nerve fibers and their endings (Rasmussen, 1938). These are the axones of special nerve cells situated in the supraoptic and paraventricular nuclei of each side in the anterior hypothalamus (Fisher *et al.*, 1938). The cells and their axones exhibit neurosecretory granules that occur all the way down into the pars nervosa. Barer *et al.* (1963) isolated these granules and showed that they contained oxytocin and vasopressin, the milk ejection and antidiuretic principles, respectively, of the pars nervosa. Oxytocin and arginine vasopressin (lysine vasopressin in pigs) are characteristic of mammalian partes nervosae. Birds, reptiles, amphibians, and Dipnoi have mesotocin and vasotocin, bony fishes (including *Polypterus*) have isotocin and vasotocin, and elasmobranchs have glumitocin and vasotocin (Acher, 1971). All these peptides have a common molecular pattern characterized by a sequence of nine amino acids with a 1–6 disulphide bridge. Differences of these peptides from each other consist of differences in the amino acids at positions 3, 4, and 8:

1	2	3	4	5	6	7	8	9	
Cys	Tyr	Ile	Ser	Asp	Cys	Pro	Gln	Gly(NH$_2$)	glumitocin
							Ile		isotocin
		Gln					Ile		mesotocin
		Gln					Arg		vasotocin
		Gln					Leu		oxytocin
		Phe	Gln				Arg		vasopressin
		Phe	Gln				Lys		lysine-vasopressin

Acher *et al.* (1973) have shown by chemical means that the nonapeptides present in the pars nervosa of *Tachyglossus* are the same as those found in the eutherian pars nervosa, oxytocin, and vasopressin. This is of great interest since oxytocin is the milk "let-down" hormone (p. 257) responsible for initiation of contraction of myoepithelium surrounding the alveoli of the mammary gland. Since the pituitaries of eutherians and *Tachyglossus* have the mammalian nonapeptides, Acher *et al.* predict that the metatherian pars nervosa will prove to have the same two hormones; doubtless that of *Ornithorhynchus* will prove to do likewise. However, 28 years ago, before pure oxytocin and vasopressin were readily available, Feakes *et al.* (1950) found that injection of a posterior lobe preparation containing oxytocin and very little vasopressin induced a fall in blood pressure in the platypus whereas the pressor fraction containing vasopressin and little oxytocin brought about a rise in blood pressure. This also happened in the lizard, *Trachysaurus rugosus*, whereas in the usual laboratory mammals the pressor fraction raises blood pressure and the oxytocic fraction has no effect.

The Adrenal Gland

BIOSYNTHESIS OF ADRENOCORTICAL STEROIDS. In contrast to the situation in the platypus the level of free corticosteroids is very low in peripheral blood; in one only out of five samples of peripheral whole blood were trace amounts of cortisol and corticosterone detectable (Weiss and McDonald, 1965). Rates of secretion of those steroids into the adrenal vein of laparotomized, anesthetized echidnas were also very low; the maximum rate for corticosterone, for examples was only 2.6 μg/100 mg adrenal/hr which is minute compared with the rates of secretion found for rat, rabbit, guinea pig, and wombat adrenals (Weiss and McDonald, 1966). Free or conjugated corticosterone, cortisol, or their tetrahydro-derivatives could not be detected in the urine of echidnas kept in metabolism cages. The echidna adrenal, however, does not lack the capacity to convert [14C] pregnenolone and [14C] progesterone to other steroids (Table 26). It is seen from the data of this Table and of Table 25 that the patterns of synthesis of steroids in the adrenals of the two genera are different. The yields of cortisol, cortisone, and corticosterone from [14C] pregnenolone amounted to ca. 70% of all the steroids formed in the platypus adrenal whereas cortisone was not detected and the other two steroids were formed in very small amounts, in the echidna adrenals. These facts are in agreement with the finding that cortisone is not detectable and cortisol and corticosterone are barely detectable in peripheral blood of echidnas, whereas they are present in considerable quantities in platypus blood. However, the echidna adrenal synthesizes considerable amounts of 17-deoxysteroids (Table 26) not synthesized to any great extent by platypus adrenal tissue.

TABLE 26

Percentage Yields of Conversion Products from [14C] Pregnenolone and [14C] Progesterone by Echidna Adrenal Homogenates[a]

Steroid isolated	From [14C] progesterone		From [14C] pregnenolone	
	E1 male	E2 male	E3 male	E4 female
Cortisol	0.2	0.1	0.1	0.5
Corticosterone	0.5	0.3	0.6	0.8
11-Dehydrocorticosterone	0.2	—	0.2	6.5
Aldosterone	0.03	0.01	0.02	0.01
11 Deoxycortisol	13.5	5.5	15.3	9.5
11-Deoxycorticosterone	42.5	50.8	38.7	26.0
17α Hydroxy-20β-dihydro progesterone	0.5	0.3	0.2	—
17α Hydroxyprogesterone	7.5	9.2	2.3	2.8
Androstenedione	3.0	4.1	6.8	3.0
Progesterone	5.2	12.0	1.4	0.5
Pregnenolone	—	—	11.3	3.0

[a] Data from Weiss (1973).

EFFECTS OF ADRENALECTOMY.   McDonald and Augee (1968), intrigued by the observation of extremely low secretion rates of steroids by echidna adrenals, decided that removal of both adrenals would shed some light on the functions of the gland. Adrenalectomy was carried out in two stages, the right being removed first and the left 7–109 days later. The authors looked for signs of hypertrophy of the remaining adrenal after unilateral adrenalectomy but found no convincing evidence that it occurred.

During the first week after removal of the second adrenal, plasma potassium rose a little for 3 days and thereafter was maintained at normal levels; plasma sodium and glucose remained within the normal range as did urinary electrolytes. The animals maintained their weight and exhibited normal activity for as long as the experiments were kept going; for two echidnas this was 20 weeks after bilateral adrenalectomy. McDonald and Augee put forward two propositions concerning these data: either adrenocortical regulation of metabolism in *Tachyglossus* is negligible and not important for survival, under laboratory conditions, or that their metabolism is extremely sensitive to the action of corticosteroids and that secretion from microscopic rests of adrenal tissue, undetected after careful scrutiny, is sufficient for normal regulation.

Subsequently Augee and McDonald (1973) demonstrated that adrenocortical secretions are essential for resistance to cold stress. Adrenalectomized echidnas at rest at TA 25°C exhibited O_2 consumption of 0.15 ml/g/hr, heart rate 70 beats/min, and TB of 32.2°C, all the same as in intact echidnas (Chapter 5). But when placed in the cold room at 5°C, six recently adrenalectomized echidnas became very much more active than intact echidnas exposed to the same conditions. However, after about 6 hours of this exposure their activity rapidly decreased as did O_2 consumption and heart rate; body temperature fell to below 15°C and the echidnas became torpid within 48 hours of being exposed to the low TA. A seventh echidna that had been adrenalectomized 12 months previously remained hyperactive and maintained its body temperature for 48 hours but by 60 hours it was torpid also. If these adrenalectomized animals were allowed to remain torpid with TB within 1°C of ambient they failed to arouse and died. Those removed to an ambient temperature of 25°C immediately after falling into torpor, recovered.

It was found by Augee and McDonald that in both intact and adrenalectomized animals plasma glucose levels fell progressively from an average value of 86 mg% before any change in TB was discernible. The average plasma glucose level of torpor was 17 mg% in intact, and 9 mg% in adrenalectomized echidnas. It was also found that injections of cortisol into fasting adrenalectomized echidnas maintained blood sugar at a high level and prevented entry into torpor. Many days later, after the effects of the cortisol had worn off, these same echidnas became torpid within 36 hours when exposed to cold again. Intravenous injections or continuous infusion of glucose was found to be just as effective as corticosteroids in preventing entry into torpor. Furthermore injection of adrenaline into an intact and two adrenalectomized echidnas indicated that both could mobilize their glycogen reserves, and glucose tolerance curves in both intact and adrenalectomized echidnas were similar to those reported by Griffiths (1965a) for normal echidnas. Thus there is no defect in the ability to adrenalectomized echidnas to mobilize or utilize glucose. Augee and McDonald, therefore, conclude that the probable cause of the fall in plasma glucose and TB in adrenalectomized cold-stressed echidnas is "failure to replenish liver glycogen reserves and inability to utilize alternative energy sources in the form of mobilized fat reserves."

An observation of considerable interest made by Augee and McDonald during the above experiments was that marked hypertrophy of the adrenals occurred in echidnas repeatedly subjected to fasting and to cold temperatures. The adrenal weights of these animals (86.8 mg/kg) were almost double those of unstressed echidnas.

Subsequently Sernia and McDonald (1977a) by the use of Sephadex LH-20 column chromatography and radioligand assay, could detect only cortisol and corticosterone in the peripheral plasma of conscious, intact echidnas in the very low concentrations of 0.07 μg/100 ml and 0.14 μg/100 ml, respectively, results substantially in agreement with those of Weiss and McDonald (1965). They also detected a diurnal rhythm in the combined plasma concentration of those steroids. The very low plasma concentrations could be increased by intravenous infusion of synthetic ACTH (Fig. 49); maximal values of 0.42 μg/100 ml for cortisol and 1.06 μg/100 ml for corticosterone were observed during infusion of ACTH at the rate of 1 iu/kg/hr. This, however, is but 1/160 the potency of that ACTH in man.

Using these techniques Sernia and McDonald (1977c) found that the metabolic clearance and production rates of cortisol and corticosterone increased in fasting echidnas repeatedly exposed to cold (4°C), while the plasma corticosteroid concentration decreased. In spite of the increase in size of the adrenals, mentioned above, the secretory capacity was apparently less since the response to ACTH was less after exposure to cold. They also measured concentration of plasma corticosteroids in a fasting echidna chronically exposed to TA of 4°C and found a slow decrease in corticosteroid concentration as TB dropped, the lowest value being attained during torpor. When this echidna was returned to room temperature there was an initially rapid increase in corticosteroid concentration followed by a gradual return to preexperimental concentrations. Sernia and McDonald (1977c) conclude from these data that specific changes in adrenocortical function are related to thermogenic activity in echidnas; doubtless these changes are the prime cause of the fall in blood glucose which is the essential factor leading to torpor when exposed to low TA.

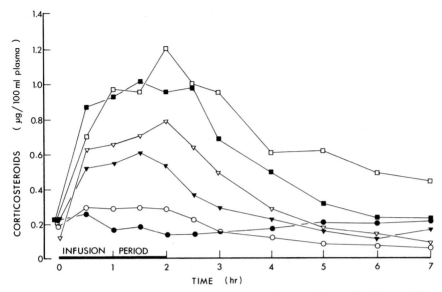

Figure 49. *Tachyglossus.* Changes in the concentration of plasma corticosteroids during and after 2-hour infusion of ACTH. ●, Saline; ○, ACTH 0.016 iu/kg/hr; ▼, ACTH 0.04 iu/kg/hr; ▽, ACTH 0.20 iu/kg/hr; ■, ACTH, 1.00 iu/kg/hr; □, ACTH 5.00 iu/kg/hr. *(From Sernia and MacDonald, 1977a.)*

METABOLIC EFFECTS OF CORTISOL, CORTICOSTERONE, AND ACTH. Following the discovery of the importance of the adrenals in maintenance of TB under conditions of cold stress, Sernia and McDonald (1977b) made a study of the effects of cortisol, corticosterone, and ACTH on indices of carbohydrate, protein, and fat metabolism in intact echidnas. Short-term infusions of each of those hormones had a slight hyperglycemic effect, and no significant effects on levels of plasma amino acids and urea. The treatments, however, did bring about considerable increases in the levels of free fatty acids in plasma. Long-term infusions over a period of days gave the same results. In the latter experiments nitrogen intake and excretion were determined but the treatment with corticoids brought about no changes. Some years ago Griffiths (1965a) reported that massive doses of cortisone (25 mg/day) injected intramuscularly into an echidna failed to induce diabetes—a finding in agreement with the above. Another experiment carried out subsequently on one animal shows just how refractory the metabolism of normal intact *Tachyglossus* is to corticosteroids: initial body weight was 1700 g, it ingested 1.77 g termite nitrogen + 10 g glucose daily for 8 days, body weight gain was 22 g/day, daily NH_3-N was 34.6 mg and daily urea–nitrogen excretion was 0.99 g. It then received 50 mg cortisol intramuscularly daily for 9 days and ingested the same diet as before quantitatively and qualitatively. During the fifth 24-hour period it excreted 830 mg glucose and during the sixth 350 mg; after that no glycosuria was detectable. NH_3-N and urea–nitrogen excretion were practically the same as before—39.4 mg and 1.01 g/day, respectively; however, body weight increase was greater—35 g/day. On days 5 and 6 the blood sugar level must have exceeded 200 mg% since the renal threshold for echidnas is about the same as for man—180 mg% (Griffiths, 1965a). However, the point is that it took the fantastic dosage of 25 mg/kg/day of cortisol, which is a powerful glucocorticoid, plus ingestion of 10 g glucose along with the food to achieve that very mild hyperglycemia.

Renin Secretion

Renin is an enzyme secreted by granular cells in the walls of juxtaglomerular arterioles of the eutherian kidney. The enzyme acts on its substrate, angiotensinogen, a serum α_2 globulin formed by the liver, splitting off a polypeptide, angiotensin I. A dipeptide is split off this by angiotensinase yielding angiotensin II which is a potent pressor substance. Angiotensin II also stimulates the adrenal cortex to secrete aldosterone resulting in Na^+ retention. This, coupled with the observation of McDonald and Augee (1968) that bilateral adrenalectomy has practically no effect on fluid and electrolyte balance in *Tachyglossus* prompted Reid (1971) to determine whether or not *Tachyglossus* has a renin-angiotensin system. It was found that the echidna kidney has a juxtaglomerular apparatus identical to the renin-secreting organ in the eutherian kidney but no plasma renin could be demonstrated in normal nor in adrenalectomized echidnas. However it was easily measurable in extracts of kidney tissue and in the plasma of an echidna depleted of sodium by treatment with frusemide. Depletion of sodium also brought about an increase in renal renin content and in plasma angiotensinogen. The renal renin content, however, was not altered by adrenalectomy. According to Reid this result together with the observation of increases in plasma and renal renin following sodium depletion supports the notion that the increase in activity of the renin angiotensin system observed in eutherians following adrenalectomy is an effect caused by sodium depletion and not by the absence of adrenal steroids. Reid also points out that what is needed right now is a study of the effect of angiotensin on the secretion of aldosterone in echidnas; apparently in the absence of the renin angiotensin system the monotreme adrenal is almost incapable of synthesizing that steroid (Tables 25 and 26) (see Appendix).

The Islets of Langerhans

The pancreas is not diffuse as in rats and rabbits but discrete as in carnivora, apes, and man—the islets, in the form of spheroidal groups of cells, being distributed throughout the exocrine tissue. The islets exhibit the usual alpha, beta, and indifferent cells found in the islets of all vertebrates. In birds, mammals, and apparently in reptiles (Miller and Lagios, 1970) the alpha cells secrete glucagon, the hyperglycemic principle, and the beta cells secrete insulin. Crude insulin preparations made from echidna pancreas exhibited a glucagon effect only (Griffiths, 1965a) indicating that more glucagon than insulin is stored in the pancreas; however, the quick fall of the blood sugar to normal levels in glucose tolerance tests (Griffiths, 1965a; Augee and McDonald, 1973) argues that a rapid release of stored insulin takes place in response to hyperglycemia. Hypoglycemia follows injection of insulin into echidnas (Griffiths, 1965a; Sernia, 1977); blood sugar levels as low as 30 mg% are tolerated without distress. Injection of insulin leads to a rise in free fatty acids and falls in amino acids and urea of the plasma (Sernia, 1977). However, anabolic effects of insulin in echidnas have not yet been detected (Griffiths, 1968).

The Thyroid Gland and Other Derivatives of the Pharyngeal Pouches

The development of the thyroids, thymus, parathyroids, carotid glands, and the postbranchial bodies have been described by Maurer (1899). He had embryonic stages 40–47 (see p. 241 for description of these embryos) and a pouch young 12 cm long available for study. At stage 40 the usual four pairs of pharyngeal pouches are present. The first pair extends between the mandibular and hyoid arches and come into contact with the auditory capsules giving rise to the tympanic cavities on either side and to the Eustachian tubes. The second pair gives rise to two outgrowths that make contact with the wall of the third arterial arch; each of these outgrowths is nipped off and transformed into ganglion tissue and along with other cellular elements forms the carotid body. On the ventral surface of the pharynx between the first and second sets of pouches a median diverticulum is present; when freed from the pharynx this will form the thyroid gland. The third pair of pouches gives rise to two sets of outgrowths. These are the anlagen of paired thymus and parathyroid glands; later they will break away from the pouches and migrate caudally along with the thyroid anlage. By stage 47, i.e., just after hatching, this complex of glands comes to lie in the thoracic cavity posterior to the diaphragm and attaches itself to the trachea. The thymus glands are well developed at this stage (Prof. G. Schofield, personal communication). The complex is finally joined by another pair of parathyroids derived from the fourth pair of pharyngeal pouches. In the adult echidna all these glands are intermingled with fat so that the thyroid cannot be distinguished as a separate entity (Augee, 1969) especially as it is also spread in a diffuse manner in the region of the aortic arch and innominate arteries. This makes it impossible to remove the thyroid completely and any attempt to remove it also involves removal of parathyroid and thymus tissue. Augee noted that the thyroids of echidnas that had been in torpor were hypertrophied so that they could be distinguished as discrete glands. Histologically the thyroids of normal echidnas exhibited the usual collection of follicles lined by a single layer of epithelial cells surrounding a central mass of colloid, the follicles varying considerably in size from small to large. However, Augee detected a trend towards a greater number of large follicles in the thyroids of the echidnas killed during or following torpor.

M. L. Augee and A. J. Hulbert (personal communication) are at present studying metabolic rates in echidnas "thyroidectomized" by treatment with radioactive iodine.

Glands of the Immune System

Anatomy

The lymphoid tissue in eutherians is organized for the most part into lymph glands. Each gland has a capsule and trabeculae dividing the gland into a number of nodules each filled with lymphocytes; at the center of each nodule is a germinal region of pyroninophil cells undergoing mitosis and forming new lymphocytes. Each lymph gland is supplied with an afferent lymph vessel that breaks down into a number of sinuses bathing the lymphocytes with lymph, and efferent lymphatics that convey lymphocytes and lymph away from the gland. In *Tachyglossus* the nodules are not grouped into glands but occur as entities scattered in the chest, neck, pelvic region, mesenteries, pharyngeal region, and some suspended in the lumina of large lymphatic vessels, an arrangement reminiscent of the sinuses in the eutherian lymph gland. Each nodule exhibits a cortical zone of small lymphocytes and a germinal center of pyroninophil cells.

The spleen consists of two long flat straps and where they meet a third strap with a club-shaped appendage is found giving the organ a triradiate shape (Cuvier, quoted by Owen, 1839–1847). Basir (1932) found that the spleen is invested with a capsule of fibrous and smooth muscle tissue and from the capsule trabeculae invade the splenic pulp. The latter consists of a reticulate syncytium forming a loose stroma filled with erythrocytes. The arteries entering the pulp are enclosed in connective tissue sheaths which become invaded by small lymphocytes to form the white pulp or Malpighian corpuscles. The red pulp consists of reticular cells, macrophages, and erythrocytes. According to Basir there are no germinal centers in the white pulp but Diener and Ealey (1965) speak of marginal zones and of germinal centers that contain pyroninophil cells in the Malpighian corpuscles.

Each thymus gland consists of highly vascular lobules each differentiated into cortical and medullary zones—the cortex exhibiting densely packed small lymphocytes among which are dispersed a few reticulocytes. In the medullary zone, however, there are few lymphocytes, many reticulocytes, and many Hassal bodies—rounded acidophilic structures of concentrically arranged cells. Apparently the thymus is large and remains well developed in adult echidnas since

Diener and Ealey say they studied the glands in echidnas selected from a colony used for physiological studies.

By far the most interesting of the glands of the echidna immune system are those associated with the gut. There is an appendix about 2.5 cm in length situated at the commencement of the large intestine; Diener and Ealey (1965) and Diener *et al.* (1967b) found it contained numerous lymph nodules and they also found Peyer's patches in the gut. Subsequently Schofield and Cahill (1969) confirmed Klaatsch's (1892) observations of tubular submuscosal extensions of intestinal glands into lymphoid aggregations in ileum and appendix and also the same structures in colon and cloaca. Many of the submucosal lymphoid nodules studied in all four regions mentioned exhibit a cortex and a pyroninophil germinal center; in ileum and colon the nodules are macroscopically visible as Peyer's patches. In the cloaca, however, the submucosal nodules do not form conspicuous elevations on the surface of the mucosa. The submucosal lymphoid nodules in ileum, appendix, colon, and cloaca all exhibit invasions by compound elongated intestinal glands which extend into the central area of the subjacent lymphoid tissue. This intimate association of lymphoid tissue and gut glands has been termed a lymphoepithelial gland by Schofield and Cahill who draw attention to the fact that these glands bear a strong resemblance to those found in the bursa of Fabricius in very young birds, but not in adults. Lymphoepithelial glands, however, could not be detected in the gut of a newly-hatched echidna (Prof. G. Schofield, personal communication).

Immune responses of the lymphoid system to injection of ^{125}I-labeled flagellar antigen of *Salmonella adelaide* were described in a preliminary report by Diener and Ealey (1965) in which differences from the responses of eutherian immune systems were emphasized; however, in their full report (Diener *et al.*, 1967b) it can be found that there are many similarities. Following injection of labeled antigen biopsy samples of lymph nodules were taken at intervals of 5 hours to 28 days. The uptake of antigen was found to be erratic, some nodules showing considerable uptake while others a few millimeters away showed none. At first the antigen, in those nodules taking it up, was localized at the periphery, but 5–7 days later it was found in varying degrees of concentration in the central proliferative area. The antigen trapping in the peripheral area is associated with a network of reticular fibers—structures identical to those responsible for the trapping of antigen in the lymph glands of rats (Miller and Nossal, 1964). Another resemblance to the rat lymph nodule is tbe accumulation in some nodules of antigen to form, in an eccentrically located germinal center, "a crescentic cap, strikingly similar to the pattern reported in the secondary follicles of the rat lymph node."

The distribution of cell-proliferative activity in the lymph nodule of the echidna was studied with the aid of [³H] thymidine injected into antigenically stimulated animals. It was found that there were far more labeled cells in the germinal center than in the periphery of the nodule and that macrophages (tingible bodies) also contained the label indicating, the authors say, that they contain products from cells that were actively synthesizing DNA at the time of thymidine injection.

Antigen was also located in lymphoid nodules in spleen, appendix, Peyer's patches, and thymus including Hassal bodies in the latter organ.

The serological response to the flagellar antigen was production of antibody in varying amounts from echidna to echidna. The best responses were obtained by injecting antigen into the jugular region where the main mass of lymph nodules is situated. Each echidna was injected twice, once at the start of the experiment and then 7 weeks later. All echidnas gave a good primary response (Table 27) and three out of the five animals gave a better secondary than primary response (1, 2, and 3 in Table 27). However, none of those secondary responses were the result of full replacement of mercaptoethanol (ME)-sensitive antibody by ME-resistant antibody as happens in the rat. In one echidna there was no anamnestic response at all. Other experiments showed that the occurrence of the anamnestic response was erratic and that it was much smaller in magnitude than in eutherians. Diener et al. (1967a) adduce evidence that echidna ME-sensitive antibody is analogous

TABLE 27

Primary and Secondary Immune Response of *Tachyglossus* to 1 mg of Flagella from *Salmonella adelaide*, Injected into the Jugular Region[a,b]

Weeks after primary immunization	Animal number									
	1		2		3		4		5	
	T	MER	T	MER	T	MER	T	MER	T	MER
1	<5		<5		<5		<5		<5	
2	480	120	30	30	60	60	60	60	320	320
3	1200	140	960	30	1280	1280	1280	960	2560	2560
4	9600	3200	1600	100	6400	1600	3200	3200	25600	12800
5	1600	800			1600	800	1600	800	12800	4800
6	400	400	6400	6400	CH		600	100	12800	1200
7	CH		3200	1200	3840	60	CH		6400	1200
8	2400	400	CH		1280	160	400	100	CH	
9	25600	1200	6400	3200	3200	1280	1600	300	2400	2400
10	12800	6400	9600	1600	25600	12800	2400	1600	800	300
11			12800	6400	25600	12800	2400	200	1600	1600
12			4800	3200	12800	600	4800	1600	3200	50
13							3200	3200	800	800

[a] Data of Diener et al. (1967a).

[b] CH, challenging injection; T, total antibody titer; MER, mercaptoethanol resistant antibody titer.

to the IgM immunoglobulin and ME-resistant antibody to the IgG immunoglobulin of eutherians.

The anamnestic response in echidnas is apparently no better than that found in reptiles. In the rhynchocephalian reptile, *Sphenodon punctatus* maintained at 20°C, secondary responses to *Salmonella adelaide* flagellar antigen consisted of a return to tbe level of titers observed during the primary response (Marchalonis *et al.*, 1969) and the antibodies to this antigen were confined to the IgM class only. However, it has been shown by Wetherall and Turner (1972) that immunological responses in another reptile, the viviparous lizard *Tiliqua rugosa*, depend on the type of antigen used: anamnestic responses were observed with two of the three antigens used, two classes of antibody IgG and IgM were formed in responses to two of the antigens but only IgM class of antibody was formed in response to another antigen. Moreover the kinetics and magnitude of the responses were dependent on body temperature. It is not unlikely, therefore, that the immune response of *Sphenodon* could equal that of *Tiliqua* if the former were maintained at TB of 25°–30°C as was the case for the experiments giving the "best" results with *Tiliqua*. The point of the above, however, is that the immune response of *Tachyglossus* is no better than that of a reptile, *Tiliqua*.

A Note on Transfer of Passive Immunity

In a female carrying a 40-day-old pouch young antibodies to red cells of *Macropus giganteus* were raised (M. Griffiths and D. W. Cooper, unpublished). Two injections of erythrocytes were given during a period of 19 days; on the 19th day good titers of antibody were present in both blood and milk of the female but none could be detected in the blood of the pouch young, now 59 days old. This is merely a start on the problem but it indicates that if passive immunity can be acquired through the milk experiments will have to be carried out on females carrying pouch young much less than 40 days old.

7

Special Senses, Organization of the Neocortex, and Behavior

ORNITHORHYNCHUS

Olfaction

THE OLFACTORY ORGANS. The essential organ of olfaction in the platypus, or for that matter in all vertebrates, is the collection of first-order neurones in the olfactory epithelium lining the turbinals of the nasal passages and capsules. Olfaction would not seem to be of prime importance to the platypus since the olfactory epithelium is cut off from outside stimuli by the narial valves when the animal is under the water feeding (p. 3). However, sense of smell may be of some importance in courtship and in the reflex induction of milk ejection when feeding young in the breeding burrow (p. 257). The notion that olfaction is not of prime importance is supported by the fact that the turbinals are fewer in number and far smaller in extent than in *Tachyglossus* which almost certainly uses the sense of smell for detection of prey. In *Ornithorhynchus* there is a maxillary turbinal comparable in size to that in *Tachyglossus* but the nasal and ethmoidal turbinals are relatively tiny. As well as those organs of olfaction, monotremes also exhibit Jacobson's organ found in reptiles and many other mammals. It consists of a pair of pouches, originally formed from the nasal capsules, which open into the anterior part of the mouth cavity. They probably serve to smell food in the mouth.

HISTOLOGY OF THE OLFACTORY BULBS. The first-order neurones in the olfactory epithelium are directly exposed to the stimulus drawn into the nasal passages and capsule. The fibers (axons) from the neurones ensheath the olfactory bulbs and form the olfactory nerve; there are no synaptic connections between adjacent fibers so that the sensory message conveyed by each individual fiber remains undisturbed until it reaches the glomerular relay. Here the fibers from the olfactory sense cells discharge into the dendrites of the mitral cells and the tufted cells. Fibers from many olfactory cells connect with the brain dendrite of a single mitral or tufted cell. This arrangement (direct stimulation of the receptor and summation of energy from many olfactory fibers on a single neurone of the second order) helps to explain the extremely low thresholds of olfactory stimuli.

In the metatherian and eutherian olfactory bulb the mitral cells form a conspicuous narrow layer of densely packed cells. The perikaryon of each cell is large (about 20 nm in diameter) and its shape resembles a bishop's mitre. The tufted cells, which also are large, are found midway between the glomeruli and the mitral cell layer. The dendritic processes of these tufted cells are frequently connected to more than one glomerulus and thus to a larger spectrum of sensory messages from the receptor neurones than the mitral cells are. The axons of the mitral and tufted cells together pass caudad at the lateral and medial olfactory tracts to various centers in the forebrain. In addition to the above two types of cell, there is a third, peculiar type which has no axon but has several dendrites— the granule cells. These are formed into a granule cell layer; their dendrites divide into many small branches that form synaptic junctions with the lateral dendrites of the mitral cells. In the platypus olfactory bulbs, in the place where one would expect to find a well-defined layer of mitral cells, there is an indistinct line made up of a few granule cells (Switzer and Johnson, 1977). There are, however, tufted cells with large perikaryons lying between the granule cells and the glomeruli. The bulb in *Tachyglossus* also lacks a mitral cell layer.

The monotreme olfactory bulb is not alone in lacking a well-defined mitral cell layer; the bulbs of various species of reptiles and birds exhibit every gradation from well-defined layers of mitral cells through more scattered layers to those with a diffuse broad band of tufted cells. However, in all vertebrates studied so far (Andres, 1970; Rudin, 1974; Switzer and Johnson, 1977), the bulbs have diffusely distributed cells with large perikaryons which may be considered to be tufted cells, and only certain bulbs have the mitral cell monolayer. In view of this Switzer and Johnson suggest that the mitral cells are only a layered subset of the tufted cells. Since metatherian and eutherian bulbs have a mitral monolayer and the monotremes do not, the question arises what olfactory *tours de force* can the former carry out that the latter cannot?

OLFACTORY PATHWAYS IN THE BRAIN.   Hines (1929) states that the olfactory tract from the bulbs follows much the same pattern as they do in other mammals but that she was unable to trace all pathways with certainty. The structures and tracts appear to be as follows: the olfactory tract divides into medial and lateral tracts, the medial going to the hippocampus via the fornix and mamillary peduncular tract to the mammillary body thence to lower motor centers in the medulla oblongata. Medial tract fibers also pass to the habenular ganglion and the peduncular system via the fasciculus retroflexus. The lateral tract fibers reach the pyriform cortex, hippocampus, and amygdala (lateral olfactory area) and then pass via the fornix to the mammillary body. Thus the discharge into the motor systems is similar to that of the medial olfactory tract. These pathways are similar to those found in other mammals.

Organs of Taste

The tongue of *Ornithorhynchus* exhibits two small slits on the dorsal surface of its rounded hind portion just posterior to the two keratinous teeth. These slits lead down to deeply seated circumvallate papillae (Oppel, 1899), the organs of taste characteristic of mammals. Laterally at the posterior end of the tongue two foliate papillae are located on each side. As far as I know the structures of these vallate and foliate papillae have not been studied but they are known for the tongue of *Tachyglossus* (p. 183).

Vision

ANATOMY OF THE EYE. Walls (1942) remarks "The eye of *Ornithorhynchus* has been described once only, by Gunn in 1884 from material preserved in whisky by a Mr. Sinclair who clearly took his science very seriously." Walls has given another description based on material sent to him by Dr. K. O'Day of Melbourne who, apparently, used another fixative. The eye is spherical and very small, only 6 mm in diameter. Both lids exhibit tarsal plates and there is a nictitating membrane; lacrimal and Harderian glands are present. The sclera which surrounds the posterior two thirds of the eye is quite remarkable in that it consists of fibrous tissue underlain by a layer of cartilage; a cartilaginous cup is unknown in the eyes of metatherians and eutherians. The cartilaginous cup is 400 μ thick fundally but it tapers to 25 μ at its terminal margin. The extrinsic muscles are inserted onto the sclera and their origins have the same relationships as they have in other mammals: the four rectus muscles, superior, inferior, medial, and lateral take their origins at the rear of the orbit; distally they become tendinous at their insertions onto the sclera. As far as the superior oblique is concerned the platypus, according to Walls (1942), exhibits that fantastic arrangement found only in the mammals: the muscle passes forward from its origin with the recti, becomes tendinous, and passes through a pulley at the anterior nasal wall of the orbit; it is then sharply inflected medially and is inserted onto the dorsal surface of the eyeball. Göppert (1894) and Franz (1934) describe the same arrangement of the superior oblique in the monotremes. Saban (1969b), without reference to Walls in this connection but acknowledging Göppert and Franz, flatly denies that the superior oblique passes through a pulley and refers to Göppert and Franz in these terms "Ces auteurs semblant avoir confondu l'oblique supérieur avec le retractor de la nictitante qui correspond à cette description.'' Walls specifically states that the pulley in *Ornithorhynchus* is chondroid rather than soft as in the echidnas. The inferior oblique arises at the nasal side of the anterior part of the orbit and passes medially to be inserted on the underside of the eye ball.

Movement of the upper lid is brought about by contraction of the m. palpebralis lateralis and the m.p. medialis; the lower lid is moved by the m.p. inferior while the nictitating membrane has its own motor muscle, the m. rectractor membranae nictitantis (Saban, 1969b).

Internal to the cartilage layer is a thin (50 μ) highly vascular pigmented chorioid, which has capillaries of large diameter. The chorioid is continued forward to form the unveal portion of the ciliary body which exhibits a number of rather thick processes. At the anterior end of the ciliary body the processes are joined to each other by membranes forming an annular ring—the ciliary web. Walls says there is no trace of a ciliary muscle and that the web ''is a decidedly mammalian character shared by many marsupials and placentals but by no sauropsidan.'' The iris is simple, consisting of little more than a prolongation of the two retinal layers and a few small blood vessels attached loosely to the anterior face; a dilator muscle is not present but there is a very well-developed sphincter muscle around the aperture of the iris.

The lens is relatively flat having a flatness index (diameter divided by thickness) of 1.38–1.50 but it is not nearly as flat as that of *Tachyglossus* (p. 184). It is suspended by zonule fibers arising in the ciliary body and web. Walls says ''no monotreme has any demonstrable accommodation... It is not known whether *Ornithorhynchus* approaches emmetropia in either air or water, but the implications are that the eye is better adjusted to the latter medium.'' This would seem to be of little importance since platypuses close their eyes when under water.

The retina in *Ornithorhynchus* is thin and avascular; there are three rows of outer nuclei, four of inner, and a single row of ganglion cells. Rods and cones are present, the cones being double or single and both kinds exhibit an intracellular oil droplet. The retinofugal fibers pass into the brain as the optic nerve which, as Hines (1929) points out, is very small. Nevertheless Bruesch and Arey (1942) found that the number of fibers therein was between 31,000 and 33,000; they suggested that this high number was achieved by close packing of very fine fibers. The number found is about double that for the *Tachyglossus* optic nerve.

PRIMARY OPTIC PATHWAYS. The path of the retinofugal fibers within the central nervous system has been traced by Campbell and Hayhow (1972) who used the techniques of unilateral eye enucleation; after a period of 17–18 days postoperative survival, the brains were sectioned and the tracts identified therein by the presence of degenerating axons stained by the method of Nauta and Gygax (1954) or of Fink and Heimer (1967).

Decussation in the chiasma was found to be practically complete, a few fibers only passing to the ipselateral side of the brain. The optic tracts course lateral-caudally on the ventral surface of the diencephalon and then pass dorsally. En route, groups of fibers leave the dorsal aspect of the proximal part of the contralateral tract and pass medially and caudad through the ventral diencephalon to the rostral tegmentum of the midbrain. Here they terminate in a nucleus designated by Campbell and Hayhow, the medial terminal nucleus of the accessory optic system.

The rest of the fibers of the optic tract pass into the lateral geniculate system consisting of two nuclei disparate in size and location which are designated LGNa and LGNb by Campbell and Hayhow (1972). Most of these fibers curve dorsally round the posterior margin of the nucleus ventrolateralis thalami and terminate in the prominent lateral geniculate nucleus—LGNa. The rest of the fibers that do not terminate here pass over the dorsal surface of the nucleus and form two diverging tracts; the larger of these runs medially along the anterior margin of the superior colliculus and is known as the superior brachium. Some of the fibers passing in the brachium terminate on a group of scattered cells embedded in the general mass of the fibers. This group, from its situation, resembles the nucleus of the optic tract found in the brachiums of eutherians and metatherians. The rest of the fibers in the brachium turn caudally over the surface of the colliculus forming the stratum fibrarum zonalium; the fibers terminate immediately below in the stratum griseum.

As mentioned above, the brachium constitutes the larger of two tracts passing through LGNa without terminating; the smaller passes forward on the dorsal surface of the nucleus ventrolateralis thalami. The fibers of this tract terminate in a rather diffuse series of cells designated LGNb which is much smaller than LGNa and, of course, receives far fewer fibers. These fibers to LGNb seem to be, according to Campbell and Hayhow, bilateral but the contralateral projection is by far the larger; there is also ipselateral projection of retinofugal fibers to LGNa. The above findings will be compared with those found for the primary optic tracts in *Tachyglossus* (p. 186).

No experiments have been carried out to determine whether or not platypuses are capable of visual discriminations or to determine the dioptric characteristics of the eye, as they have been for *Tachyglossus*.

Hearing

ANATOMY OF THE EAR. The tympanic bone and membrane are situated on the ventral surface of the skull and the external orifice of the ear is on the dorsal surface of the head; sound waves entering at the ear hole pass down a long curved cartilaginous "ear trumpet" which is directed ventrally, medially, and finally dorsally to the tympanic bone.* The cartilage is soft and flexible and lacks the cartilaginous strips that stiffen the ear tubes of *Tachyglossus*. The ear tube is attached to the tympanic bone and so forms a "tent" over the middle ear bones (Gates *et al.*, 1974). The three cartilages and the ossification found at the posterior end of Meckel's cartilage in the chondrocranium have become tympanic, malleus, incus, and stapes of the middle ear. The tympanic is hook-shaped, the long axis of the hook lying lateromedially; the tympanic membrane suspended within the hook is oval in shape and firmly attached to it is the long manubrium of the malleus. Relative to the malleus, the incus is small and its articulation with the stapes is tiny since the latter is a long thin columella; however, its base is expanded into a circular plate that sits on the membrane of the fenestra ovalis (f. vestibuli) located in the roof of the tympanic cavity which is filled with air. Gates *et al.* (1974) state that although the tympanic, malleus, and incus are firmly connected to one another, they can be separated from one another without difficulty, unlike those of *Tachyglossus* (p. 189). Although the ear ossicles of the platypus (and of the echidna) are very like those of the wombat and koala (Gregory, 1947), in fact most marsupials have a collumelliform stapes (p. 190) as does the pangolin *Manis javanica*. Gates *et al.* (1974) point out that the middle ear of the platypus is not firmly anchored to bony structures but "floats" in a cavity surrounded by connective tissue and muscle. On the other side of the membrane in the oval window is a cavity, the vestibule, leading to the bony labyrinth and the inner ear. We are indebted to Pritchard (1881) for his account, a classic of 19th century anatomical description, of the structures of that organ:

The cochlear portion of the bony labyrinth consists of a curved tube about 6.3 long and 1.3 mm in diameter projecting from the vestibule and embedded in the petrosal bone. The tube terminates in a slightly enlarged rounded extremity. The enlargement is due to the coiling up of the anterior extremity of the membranous labyrinth showing a tendency to a spiral form. Thus as Pritchard puts it, "the ductus cochlearis of the *Ornithorhynchus* may be likened to that of a typical mammal which has been unwound except at its extreme apex." The cochlea is divided into three compartments: the scalae tympani, vestibuli, and media (ductus cochlearis). The vestibule and scalae vestibuli and tympani are filled with perilymph which also bathes the outside of the membranous labyrinth. Near the apex of the cochlea the scala vestibuli communicates with the scala tympani by means of an oval hole—the helicotrema.

The scala media is filled with endolymph since it is an extension of the membranous labyrinth. It is connected posteriorly to the sacculus "by means of a circular tube with delicate membranous walls in a manner that I have not com-

---

*There is no tympanic bulla in the monotremes nor is there a bony external auditory meatus.

pletely made out." Doubtless Pritchard is talking about the ductus reuniens described by Alexander (1904) in the echidna membranous labyrinth (Fig. 57). The scala media is triangular in cross section, the base of the triangle being the membrana basilaris which separates the scala media from the scala tympani; the second side of the triangle is formed by Reissner's membrane separating the scalae media and vestibuli while the third side of the scala media is the stria vascularis exhibiting numerous blood vessels running parallel to the long axis of the cochlea.

Lying along the whole length of the membrana basilaris is a true organ of Corti closely resembling that of the eutherian and metatherian cochlea and emphatically unlike the compact auditory papilla of the sauropsidan cochlea. The organ of Corti consists of two rows of rods separated from each other proximally but fused distally so that the two rows and the basal membrane on which they stand form a tunnel triangular in cross section. On each side of the rods rows of hair cells are arranged, two rows to the center side of the center rods and three to the inner side of the inner rods. The hair cells are elongated, have rounded nuclei, and bear distally four or five bristles or hairs. A reticulate membrane covers the upper surfaces of the rod heads and hair cells, the hairs passing through the interstices of the membrane. Passing from the membrana reticularis to the membrana basilaris is a series of thick fibers or trabeculae each attached to the basal membrane by an expanded foot plate. Between the trabeculae and below the level of the hair cells are a number of nuclei, apparently not surrounded by cell walls—Deiter's cells. The inner and outer boundaries to the organ are formed of columnar cells (the supporting cells) and lateral to these two longitudinal grooves run along the whole length of the organ, the marginal and secondary sulci [the internal and external sulci in Alexander's (1904) terminology for the echidna organ of Corti]. From the top of the marginal sulcus a shelflike structure projects out into the scala media and overhangs the organ of Corti; this is the tectorial membrane.

The organ of Corti does not extend quite to the end of the scala media; distal to its termination the ductus cochlearis exhibits a construction, the isthmus lagena, which opens into an expanded chamber, the lagena. Within the lagena is a macula similar to the maculae acusticae of the saccule and utricle; this lagenar macula is made up of nerve fibers and columnar epithelial cells bearing at their distal end hairs or cilia which project out into a mucoid mass in the lagena chamber. Pritchard surmised that in suitably fixed cochleas this mass would exhibit otoliths and Alexander (1904) did indeed demonstrate their presence in a "gelatinous mass" found in the lagena of *Tachyglossus*.

The chief interest of the lagena and its macula lies in the fact that it has not been found in the cochleas of other mammals but it does occur in those of birds and reptiles. Thus once again we have an instance of that curious intermingling of reptilian and mammalian characteristics in a prototherian organ: a cochlea with incipient coiling intermediate between the straight cochlea of the Mesozoic mammal Triconodon (Simpson, 1928) and the spiral cochlea of the metatherian and eutherian ear, with a lagena and lagenar macula, yet with a true organ of Corti, identical to that found in the mammalian cochlea for the conversion of sound waves to nerve impulses in the cochlear division of the eighth nerve.

The acoustic nerve consists of fibers from the bipolar cells of a ganglion located in modified bone tissue lying lateral to the scala media; this set of fibers passes from the ganglion to the dorsal and ventral cochlear nuclei in the medulla oblongata. The fibers from the other poles of the ganglion cells pass to the lower lip of the marginal sulcus through a single row of holes, the habenula perforata, and so to the hair cells of the organ of Corti.

Denker (1901) and Simpson (1938) have made, inter alia, further observations on the osteography of the periotic bone of *Ornithorhynchus* which enable one to round-off the description of the inner ear: as mentioned above the scalae vestibuli and tympani are filled with perilymph and communicate with one another through the helicotrema. The perilymph apparently is elaborated in the sub-arachnoid space and is passed to the scala tympani through a channel in the bony labyrinth, the cochlear aqueduct. The perilymph is retained within the system by the membrane in the oval window and by the membrane in another window at the base of the scala tympani—the fenestra rotunda or round window that looks through into the tympanic cavity as does the oval window; the tympanic cavity is filled with air.

Although the vestibular labyrinth and the scala media are filled with en-dolymph the areas of secretion and resorption of this fluid are not definitely known even for eutherians and man (Schuknecht, 1970). It is presumed to be secreted in the scala media by the stria vascularis and by an entirely different tissue in the vestibular labyrinth. Apparently the endolymphatic sac is an organ for resorption of endolymphs in both parts of the labyrinth. This sac in *Or-nithorhynchus* (Denker, 1901) is situated on the endocranial surface of the peri-otic and communicates with the saccule and utricle of the membranous labyrinth by the ductus endolymphaticus. The endolymph of the saccule and scala media are in communication through the ductus reuniens, that of the utricle passes to the three semicircular canals. The arrangement of the parts of the membranous labyrinth figured by Denker is much like that shown in Fig. 57 for *Tachy-glossus*.

PERIPHERAL AUDITORY FUNCTION. From what is known of middle and inner ear function in eutherians and marsupials it can be assumed that the follow-ing holds good for *Ornithorhynchus*: sound waves pass down the cartilaginous ear trumpets and set up vibrations in the tympanic membrane which are con-ducted by the malleus, incus, and stapes to the membrane in the oval window. These waves are transmitted through the perilymph and since the membrane in the round window is elastic it can bulge into the tympanic cavity (filled with air); thus mechanical movement of the stapes produces corresponding movement in the perilymph; since the scala media is surrounded by perilymph these movements are transmitted via the endolymph to the tectorial membrane in which the cilia of the hair cells of Corti's organ are embedded. It is probable that movements of the perilymph can also be transmitted directly to the organ of Corti since there is evidence that the tunnel is filled with perilymph (Schukneckt, 1970). Whatever the mechanism involved, deformation of the cilia of the hair cells leads to generation of minute electrical potentials. These potentials, known as cochlear microphonics (CM), follow exactly sound pressure vibrations and these can be measured by an electrode placed on the round window. Gates *et al.* (1974) have recorded CM at the round window in *Ornithorhynchus* in response to pure tones of frequencies between 0.5 and 20.0 kHz. At selected frequencies between these two limits the sound pressure level (SPL) was increased until a potential of $1.0 \mu V$ was achieved at the round window. This SPL was defined as the threshold for the frequency. From the plot of sound intensity (expressed as

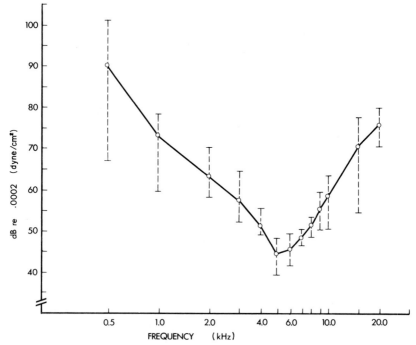

Figure 50. *Ornithorhynchus.* The 1.0-μV cochlear microphonic threshold curve. Best frequency lies at 5 kHz with high frequency rolloff of 20 dB/octave and low frequency rolloff of 15 dB/octave. *(From Gates et al., 1974.)*

decibels) as a function of frequency.(Fig. 50) it can be seen that the best sensitivity occurred at 5 kHz. The general shape and the location of the most effective frequency on the curve is strikingly like that for the echidna CM audiogram (Johnstone, cited in Griffiths, 1968) except that the thresholds or the potential elicited at the echidna round window was 10 μV at the same sound pressure levels applied in the platypus experiments. Reasons for this difference will be discussed in the section on peripheral auditory function in *Tachyglossus* (p. 193).

Gates *et al.* (1974) also measured amplitude of CM as a function of tone intensity at three different frequencies: 2.0, 6.0, and 10.0 kHz. At all frequencies the CM was a smooth and linear function of sound pressure (Fig. 51). This is very different from the CM as a function of SPL exhibited by the cochleas of lizards: instead of a smooth curve, it consists of a series of peaks and troughs (Wever *et al.*, 1963) but it is identical to that of the marsupial *Didelphis virginiana* (Fernandez and Schmidt, 1963) and to those of eutherians.

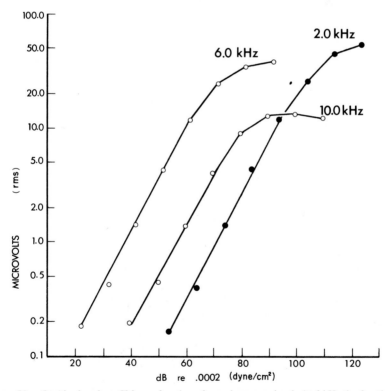

Figure 51. *Ornithorhynchus.* CM as a function of sound pressure level. At 6 kHz the function is linear over a range of 40 dB with asymptote near 90 dB SPL. *(From Gates et al., 1974.)*

THE AUDITORY PATHWAYS IN THE BRAIN.   Efferent fibers from the ganglion into the bipolar cells (the spiral ganglion of eutherians) pass as the cochlear nerve and terminate in the dorsal and ventral cochlear nuclei in the medulla oblongata. From here on opinions differ as to the pathways taken. Hines (1929) has it that efferent auditory fibers from the cochlear nuclei pass to the superior olive and from here secondary fibers pass to the opposite side of the brain via a decussation known as the corpus trapezoideum, and travel rostrally as the lateral lemniscus to the way-station in the thalamus, the medial geniculate body. Dorsal and ventral cochlear fibers also pass directly to the lateral lemniscus via the striae medullares acusticae. Other efferent auditory fibers proceed from the olivary complex to the inferior colliculus of the midbrain and efferent fibers from the colliculus terminate in the medial geniculate body. The main interest of all this stems from the fact that in mammals there are three auditory decussations—the dorsal of von Monakow (the striae medullares acusticae), the intermediate of Held, and the corpus trapezoideum (the major decussation), and that Hines identified an auditory decussation in *Ornithorhynchus* as a corpus trapezoideum. Abbie (1934) flatly denies that *Ornithorhynchus* has the trapezoid body and says that there are only two auditory decussations as in the Sauropsida, i.e., the dorsal of von Monakow and the intermediate of Held. Abbie suggests that Hines mistook the intermediate of Held for the corpus trapezoideum. He says there is, however, a rudimentary trapezoid body in which a few auditory fibers decussate in the medulla oblongata of *Tachyglossus* (p. 194), and he holds the view that the corpus trapezoideum has become the main auditory decussation in higher mammals as a result of development of long coiled cochleas.

Tactile Organs of the Muzzle

Home (1802a) remarked, ''The olfactory nerves are small and so are the optic nerves; but the fifth pair which supplies the muscles of the face are uncommonly large. We should be led from this circumstance to believe, that the sensibility of the different parts of the bill is very great, and therefore it answers to the purpose of a hand, and is capable of nice discrimination in its feeling.'' Eighty-three years later Poulton (1885) had occasion to examine the anatomical basis of the ''nice discrimination.'' The skin of the muzzle and frontal shields exhibits a large number of pores which are the openings of sweat glands. Associated with these are rodlike sensory organs also distributed over the entire surface of the muzzle and shields but in greater numbers at the distal end of the muzzle (Bohringer, 1976). Each rod is embedded in the dermis perpendicular to the surface and at its base are located 2–6 Pacinian bodies. Wilson and Martin (1893b) confirmed and extended Poulton's observations: the rods are enclosed for half their length by an invagination of the epidermal cells, the deeper half is surrounded by the dermal connective tissue arranged in a circular manner to form a definite sheath. The rod itself is made up of three concentric layers; the outermost consists of polyhedral cells with their long axes horizontally. The second layer is made up "of narrow imbricated nucleated cells arranged not unlike the tiles of a roof, at a more oblique angle than those of the laminae forming the core and completely encircling this latter." The core is made up of superimposed elements in the shape of truncated cones and each element is in turn made up of three or four nucleated cells which are keratinized; in longitudinal section the core appears obliquely striated since the cells of the cones are arranged at about an angle of 45° to the long axis of the rod.* Wilson and Martin (1893b) also observed large Pacinian bodies, three times larger than those at the base of the rod, in the dermis a short distance from the rod. R. C. Bohringer (personal communication) has studied the ultrastructure and innervation of the Pacinian bodies: they have a multilamellated capsule and each is innervated by a single myelinated nerve fiber that loses its myelin sheath on passing into the capsule. The nerve fiber exhibits mitochondria arranged around the periphery, neurofilaments in the center, and numerous vesicles. The capsule itself has 12–20 lamellae between which a collagenlike substance is found. The nuclei and the organelles of the cells on the outermost lamellae resemble those of Schwann cells; numerous pinocytic vesicles are present on all lamellae.

The rods themselves exhibit free nerve endings and in intimate contact with the keratocytes at the base of the rod are located receptor organs different from the Pacinian bodies. These were called lenticular bodies by Wilson and Martin (1893b) and they have been identified by Bohringer as Merkels discs or cells. There are 4–6 of these, each associated with a nerve ending at the base of the rod. The Merkels cell contains a bi- or multilobed nucleus and a cytoplasm packed with granules. The cells are attached to the surrounding keratocytes by desmosomes and each Merkels disc is associated with a nerve ending.

A leash of myelinated fibers proceeds from each end-organ complex (Pacinian bodies plus Merkels cells) which combine with those of other end-organ complexes to form the nine major trigeminal nerve bundles described by Bohringer (1977) (also see p. 9).

The end organs described above are identical to those found in the naked rhinaria of other mammals. The arrangement of these organs at the base of a cellular rod also occurs in the nose of the mole (Halata, 1972) and in the beak of *Zaglossus* (Kolmer, 1925); as far as I am aware, tactile organs have not been studied in the snout of *Tachyglossus*; they have, however, been studied physiologically (p. 195).

---

*Poulton, incidentally, thought these rods were modified hairs, but the above description shows they have nothing in common with hair.

Poulton (1885) suggested that the rods in the muzzle communicate pressures to the receptor organs lying below. Experiments about to be described substantiate that notion.

The Organization of the Sensory and Motor Areas of the Neocortex

Bohringer and Rowe (1977) have mapped skin, vision, and auditory representation on the cortex of platypuses anesthetized with dial-urethane; they have also mapped motor areas of the cortex. Since there is an element of doubt concerning the validity of some of the results on cortical localization in other mammals due to problems inherent in the technique used (see p. 197), the very elegant techniques used by Bohringer and Rowe will be described. Short latency positive-going evoked potentials, only, were recorded for mapping cortical projection foci since these are considered to arise from the excitatory action of thalamo-cortical afferents on cortical neurones. In addition recordings were made extracellularly with a microelectrode from individual neurones at varying depths in the cortex. The receptive fields of these single neurones were delineated by gentle mechanical stimulation of the skin. The receptive field thus located was sketched on photographs of the region of the body involved. Both the evoked potentials and the microelectrode recordings were made on a grid pattern over the surface of the cortex at intervals of 1–2 mm except where the presence of surface blood vessels necessitated a change of pattern. The stimuli for the evoked-potential work were a single-pulse electrical stimulus applied with needle electrodes inserted into the skin, or a short tap stimulus of definite duration and amplitude. The mechanosensitive properties of the neurones were determined with two kinds of stimuli: steady indentations to the skin (0.5–1.5 sec duration, 30 m/sec rise time, 500 μm amplitude) and sinusoidal vibration (5–300 Hz) superimposed on a step indentation. Visual stimulation consisted of a 0.74 m/sec-electronic flash delivered 30 cm from the eye at 2-minute intervals. Auditory stimuli were clicks of 2 m/sec duration given every 10 seconds from a loud speaker placed near the ear opening.

For the mapping of the motor area of the cortex stimuli consisting of trains of pulses (60 Hz, 2 mA maximum intensity, 200 μsec pulse interval) were applied for 3 seconds every 2 minutes.

Areas on the distal part of the hindlimb, distal forelimb, and anterolateral borders of the muzzle were selected for determination of the somatosensory cortex. Responses to stimulation of the contralateral muzzle were recorded over a very large area of the hemisphere (Fig. 52). At the same places (the dots in the figure) recordings were also made following stimulation of the contralateral fore and hindlimbs but only the responses to stimulation of the muzzle are shown in the figure. Isopotential contour maps based on amplitudes of the positive compo-

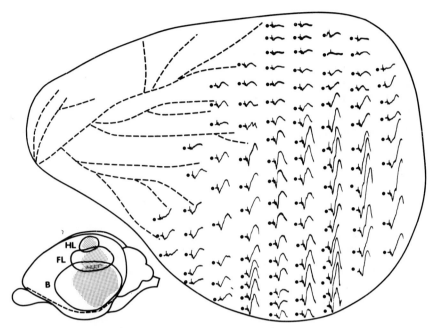

Figure 52. *Ornithorhynchus.* Evoked potentials (positively downwards) recorded from the cortical surface following bipolar stimulation of the anterolateral margin of the contalateral muzzle (iu, 100 µsec pulse). Each recording was made from the position indicated by the dot at the left of the trace. Dotted lines indicate positions of large blood vessels. The view of the hemisphere is from the dorsolateral aspect. In the inset the stippled areas represent focal projections sites and include sites at which positive-going responses exceed 100 µV for muzzle (B), 30 µV for forelimb (FB), and 10 µV for hindlimb (HL). Zones between stippling and continuous lines include sites from which smaller responses could be recorded. *(From Bohringer and Rowe, 1977.)*

nent of the evoked potential were constructed, with the result shown in the inset to Fig. 52: the areas between stippled area and continuous line represent the regions from which smaller evoked potentials were elicited. Thus the somatosensory cortical representation shows a mediolateral sequence of hindlimb (HL), forelimb (FL), and muzzle (B), the whole somatosensory area being located in the posterior half of the hemisphere.

Recordings from single neurones confirmed that the somatosensory cortex has a mediolateral sequence of HL, FL, and B and that the responsive neurones could be detected 200 µm to 2.77 mm below the surface of the cortex. Bohringer and Rowe point out that such a distribution of responsive neurones through various layers of cortex is in agreement with results obtained from many other mammalian cortices.

A matter of great interest was the difference detected in the size of receptive fields for single neurones. Tactile receptive fields on tail and trunk were as large as 15 cm², on skin of the limbs 0.5–1.0 cm², but on the muzzle, particularly on the anterior and lateral margins, the receptive fields were minute, usually not more than 1 mm in diameter.

The neurones encountered during individual penetrations by the microelectrodes perpendicular to the surface of the cortex exhibited similar receptive field sizes and locations which indicate that the platypus cortex has a columnar organization like that of higher mammals. This was substantiated by the observation that when microelectrodes were passed through successive columns in the area of muzzle representation, regular and progressive shifts of muzzle representation occurred.

As far as body representation in the anteroposterior plane of the cortex is concerned it was found that neurones with receptive fields on the axial body regions lie posterior to those on the body apices.

In eutherian hemispheres there are two somatosensory areas (SI, SII, see p. 196); Bohringer and Rowe found no evidence of an SII on the dorsolateral regions of the platypus cortex; however, the microelectrode technique allowed them to explore the ventral region of the neocortex buried deep in the skull. This was carried out by driving the microelectrode from the dorsal surface down to the ventral surface of the hemisphere. During the course of the penetration neurones were tested for responsiveness to tactile stimuli applied to the surface of the body. In the ventral region of the cortex, spontaneously active neurones were encountered but no peripheral receptive field could be detected. Penetrations through lateral regions of the cortex revealed a continuous representation of the muzzle dorsally round to the rhinal sulcus; there was no evidence, however, of an SII.

The mechanosensitivity of a series of cortical neurones receiving impulses from the muzzle were studied using the controlled and reproducible mechanical stimuli listed above. Nearly all the neurones studied exhibited spontaneous activity of between 2 and 5 impulses/sec. In response to an indentation stimulus (0.5–1.5 sec duration) at their receptive field all the neurones exhibited rapidly adapting discharge patterns but they responded only to the "on" and "off" phases of the stimulus. When tested with trains of sinusoidal vibration up to 1 sec in duration applied to their receptor fields, some of these neurones were insensitive at all frequencies applied (5–300 Hz). Others were responsive to low frequency sinusoidal vibration; the responses of these particular neurones to stimuli of different frequencies were determined with the result shown in Fig. 53. At frequencies of 10–20 Hz their activity was phase-locked to the applied stimulation, at higher frequencies, e.g., 60 Hz, there was a response to the first cycle of vibration but no other response followed. Although Bohringer and Rowe have described the above as the properties of cortical neurones they do not mean to imply anything about where functional properties of those neurones are set or determined; some of the differences in behavior of the cortical cells may reflect

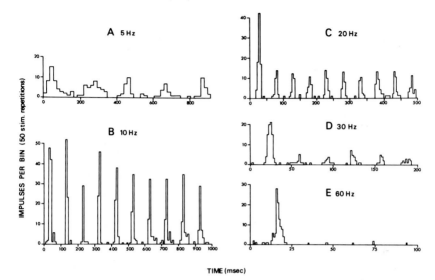

Figure 53. *Ornithorhynchus.* Poststimulus time histograms constructed from the accumulated responses of a cortical neurone to 50 repetitions of the 1-second duration vibration train at five different frequencies (A–E). The analyses started at the onset of the sinusoidal vibration train. The period of analysis was different for each frequency of vibratory stimulation (see abscissae), but in each case was divided into a number of equally spaced time segments or bins, the height of each indicating the number of impulses accumulated during the 50 repetitions of the stimulus train. Bin widths were 18.18 msec in A, 9.07 msec in B, 4.54 msec in C, 1.83 msec in D, and 0.83 msec in E. *(From Bohringer and Rowe, 1977.)*

differences in the behavior of receptors in the muzzle. Recording from afferent trigeminal fibers from the muzzle, as has been carried out with fibers from the snout of *Tachyglossus* (MacIntyre and Kenins, 1974), may determine the matter.

The visual and auditory stimuli evoked potentials at two partially overlapping areas on the dorsal surface of the caudal pole of the hemisphere; the center of the visual area is dorsal to that of the auditory.

Movements of contralateral forelimb and muzzle could be elicited by stimulation of the cortex (Fig. 54). Flexion (upgoing arrows in the figure) and extension (downgoing arrows) could be induced in the forelimb, and opening, closing, and lateral movements (upgoing, downgoing, and horizontal arrows, respectively) of the muzzle could also be observed. There is considerable overlap of motor and somatosensory areas in the cortex (inset to Fig. 54). However, in none of the experiments could movements be elicited in the hindlimb or in the tail.

The outstanding feature of this cortex, as pointed out by Bohringer and Rowe, is the enormous area connected with inputs from the muzzle, which is consistent with the platypus' reliance on tactile sensory information since the eyes, ears,

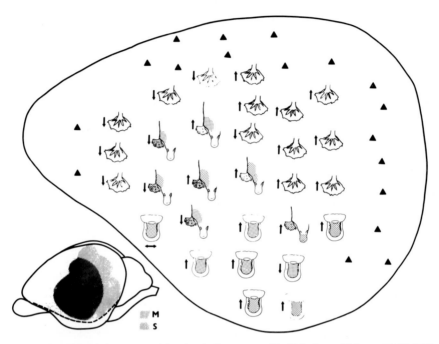

Figure 54. *Ornithorhynchus*. The figurines indicate areas of the body from which movements were elicited by stimulation of the cortex. Upgoing arrows indicate flexion of forelimb or muzzle opening, downgoing arrows denote forelimb extension or muzzle closings. Horizontal arrows denote lateral movements of mandible. Triangles mark points from which no movements could be elicited. The inset shows the relative positions of motor (M) and somatosensory (S) cortical areas. View of hemisphere from dorsolateral aspect. *(From Bohringer and Rowe, 1977.)*

and olfactory organs are cut off from the outside world when the animal is submerged and feeding. Bohringer and Rowe, however, make the points "There was no suggestion from our study, nor is it probable, that individually, these neurones whose receptive fields were punctate and often less than 1 mm in diameter, could signal the direction of the stimuli moving across the bill surface. However, the highly ordered somatotopic organization within the bill area suggests that the direction of moving tactile stimuli could be coded by a population response based on the sequence in which adjacent cortical cell columns underwent peripheral-induced activation. The speed as well as the direction of the moving stimulus could be encoded by the rate at which neural activation proceeded through successive columnar arrays."

A comparison of this cortex with those of *Tachyglossus* and other mammals will be discussed in this Chapter.

Behavior

PLAY IN THE YOUNG; VOCALIZATION; LARYNX. One needs to read only George Bennett's (1835) charming description of two nestling platypuses at play to be absolutely convinced that the monotremes are mammals. With the help of a party of aborigines armed with hard-pointed sticks, he dug out a breeding burrow that proved to be 35 feet long and to be inhabited by two young platypuses. They were fully furred, "growled exceedingly at being exposed to the light of day," and measured 10 inches each from the extremity of the muzzle to the tip of the tail. The mother was also taken, later, outside the burrow. Bennett relates "The eyes of the aborigines both young and old, glistened, and their mouths watered, when they saw the fine condition of the young Mallangongs. The exclamations of 'Cobbong fat' (large or very fat) and 'Murry budgeree patta' (very good to eat) became so frequent and so earnest, that I began to tremble for the safety of my destined favourites." However, he managed to get them home and kept them for 5 weeks during which time he made many observations of their habits. Not the least interesting was the following:

> One evening both the animals came out about dusk, went as usual and ate food from the saucer, and then commenced playing one with the other like puppies, attacking with their mandibles and raising the fore paws against each other. In the struggle one would get thrust down and at the moment when the spectator would expect it to rise again and renew the combat, it would commence stretching itself, its antagonist looking on and waiting for the sport to be renewed . . . Sometimes I have been able to enter into play with them, by scratching and tickling them with my finger: they seemed to enjoy it exceedingly, opening their mandibles, biting playfully at the finger, and moving about like puppies indulged with similar treatment.

They also like to play in shallow water exhibiting the same antics: running about, wrestling, seizing one another with the mandibles, and rolling over in the water, followed by intervals of grooming.

Bennett frequently refers to awful growls coming from the sleeping platypuses when disturbed. Adults make the same noises but I have detected a querulous note in the growling from time to time—a blending of the growl of a puppy and the noise of a broody hen. Bennett states that one evening when the two young ones were running about his room one of them emitted a squeaking noise, the other, hidden behind some furniture, immediately answered with the same noise and the first animal ran at once in the direction from which the call came, and located its playmate.

These noises are made with a larynx supported by a perfectly mammalian laryngeal skeleton; the stylohyals at their attachments to the cristae paroticae have remained cartilaginous but their proximal portions have now differentiated into two parts, ceratohyals and hypohyals, which are bony and are joined to the ossified median ventral body of the larynx, the basihyal copula (the fused basihyals). The other elements of the laryngeal skeleton retain the evidence of their branchial origin: paired thyrohyal cartilages arise laterally from the basihyal copula and pass dorsally around the larynx; attached to the caudal end of the basihyal copula is a crescent-shaped thyroid copula from which arise paired anterior and posterior laryngeal cartilages. Posterior to the thyroid copula is a large cricoid cartilage.

The laryngeal skeleton of the monotremes is primitive in that two pairs of laryngeal cartilages arise from the thyroid copula whereas in the Eutheria there is apparently only one pair, but these are formed by the fusion of two pairs of laryngeal cartilages. Goodrich (1958) points out that a relic of the dual nature of the laryngeal cartilages is apparent in the laryngeal skeleton of the Metatheria in the form of a foramen on either side in the fused anterior and posterior laryngeal cartilages.

COURTSHIP.   Strahan and Thomas (1975) have made an extensive series of observations on courtship behavior exhibited by a pair of platypuses kept at Taronga Park Zoo, Sydney. The period over which observations were made was 140 days and all of the behavior took place in the water of the swimming tank provided for the captives.

For most of the year the male and female took no notice of each other but in August (which is the start of the breeding season for wild platypuses, p. 220) the female became interested in the male and carried out the following repertoire of movements:

1. The frontal approach; the male is resting on the surface of the water when the female swims to him and rests with her muzzle touching his; this may last as long as 10 minutes.

2. Side-passing; the female swims to the male as in the frontal approach and then passes alongside rubbing against his flank.

3. Ventral-passing; the female swims to the male and rolls over just before meeting him and passes underneath so that her ventral surface rubs against his as she passes beneath him upside down.

4. Under-passing; the female approaches from one side as the male rests on the surface, dives under his body and surfaces.

5. Circling; the female swims in a tight circle on the surface of the water and the male may follow and hang onto her tail with his jaws. This act also occurs after ventral- and under-passing and when it takes place the male is towed along passively or he may assist with swimming motions of his forelegs.

The repeated grasp of the tail in the jaws leads to loss of hair from the female's tail. At times tail-holding is followed by the male climbing onto the female's back and grasping the nape of her neck in his jaws. A variant of this behavior was observed—the male climbs onto the female's back but curls his body down and around until his tail is directed forwards and lies closely appressed to the ventral surface of the female. In all probability this is the position for copulation (see below). Verreaux (1848) observed the neck-grasping behavior in a pair of platypuses living in the Derwent River in Tasmania; the time of the year was September. In 1944 Fleay observed tail-holding and circling exhibited by a pair of platypuses kept in captivity; this behavior occurred at various times between September 14th and October 11th. A presumed mating took place during the latter day. Of this Fleay says, "the male animal doubled his body under while maintaining his grip on the female's tail with his bill." They remained interlocked for 10 minutes.

During the next 10 days the female made a new entrance to a burrow she had made in a mound of earth provided for her in her quarters. On October 23rd she became very active collecting leaves and swimming with them to the burrow. This was achieved by packing the leaves between her venter and her tail which was bent down and forward when required. The load was compacted by backward kicks of the hind feet so that the leaves would not be lost during the swim back to the burrow. These leaves were undoubtedly used to form a nest (see p. 218). She retreated to the burrow on the night of October 25th and did not reappear until October 31st. All this activity was followed, apparently by egg laying, incubation, hatching, and rearing since a very young nestling about 9 inches in length was observed on January 3rd; it was assumed to be 8½ weeks old. How and when incubation and hatching occurred is not known since Fleay made no attempt to disturb the female in her burrow.

The courtship behavior of Strahan and Thomas' platypuses was also followed by egg-laying. Between days 85 and 112 of their observations the intensity of the courtship rituals increased and on day 104 the female was seen to carry leaves from the bottom of the tank to the exit platform. On day 113 two eggs were found in her nesting box. The time of the year, December 13th, however, was later than the time of Fleay's presumed egg-laying.

EVIDENCE OF PARADOXICAL SLEEP.   Allison *et al*. (1972) could not detect by electrophysiological methods, convincing evidence of PS in adult *Tachyglossus* (p. 205) and implied that this lack of PS may be a monotreme character of some phylogenetic significance. Bennett (1835), however, recorded the interesting fact that when his two juveniles were asleep they often appeared to dream of swimming, "Their forepaws in movement as if in the act." This would seem to be similar to the rapid eye movement sleep of dogs and cats. Perhaps the juvenile platypus would be a better experimental animal for demonstration of PS in monotremes for at least three reasons: they are available in February and March; they would not have to be kept in the laboratory for long; and they go to sleep readily.

TACHYGLOSSUS

Olfaction

Unlike the olfactory organs in *Ornithorhynchus* those of *Tachyglossus* are very well developed. The cribriform plate, which is very large, extends a long way forward of the transversely placed mesethmoid so that it lies horizontally; it is pierced by a large number of pores through which the numerous branches of the olfactory nerves pass from the nasal epithelium to the olfactory bulbs. The area covered by the olfactory epithelium is enormous; there are seven vertical endoturbinals dependent from the cribriform plate, a large number of ectoturbinals (ethmoturbinals) as well as sets of nasoturbinals and maxilloturbinals.

The bulbs, as in *Ornithorhynchus* lack the layer of mitral cells (Switzer and Johnson, 1977). It can be gleaned from Ziehen's (1908) and Abbie's (1934) descriptions that the organization of the olfactory system in the *Tachyglossus* brain follows the usual mammalian pattern but there are one or two matters of interest: efferent fibers leave the anterior olfactory nucleus in two sets, the lateral and medial olfactory striae. The former pass to the hippocampal cortical center either directly or indirectly via the lateral olfactory area. From the hippocampus olfactory impulses pass by the fimbriafornix and stria medullaris to the habenular ganglion, the thalamic way-station for projection of olfactory impulses to lower centers. From the habenular ganglion efferent fibers pass in the very large tractus habenulo-peduncularis to the interpeduncular ganglion which forms a very prominent protuberance on the ventral surface of the midbrain between the cerebral peduncles. Efferent fibers from the ganglion radiate to visceral motor centers in the medulla oblongata. The tractus habenulo-peduncularis forms part of a much larger tract, the fasciculus retroflexus, which also has ascending fibers linking the nucleus lateroventralis of the thalamus with the habenular ganglion. Thus the trigeminal end-station in the thalamus is connected to the olfactory system linking tactile sensibility of the head and snout with smell (Abbie, 1934).

The fibers of the medial olfactory striae pass to a medial olfactory area whence secondary fibers pass to the hippocampus, the habenular ganglion, and to olfacto-visceral centers in the tuber cinereum (hypothalamus). It is possible that *Tachyglossus* has an olfacto-tegmental tract linking the medial olfactory area directly with secretory and motor centers in the medulla oblongata.

No physiological or behavioral studies of olfaction have been carried out on echidnas but it would seem that a sense of smell will prove to be of importance to echidnas in at least three ways: (1) Echidnas are solitary animals for most of the year but they do manage to meet up with other echidnas at the breeding season (p. 204); very likely olfaction plays a part in bringing them together; (2) olfaction, in all probability, plays a part in detection of the ants and termites that constitute the diet; (3) sense of smell is probably the one special sense that an echidna has at the time it emerges from the egg since fully differentiated olfactory sense receptors are at that time present in the nasal chamber (p. 280) but the malleus is still part of Meckel's cartilage and the eyes are covered with a thin layer of epithelium resembling skin. Echidnas exhibit Jacobson's organ opening into the anterior end of the buccal cavity.

Organs of Taste

At the hind end of the "dental" pad of the tongue in the midline two slits forming a V, are located. These lead down into deeply seated circumvallate papillae in the trenches of which taste buds occur (Oppel, 1899). Glands, known as the serous glands of von Ebner, discharge into the trenches around the papillae. As well as these papillae sets of foliate papillae are located laterally and posterior to the dental pad, these are also furnished with taste buds; a difference between foliate and circumvallate papillae is size, foliate being smaller. Circumvallate papillae are found only in the tongues of the Mammalia.

Vision

ANATOMY OF THE EYE.   Gates (1973) has reviewed, inter alia, the literature on the structure of the eye in *Tachyglossus* and it appears that the subject is bedeviled by inadequate description of techniques and findings, and by contradictions in accounts by different authors and even by the same author. The following description is based on those of Gresser and Noback (1935), Walls (1942), O'Day (1952), Prince (1956), Duke-Elder (1958), and on Gates' criticisms of those descriptions: the eye is roughly spherical, with a horizontal diameter of 9.5 mm and a vertical diameter of 9.0 mm. Unlike the eye of *Ornithorhynchus* that of *Tachyglossus* has no nictitating membrane and only the lower lid is furnished with a tarsal plate; however, Harderian and lacrimal glands are present. The ultrastructure of the latter has been described in great detail by Atkins and Schofield (1971) but it is not apparent from their study whether there is anything unusual about them.

The surface layers of the cornea are heavily keratinized as they are in aardvarks and some armadillos that eat ants. The cartilaginous cup is present just below the sclera (Fig. 55) but it is much thinner (only 27 μ thick in the region of the optic nerve) than in the eyes of *Ornithorhynchus* and *Zaglossus*.

The six extrinsic muscles have the same relationships as they have in *Ornithorhynchus* but the superior oblique exhibits an extra slip of muscle taking its origin at the anterior nasal aspect of the orbit and its insertion onto the sclera near the insertion of the main part of the superior oblique; this slip of muscle is sometimes seen in the human orbit. In addition to the above extrinsic muscles, four authors (Johnson, 1901; Walls, 1942; Prince, 1956; Duke-Elder, 1958)

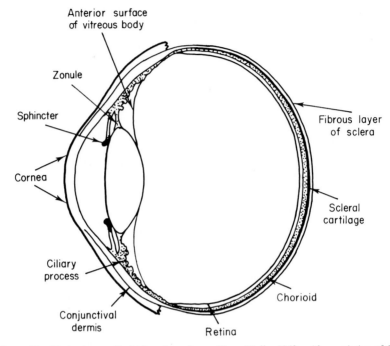

Figure 55. *Tachyglossus.* Sagittal section of eye. *(From Walls, 1942; with permission of Cranbrook Institute of Science.)*

found that a retractor bulbi muscle is present. Without reference to these observations, Saban (1969a) says: "Il n'a pas de m. retractor bulbi." He does, however, say that there is a suspensor bulbi that is not found in any other vertebrate.

The chorioid, as in the eye of the platypus, is carried forward to form the uveal portion of the ciliary body. The latter consists of about 60 squat radially arranged processes interconnected anteriorly by the annular ciliary web. Gresser and Noback (1935) say there is a ciliary muscle differentiated into circular, radial, and meridional fibers. O'Day, Walls, and Prince have it that there is no ciliary muscle at all. The lens according to Walls is the "flattest of all lenses" having a flatness index of 2.75. It is suspended by zonule fibers arising in the ciliary body and the fibers are inserted compactly at the periphery of the lens. The iris is like that of the *Ornithorhynchus* eye.

The authors mentioned above agree that the retina is very thin and avascular depending on the substantial chorioid vessels for its supply of nutrients and oxygen; they are also fairly well agreed as to its general configuration: a layer of visual receptors, five layers of nuclei, a single layer of ganglion cells, and a thin layer of nerve fibers which, as Gates (1973) says, indicates a high degree of summation. There are discrepancies in accounts of the nature of the visual receptors: O'Day (1938) found a few cones in a predominantly rod retina but in 1952 he says the retina has rods only; Walls (1942) and Duke-Elder (1958) have it that there are rods only while Prince (1956) saw a few cones in a predominantly rod retina. It would seem that some echidnas might have a few cones in the retina and others have a pure rod retina.

The retinofugal fibers leave as the optic nerve, slightly to the nasal side of the back of the orbit by a single root. In cross section it presents the usual pattern of fascicles of closely packed myelinated fibers, the fascicles separated from each other by septa. Occasionally very small discrete bundles of unmyelinated fibers can be seen. The diameter of the myelinated fibers vary from 0.36–7.16 μ. Histograms of fiber diameters from the nerves of two different echidnas were remarkably similar, both distributions were unimodal exhibiting peaks in the 1.82–2.30 μ range. Estimates of the number of fibers in the optic nerves of the two animals were also remarkably similar—14,950 and 14,709, respectively. Although this number is small compared with those for "highly visual" animals such as monkeys and ducks, the optic fiber count as Gates points out is substantial and it is double that for the bat *Eptesicus fuscus*, a mammal capable of making a single visual discrimination. The echidna on this basis has a reasonable degree of visual function and this conclusion is substantiated by behavioral data to be described (p. 200).

The apparent lack of intraocular musculature has led to the notion that the echidna eye lacks a mechanism of accommodation but some retinoscopic observations by Gates suggest very strongly that the echidna eye can accommodate. Mydriasis was achieved before observations were attempted by placing a drop of homatropine hydrobromide and one of phenylephrine in the eye and observations had to be carried out on animals anesthetized with fluothane since normal ones proved to be photophobic. When the effect of the anesthetic began to wear off it was found that the eye protruded and that a retinoscopic examination could be made. The refractive changes observed during early and late recovery from anesthesia are shown below.

Echidna		Early recovery dioptres		Late recovery dioptres
(1)	left	0.00	left	+3.50
	right	+0.50	right	+4.00
(2)	left	−0.05	left	+3.50
	right	0.00	right	+3.50

These data show that a considerable change is occurring in the refractive error of the eye in a short period of time. Gates offers two explanations:

1. That Gresser and Noback (1935) are right that there are ciliary muscles in the tachyglossid eye capable of changing the shape of the lens and therefore the dioptric properties of the eye. A careful reexamination of the ciliary body will be necessary to settle that.

2. The protrusion is really elongation of the eyeball brought about by contraction of the extraocular muscles "somewhat reminiscent of the old method used to alter focus of a camera in which the photographic plate and not the lens is moved." Thus a sharp image is achieved at the retina by movement backward or forward brought about by change of shape of the eyeball. I might add that there is nothing novel in this suggestion since the idea is the basis of treatment of errors of refraction in the human eye (see Bates, 1956).

PRIMARY OPTIC PATHWAYS.   The optic nerves and the chiasma are diminutive in T. a. aculeatus but Schuster (1910) found that they were much larger in setosus. The decussation is practically complete but a few fibers may pass to the ipselateral side. The tiny optic tracts run from the chiasma laterally, posteriorly, and dorsally round the base of the brain, closely appressed to the cerebral peduncles, to the geniculate bodies. Here some terminate in the ventral nucleus of that body, some pass to the dorsal nucleus of the geniculate body, and the remainder pass in the superior brachium to the superior colliculus. However, not all the fibers of the brachium pass to the colliculus; some terminate at a small, well-defined group of cells lying within the brachium, which is equivalent to the "large-celled nucleus" of the optic tract found in other mammals. Having arrived at the colliculus the remaining fibers of the brachium appear to pass in the outermost layer of the colliculus, the stratum zonale, from where they turn ventrally to end up in the stratum opticum.

The above description was gleaned from Abbie's (1934) monograph on the brainstem of Tachyglossus. Apparently unaware of this work Campbell and Hayhow (1971) confirmed Abbie's interpretation using their technique of unilateral eye enucleation described above for Ornithorhynchus; after a period of 8–18 days postoperative survival the brains were sectioned and the two tracts identified by the presence of degenerating axons. Campbell and Hayhow's description of the tract is much the same as Abbie's. Decussation was found to be almost complete, only about 1% of the fibers passing to the ipselateral side of the brain and terminating in the apical portions of the lateral geniculate nucleus designated LGNa. The contralateral fibers leave the optic chiasma and pass as the optic tract laterally, caudally, and dorsally to terminate in the superior colliculus (via the brachium superior), in the nucleus mentioned above (LGNa), and in LGNb corresponding to the nucleus LGNb in Ornithorhynchus. However, the degeneration technique revealed the presence of four interstitial nuclei, not observed by Abbie and not present in Ornithorhynchus, associated with the optic pathways. These nuclei occur at approximately equal intervals between the chiasma and

LGNa. As in *Ornithorhynchus* there is an accessory optic system consisting of a bundle of fibers leaving the contralateral tract near interstitial nucleus I and passing medially and caudally on the ventral surface of the tegmentum and terminating in the medial ventral nucleus. Many eutherian mammals have three accessory optic nuclei, moles and bats have none; pigs, prosimians, and anthropoid primates lack the medial terminal nucleus.

Campbell and Hayhow (1971) discuss the relationship of LGNa and LGNb to the ventral and dorsal parts of the lateral geniculate nucleus of metatherians and eutherians, and to the ventral and dorsal nuclei, found in the diencephalons of reptiles receiving retinofugal fibers. In all these vertebrates the dorsal and ventral groups are close to one another and, as far as the reptiles are concerned, the dorsal nucleus sends fibers to the telencephalon. Thus it appears that the monotremes with their widely separated diencephalic nuclei receiving retinal fibers are the odd men out. However, it seems to me that the widely disparate relations of the two nuclei may be occasioned by the enormous nucleus ventrolateralis thalami displacing the two parts of the geniculate nuclei; as Campbell and Hayhow point out before a definite account of the nature of the two nuclei can be given it will be necessary to study thalamo-cortical relationships with the Nauta-Gygax technique. Such a study has been carried out for *Tachyglossus* (W. Welker and R. A. Lende, 1978) but the results, at present, cannot be reconciled with Campbell and Hayhow's findings (p. 198).

Campbell and Hayhow (1972) make a further point that the passage of the optic fibers of the brachium onto the superior colliculus via the stratum zonale is like that found in modern reptiles and is unlike that found in metatherian and eutherian colliculi where the fibers enter the stratum opticum without first traversing the stratum zonale.

Gates (1973) points out that even though only about 1% of the optic fibers decussate in *Tachyglossus*, it is possible that they come from the same area of the retina and that that area may also be "represented in the contralateral visual cortex, thus giving the animal some sort of binocular representation if the contralateral eye also contributes fibers from the same region." As well there is a small lateral divergence between the optic axes of the two eyes (Johnson, 1901) which suggests there is overlap in the visual fields of the two eyes and that there is some binocular interaction.

Hearing

ANATOMY OF THE EAR. Cartilaginous ear trumpets, similar to those of *Ornithorhynchus*, carry sound waves from the pinnas, downward, medially, and finally dorsally to the tympanic membranes located posteromedially to the glenoid depressions on the ventral surface of the skull. The trumpet is flexible but stiffened by two longitudinal strips of cartilage (not found in the trumpets of

Ornithorhynchus) from which arise a series of transverse riblike cartilages. The tympanic membrane is supported by a horseshoe-shaped tympanic bone formed by extension and curving of two of the arms of the triradiate tympanic cartilage of the chondrocranium; a vestige of the triradiate shape is retained in the adult bone. The domed tympanic cavity which of course contains air, lies dorsal to the membrane and exhibits three apertures on its dorsal surface: the fenestra ovalis (which happens to be circular in monotremes, see p. 190); the fenestra rotunda or round window; and the apertura tympanica canalis facialis (Fig. 56) which leads into the Eustacian tube communicating with the buccal cavity. There are three ear ossicles as in *Ornithorhynchus*: a columelliform stapes expanded proximally into a round footplate standing on the membrane of the fenestra ovalis; distally it articulates with the small incus which exhibits a processus longus and a processus brevis. The incus in turn articulates with a very large malleus furnished with a long anteromedially directed processus gracilis tightly attached to the

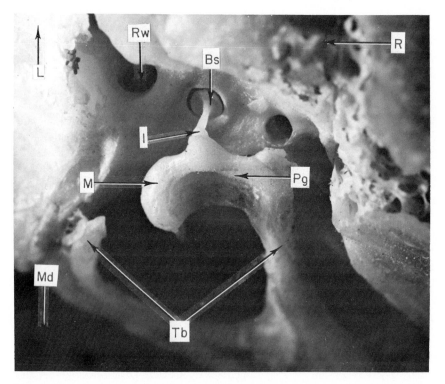

Figure 56. *Tachyglossus*. Tympanic cavity with ear ossicles *in situ*, × 9.1. Rw, round window; Bs, base of stapes in fenestra vestibuli; R, rostral; Pg, processus gracilus; Tb, tympanic bone; M. malleus; I, incus; Md, medial; L, lateral. *(From Griffiths, 1968.)*

petrosal bone. The manubrium of the malleus is also long and is attached to the tympanic membrane, while the incus is firmly ankylosed to the malleus. All this leads to the surprising conclusion that the ear bones are so firmly joined to each other and to the petrosal by the processus gracilis, that they would be resistant to movement. This is precisely what Aitken and Johnstone (1972) found experimentally (see the following section).

The organs for conversion of mechanical movement of the stapes to electrical impulses are as in *Ornithorhynchus*: the cochlea exhibits a three-quarter spiral turn and terminates in a lagena with lagenar macula. The macula consists of a proximal layer of nerve fibers and a distal one of columnar cells furnished with hairs that project into a layer of a gelatinous substance that lines the internal surface of the macula; this layer contains otoconia or otoliths. Internally the cochlea exhibits the usual three scalae; vestibuli, tympani, and media; the basal layer supports a true organ of Corti (Alexander, 1904). The relations of the remaining parts of the membranous labyrinth are shown in Fig. 57 and they are identical to those already described for *Ornithorhynchus*. In passing it may be mentioned that the semicircular canals have smooth rounded contours like those of other mammals and show no sign of the angularity of the canals of the reptilian labyrinth (Gray, 1908).

In a recent review of the anatomy of the auditory receptor organs of reptiles, birds, and mammals by Smith and Takasaka (1971), I was startled to read that I had discovered the malleus and incus of the monotreme ear: "It has been stated by Griffiths (1968) that the ossicular chain in the middle ear of the monotreme is

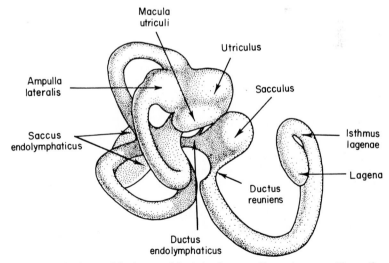

Figure 57. *Tachyglossus.* Membranous labyrinth of a stage-51 pouch young. *(From Alexander, 1904.)*

composed of three bones. Dr. Raul Hinojosa of the University of Chicago (personal communication) has recently confirmed that three ossicles are present in the spiny anteater's middle ear. However, the stapes with its single shaft more closely resembles the reptilian and avian stapes than that of other mammals.'' In fact, in that publication I quoted Doran's (1879) words ''I hardly need remind the anatomist that, as far as the ossicles are concerned, the Monotremata wear a perfectly mammalian uniform, having malleus, incus and stapes.''

As far as the monotreme stapes resembling the reptilian and avian stapes is concerned it should be pointed out that the middle ears of most marsupials and of the eutherian *Manis javanica* (pangolin) have columelliform stapes so there is nothing particularly reptilian or avian about that. Segall (1970) has also made comparisons of the auditory ossicles of *Tachyglossus* with those of marsupials and insectivores. He finds that ''a primitive form of mammalian stapes is found in the monotremes.'' This kind of stapes is defined as a columella without stapedial foramen and with a circular plate of stapes ratio 1.0 (ratio of the long diameter of the stapedial plate to the maximal width perpendicular to this). The stapes ratio of the ossicle in *Tachyglossus* and various marsupials was found to be as follows:

Tachyglossus	1.0
Dasyurus	1.1
Antechinus, Sminthopsis	1.2
Echymipera, Metachirus, Didelphis	1.3
Notoryctes, Perameles, Marmosa	1.4
Caluromys, Dendrolagus, Philander	1.5
Vombatus, Aepyprymnus	1.6
Phalanger, Petaurus	1.8
Dromiciops, Macropus	2.1

Thus the dasyurids, *Dasyurus, Antechinus*, and *Smithopsis* exhibit, along with *Tachyglossus* a ''primitive form of mammalian stapes.'' The stapes in most marsupials as already mentioned, is columelliform, i.e., it lacks a stapedial foramen but when this is present, it is invariably small. In Insectivores, however, it is a stirrup typical of the eutherian middle ear and the stapes ratio varies from 1.8 to 2.9; it would be interesting to know the stapes ratio in *Manis javanica*.

Hinojosa (cited in Smith and Takasaka, 1971) has confirmed Alexander's (1904) finding that *Tachyglossus'* cochlea has a true organ of Corti. Alexander's best preparations were from pouch young but Hinojosa has succeeded in getting good sections of the organ from an adult echidna. In cross section the organ was found to be wider than that of other mammals since there are more hair cells—three inner and six outer separated by the tunnel and often there are three tunnel rods instead of two. Cuboidal epithelium covered the external spiral sulcus and was continuous with a well-developed stria vascularis.

The hair cells were supported by phalangeal cells and tunnel rods, the latter exhibiting bundles of microtubules, tonofilaments, which are found in the tunnel rods of other mammals.

The inner hair cells were short, columnar, and had bulbous basal ends and centrally placed nuclei. The outer hair cells, however, were extraordinarily long. Electron microscopy showed that each hair cell had a row of subsurface cisternae and mitochondria just beneath the plasma membrane. Immediately below these organelles the cytoplasm exhibited longitudinally oriented fibrils and hairs which protruded from the cuticular plate. Nerve endings could be discerned proceeding to the basal ends of the hair cells. The presence of subsurface cisternae and adjacent mitochondria are characteristic of the outer hair cells of other mammals. Smith and Takasaka make the interesting point that "Perhaps the spiny anteater possesses a total number of inner and outer hair cells that is equivalent to the number found in the cochleas of some other mammals but the hair cells are crowded together on a short demilunar form rather than in a one-plus-three pattern on a long coiled ribbon."

PERIPHERAL AUDITORY FUNCTION. Johnstone (cited in Griffiths, 1968) determined the CM audiogram of *Tachyglossus* and found that its cochlea responded to a wide range of frequencies, the most sensitive frequency lying close to 5 kHz as Gates *et al.* (1974) later found was the case for the platypus cochlea. Although these two cochleas respond to a wide range of frequencies, the sensitivity is relatively poor compared with those of most, but not all, eutherian cochleas: the CM audiogram of the Mongolian gerbil is quite like that of the monotremes, with the best frequency lying between 4.5 and 5.0 kHz (Finck and Sofouglu, 1966) and the chinchilla cochlea exhibits great sensitivity for low frequencies between 1.0 and ca. 4.5 kHz (Strother, 1967). Of course references to limited range of frequencies detectable and insensitivity to high or low frequencies do not mean that hearing is defective in echidnas—it is more than likely that a cochlea most sensitive to frequencies near 5 kHz is just the thing for detecting noises emitted by termites or for detection of vibrations transmitted through bone. Johnstone (cited in Griffiths, 1968) many years ago found that a tap on the snout of the echidna elicits a large potential at the round window.

There is, however, evidence that the echidna middle ear is inferior to that of at least two other vertebrates at transmission of sound waves received by the tympanic membrane: Aitken and Johnstone (1972) have measured the velocity of stapes motion* at the fenestra ovalis. The velocity as a function of frequency at

---

*The velocity of stapes motion was measured using the Mössbauer technique. The principle involved is measurement of the degree of shift of frequency of a source of radioactivity with a very sharply defined frequency; when the source is moved the frequency shifts (Doppler effect). In these experiments Co^{57} incorporated in stainless steel foil was the source of radioactivity affixed to the moving structure—the stapes.

constant SPL of 100 dB increases as frequency is increased up to 6 kHz, but at higher frequencies it decreases (Fig. 58). Amplitude of stapes movement is largely independent of frequency up to 4–6 kHz (fig. 59) but above that frequency range, amplitude falls off. These performances are in marked contrast to those of the stapes of the guinea pig and the dragon lizard (Figs. 58 and 59). It is only at 4–6 kHz that velocities comparable to those of guinea pig and the lizard stapes can be observed. At frequencies above that peak value the curves for the lizard and the echidna are similar but they reveal that the stapes in both those animals is far less sensitive than that of the guinea pig. Thus the echidna middle ear is no better than that of a reptile in its capacity to transmit sounds of frequencies in excess of 6 kHz at the tympanic membrane and it is substantially inferior at lower frequencies.

Aitken and Johnstone argue that the low stapedial velocity in the echidna is presumably due to the fact that the malleus is tightly locked to the petrosal bone which is an integral part of the skull; they say that this circumstance is quite unlike the light suspension of the malleus in eutherian mammals. However, it is

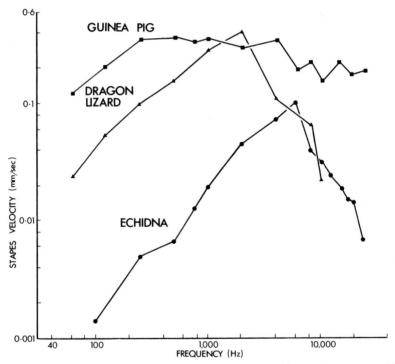

Figure 58. *Tachyglossus.* Velocity of stapes motion as a function of frequency at a constant SPL in echidna, dragon lizard (*Amphibolurus reticulatus*), and guinea pig. *(From Aitken and Johnstone, 1972.)*

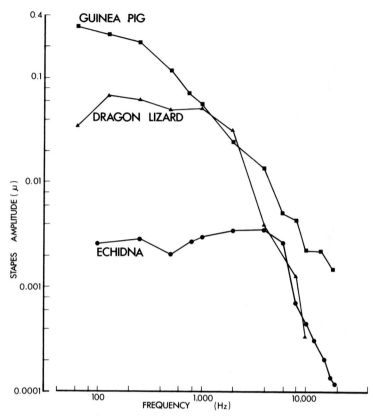

Figure 59. Comparison of mean amplitude function of the stapes for echidna, dragon lizard, and guinea pig. *(From Aitken and Johnstone, 1972.)*

this very tight attachment of the ossicular system to the periotic that makes it suitable for transmission of bone-conducted sounds to the inner ear so the union of tympanic bone to the processus gracilis of the malleus, the ankylosis of incus to malleus, and the attachment of malleus to the petrosal may all be regarded as anatomical specializations for a specialized type of hearing. Along with those specializations, however, comes an electrophysiology of the cochlea identical to that of other mammals: the CM of the echidna cochlea are smooth functions of sound pressure as they are in other mammals, including *Ornithorhynchus* (p. 171). There is one difference, however, from the CM of *Ornithorhynchus*: the same SPL applied to the ears of both genera elicits a much greater potential at the round window of the echidna than it does in *Ornithorhynchus*; Gates *et al.* (1974) say this may be the result of the differences in sizes of incus and malleus in the two genera—those of the platypus are much smaller than those in the echidna middle ear, and furthermore the platypus middle ear is not anchored to bone as it is in the echidna. These two anatomical differences may lead to less effective mechanical stimulus of the organ of Corti in the platypus with concomitant generation of smaller potentials.

There are potentials generated in the inner ear other than CM; it would be of interest to compare these in platypus and echidna but at present we have data only from the echidna, but these are of considerable interest in themselves. The potentials in question are the dc resting or endocochlear potential (EP); the summating potential (SP), a steady potential in the scala media related to average sound pressure; and the action potential (AP) in the cochlear nerve. In the eutherian cochlea EP is about 80 mV, in that of birds it is about 45 mV, and only 2–20 mV in reptilian cochleas. In mammals this potential is decreased by anoxia but reptilian EP is anoxia-insensitive (Schmidt, 1963; Schmidt and Fernandez, 1963). Johnstone (*fide* Griffiths, 1968) found that, in the cochlea of *Tachyglossus*, EP is ca. 80 mV and that it is anoxia-sensitive; SP is 400 μV and AP 100 μV, values all very close to those recorded for eutherian cochleas (Fernandez and Schmidt, 1963; Schmidt, 1963; Schmidt and Fernandez, 1963). They are very different, however, for those recorded for reptiles: EP 2–20 mV, SP 100 μV, and AP 20 μV (Johnstone and Johnstone, 1965).

THE AUDITORY PATHWAYS IN THE BRAIN.   The fibers forming the cochlear division of the eighth nerve, presumably from the echidna equivalent of the spiral ganglion, divide into two branches, one going to the small dorsal cochlear nucleus and the other to the large ventral one, both in the medulla oblongata (Abbie, 1934). From both these nuclei efferent fibers pass to the superior olive of the same side, a very few pass to a rudimentary corpus trapezoideum which exhibits a small collection of cells lying among the fibers—a foreshadowing of the nucleus of the corpus trapezoideum of higher mammals. The remainder of efferent fibers from the dorsal and ventral cochlear nuclei pass to the opposite side of the brain in the decussation of von Monakow (striae medullares acusticae). Secondary, auditory fibers from the superior olive pass in the intermediate or decussation of Held. Auditory fibers from all three decussations, corpus trapezoideum, striae medullares acusticae, and the intermediate pass rostrally in the lateral lemniscus. Thus the echidna auditory pathways are more ''mammalian'' than those of *Ornithorhynchus* (p. 172) since a corpus trapezoideum, albeit rudimentary, is present.

At the level of the mesencephalon the auditory fibers of the lateral lemniscus divide and form two branches, one to the inferior colliculus while the other continues to the medial geniculate body of the thalamus. The fibers to the inferior colliculus terminate in the lenticular nucleus and from this nucleus efferent fibers pass also to the dorsal nucleus of the medial geniculate body. Efferent fibers from this nucleus pass to the auditory area located at the posteroventral surface of the neocortex (see below). The echidna thus exhibits that cortical recognition of hearing characteristic of the auditory function of higher mammals.

Tactile Functions of the Snout

The structure of the sense organs in the snout of *Tachyglossus* has not been described but Kolmer (1925) has given an account of them in *Zaglossus* (p. 207). However, some experiments carried out by McIntyre and Kenins (1974) show that the snout of *Tachyglossus* is richly endowed with sense organs. By recording from fine filaments dissected from trigeminal nerve branches exposed in the orbit they examined the effects of various stimuli at the snout. Almost all the receptors were mechanosensitive and were little, if at all, influenced by thermal stimulation; some of the mechanosensitive receptors had extremely low thresholds. About half the receptors tested were phasic in behavior and the rest of them adapted slowly; the most sensitive phasic receptors were readily entrained by weak vibratory stimuli at frequencies between 50 and 300 Hz.

McIntyre and Kenins conclude that the snout is well suited for signaling information during the seeking of food. It is not so sensitive, however, that it cannot be used as a tool for the particularly rough job of breaking open termite-ridden logs to get at its prey (see CSIRO, 1969). An electron microscopic study of these sense organs in *Tachyglossus* would be of great interest for comparison with the receptor organs in the delicately sensitive muzzle of *Ornithorhynchus*.

The presence of a large population of sense organs in the snout offers an explanation of the presence of huge trigeminal nerves in the echidna. The snout, however, although well represented on the cortex, has nothing like the enormous representation of the muzzle found on the platypus cortex (see below).

The Organization of the Sensory and Motor Areas of the Neocortex

The organization of the auditory, visual, and somatosensory cortex in *Tachyglossus* has been studied by Lende (1964) who used the evoked-potential method only. This showed that all the sensory areas were clustered at the caudal pole of the cortex (Fig. 60). The fields were adjacent but separated by reasonably constant sulci. The auditory area lay furthest caudally; above and rostral to this was the visual area and the somatosensory area lay below the visual and rostral to the auditory areas. The somatosensory area extended only halfway up the lateral surface of the cortex but within its confines the pattern of somatotopic representation is like that of the platypus or for that matter, like that of all mammals: the caudal parts are represented dorsally (medially in the platypus and other mammals) and the anterior body regions are represented progressively ventrally in the

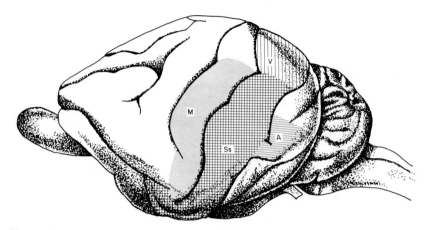

Figure 60. *Tachyglossus.* Somatosensory and motor (Ss), motor (M), visual (V), and auditory (A) areas of the cortex. *(From Lende, 1969.)*

cortex, but representation of the snout is far less than the representation of the muzzle in the platypus cortex.

Two motor areas were detected: one overlaid exactly the somatosensory area "and approximated nicely that pattern of localization." The other motor area was located in a gyrus immediately rostral to the sensory area but it exhibited no sensory counterpart. Anterior to these motor and sensory areas is a great expanse of "silent" cortex, the function of which is quite unknown.

The segregation of motor and somatosensory areas at the extreme caudal pole of the *Tachyglossus* neocortex has not been encountered in any other cortex. In the cortices of eutherians like the rat and the rabbit there are four discrete areas all with somatosensory and motor representation of varying degrees but in each area one system predominates. Two of the areas have lower sensory thresholds and have been termed somatic sensory-motor area I (SmI) and somatic sensory-motor area II (SmII). The other two areas have lower motor thresholds and are therefore called MsI and MsII. The sensory and the motor properties within each area coincide with and mirror the arrangements of the peripheral entities: caudal parts are represented medially and the anterior body regions are represented progressively laterally and ventrally.

Not all eutherian cortices are the same however; the sheep apparently has only one somatosensory area, SII (Johnson *et al.,* 1974), and the cortex of the hedgehog, studied by Lende and Sadler (1967), exhibits MsI rostrally with SmI just behind it. SmII is located inferior to SmI but MsII could not be detected. Adjacent areas overlapped and SmI was overlapped largely by the posteriorly located auditory and vidual cortices. There were no "silent" areas in this cortex. Lende (1969) admits the possibility that all this overlap may be an artifact, the result of the evoked-potential method applied to a very small cortex—it is only 13 mm long.

The organization of the marsupial cortex as exemplified by those of the American opossum *Didelphis virginiana* and an Australian wallaby *Thylogale (Macropus) eugenii* is also quite different from those of *Tachyglossus*, the platypus, and of eutherians. The motor area overlaps completely and coincidentally the somatosensory area.* Lende (1963, 1969) found no indication that either the sensory or the motor function predominated so he called this portion of the cortex the sensory-motor "amalgam." It is quite discrete and occupies a central area of the cortex extending from the rhinal sulcus up to and over the dorsal surface to the medial surface of the hemisphere. A small SmII is detectable in the opossum but not in the wallaby. The orientation of representation of the body within the amalgam is the same as in the platypus, eutherians, and *Tachyglossus*. Posterior to the amalgam at the caudal pole of the cortex lies the visual cortex dorsally and the auditory cortex ventrally. Thus the somatosensory areas in all mammals studied so far exhibit a mediolateral organization and the visual and auditory areas lie at the dorsal and ventral surfaces of the occipital pole of the cortex, respectively. The cortex of *Tachyglossus* appears to be unique in that somatosensory and motor representations do not extend up to sagittal region; the motor and somatosensory areas are grouped along with the visual and auditory cortex at the caudal pole of the hemisphere; and in that there is "relatively more 'frontal' cortex than in any other mammal, not excluding man" (Lende, 1969). Lende also recommends that further work should be done on this cortex, and indeed it would seem that a study using the microelectrode method would be necessary for comparison with the cortex of *Ornithorhynchus*.

Allison and Goff (1972) using the evoked-potential method in anesthetized and unanesthetized animals with chronically implanted electrodes have confirmed that the somatosensory area of the *Tachyglossus* cortex lies just ventral to the visual cortex and that in the somatosensory cortex low threshold stimulation elicits motor effects. They say, therefore, that the somatic area is better described as a sensorimotor area. They also confirmed that there is a separate purely motor cortex anterior to the sensorimotor area, and found some evidence for the presence of polysensory or association cortex medial to the motor and visual cortex. The late positive waves recorded had the characteristics of polysensory responses found in the cortex of other mammals, i.e., they were recorded in a nonprimary cortical area, were surface positive waves followed by a negativity, and were distinct from primary evoked responses by their longer latency, broader wave form and greater amplitude variability. It would be necessary to check on these characteristics in the echidna with single unit analysis.

*Apparently the sloth, *Bradypus tridactylus,* exhibits complete overlap of motor and somatosensory area (Saraiva and Magalhaes-Castro, 1975).

Allison and Goff also demonstrated, physiologically, interconnection of the two hemispheres: two echidnas were implanted with three pairs of stimulating electrodes placed in white matter just beneath frontal and motor cortex of the right hemisphere. In both animals stimulation of these electrodes evoked a short latency response in the homotopic area of the contralateral hemisphere. Interhippocampal evoked responses were also recorded. Allison and Goff, through lack of animals, were unable to determine whether or not the interneocortical and interhippocampal responses were mediated by the anterior and hippocampal commissures, respectively.

Thalamocortical and Spino-Cortical Relationships

W. Welker and R. A. Lende (1978) have carried out a series of selective partial ablations of somatosensory, auditory, visual, motor, and several different portions of the "silent" frontal neocortex. After 2 months survival time the brains were removed, processed histologically, and the location and extent of retrograde degenerative changes identified. Although some of the results are equivocal it was established that the "principle of adjacency," which holds good for eutherian brains, also holds good for the echidna brain, that is to say ablation of adjacent cortical areas resulted in degeneration of adjacent thalamic areas. This was particularly noticeable if the two adjacent lesions in the cortex were separated by a sulcus.

Removal of the somatosensory area led to degeneration of a thalamic nucleus called ventro-basal (VB) by Welker and Lende. Unfortunately these authors do not refer to Abbie's (1934) description of the thalamic nuclei so one is not sure if the two descriptions apply to the same nuclei. Thus Welker and Lende find that ablation of the separate motor area in the cortex results in degeneration of the ventrolateral nucleus, which, if it is the same as Abbie's nucleus ventralis lateralis is a way station for trigeminal impulses which presumably go to the somatosensory area; perhaps Welker and Lende's VB is Abbie's nucleus ventralis lateralis. Furthermore ablation of the visual cortex led to degeneration at a locus in the thalamus that did not correspond to the LGNa of Campbell and Hayhow (1971) in which those authors found most of the optic fibers terminate. LGNb, however, did exhibit degeneration; Welker and Lende state that their data "do not suggest an explanation of these basic discrepancies between the two studies." The ablation of various regions of the silent cortex led to the interesting conclusion that adjacent frontal cortical regions receive inputs from adjacent dorsomedial thalamic regions. These regions appear to consist of a relatively homogeneous mass of sparsely packed medium-sized neurones.

Lesions have also been placed in the cortex by Goldby (1939) and Draper (1971) with the object of tracing the extent of the pyramidal or cortico-spinal tracts in the spinal cord. Cortico-spinal tracts are entities found only in mamma-

lian brains and they constitute a direct connection between the motor cortex and
the lower reaches of the spinal cord. Abbie (1934) thought that in *Tachyglossus,*
after they descended from the cortex, they appeared as tracts on the ventral
surface of the medulla, decussated at the posterior end of the bulb, and terminated at the ventral horn cells of tbe first cervical segment. Goldby (1939),
however, by destroying a portion of the cortex near the midline between sulci α
and β and following the course of the degenerating fibers, found that they pass
caudally in the cerebral peduncles. At the level of the pons they decussate and
pass caudally in the Zonalbündel of Kölliker to the dorsal part of the lateral
column of the cord to at least the 24th spinal root. Goldby used the Marchi
method for tracing degenerate fibers and obtained only one good preparation. Dr.
Ed Draper has carried out a series of cortical ablations in *Tachyglossus* and has
had success in tracing degenerating fibers with the Fink and Heimer (1967)
modification of the Nauta silver impregnation technique; he has kindly allowed
me to quote the results of his unpublished study (Draper, 1971).

The lesions, as in Goldby's study, were placed in the dorsal half of the gyrus
between sulci α and β and survival times ranged from 4 to 35 days postoperative.
In preparations of tbe brain made after those periods of survival there was no sign
of contralateral degeneration in the internal capsule, thalamus, corpus striatum,
nor in other diencephalic centers. The highest level in the brain stem at which
degeneration can be ascribed to that of cortico-spinal fibers is in the ventromedial
portions of the cerebral peduncles. These fibers can be traced back on the ipselateral side until they enter the rostral border of the pons. At this level the fibers
from the peduncle turn sharply medially and ventrally; some terminate in pontine
nuclei, the rest cross the midline as a discrete compact tract, passing to the
ventrolateral pons and then caudad. In the medulla at the level of tenth cranial
nerve the cortico-spinal tract as it now may be called passes caudally and ven-
trolaterally, i.e., so superficially that it is barely covered by a few external
arcuate fibers. At all levels in the medulla some degenerating fibers could be seen
leaving the tracts and terminating in medullary nuclei. At cervical levels in the
spinal cord the cortico-spinal tract is seen to occupy the entire dorsal half of the
lateral funiculus. The tract here is very compact, there is no degeneration in the
ventral funiculi and all of it is in the decussated cortico-spinal tract lying contralateral to the cortical lesion. The tract continues down the spinal cord and at
the 24th and 25th spinal segment levels it is relatively the same size at higher
levels, occupying the dorsal half of the lateral funiculus; it continues to the
caudal limit of the cord which here is only 1 mm in diameter. Thus Goldby and
Draper are in agreement about the extent of the cortico-spinal tract but the
puzzling thing about both studies is that the lesions were placed near the midline
in the gyrus between sulci α and β which according to Lende is well outside the
motor and the sensorimotor areas, which according to Allison and Goff, could be
association cortex. Perhaps the motor cortex extends further to the midline at a

deeper level in the cortex, part of which was removed by the ablation techniques of Goldby and Draper.

A study comparable to the above has not been carried out on the platypus; Hines (1929), however, describes pyramidal (cortico-spinal) tracts in that animal in these terms: "The descending motor systems from the cortex are not noticeably large. The pyramidal tract (Koelliker, 1901) does not penetrate further than the cervical region of the cord and occupies only a small area upon the ventral surface of the medulla." Abbie (1934) agreed with Hines but he also thought that the pyramidal tracts in *Tachyglossus* were small and terminated at the first cervical segment; however, as we have seen they pass all the way down the cord so it is quite likely that the platypus will prove to have extensive cortico-spinal tracts.

Behavior

Vision and Discrimination

Gates (1973) has shown that *Tachyglossus* is very good at learning to make visual discriminations. The method used was to train the echidnas to make simple two-choice discriminations in apparatus consisting of a start box, runway, and two goal boxes fitted with inwardly opening doors, i.e., they could be pushed open by the echidna. In each of the goal boxes a reinforcement reward (food) could be offered; the doors bore the stimulus display panels on their outside surfaces. As an example of the procedure, an experiment is to determine whether or not echidnas can make a choice between a black and white square: for a correct choice the animal was rewarded with a 30-second access to food, for an incorrect choice the food was withdrawn—incorrect choices were counted as errors. The criterion of learning was that each had to choose the correct panel 18 times on any one session for six daily consecutive sessions. The experiments were controlled for olfactory cues and for inclination to develop a position habit. From the results shown in Fig. 61A,B it is apparent that all three animals involved quickly learned to discriminate between black and white, and that there was a rapid drop in the rate of error over the period of the six sessions. Gates points out that Sutherland and Mackintosh (1971) obtained learning of an easy discrimination in rats in about the same number of trials required for the echidnas.

Similar experiments were carried out in which echidnas were confronted with other discrimination tasks: vertical-horizontal, triangle-circle, and triangle-inverted triangle discrimination. It was concluded that echidnas can make brightness discriminations, discriminations between differences in orientation, and discriminations between complex shapes; they also have the capacity to make discriminations that demand visual acuity. Gates suggests that echidnas have visual acuity capability which is probably equivalent to that of the hooded rat.

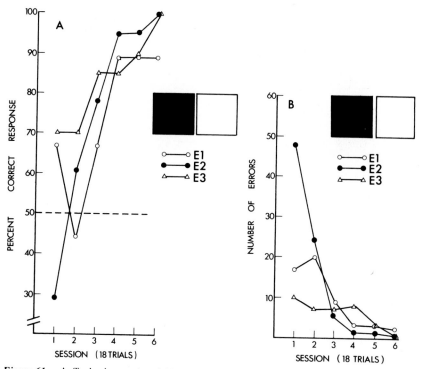

Figure 61. A, *Tachyglossus.* Acquisition of black-white discrimination by three echidnas. B, error rate during acquisition of the above. *(From Gates, 1973.)*

Gates also found evidence of interocular transfer of discrimination. Essentially the experiments were the same as the black-white discrimination task described above except that one eye was occluded with tape. Figure 62 shows that the four animals involved attained the criterion of discrimination within 100 trials with the first eye and on transferring the occluding tape to the other eye they reached criterion in 40 trials; performance of the echidnas with both eyes occluded dropped to chance levels indicating that vision was being used to make the discrimination. Similarly it was shown that interocular transfer of vertical-horizontal and of oblique stripes discriminations took place. Gates offers two explanations of how this may occur in the absence of a corpus callosum: there is evidence (Ebner 1967; Heath and Jones, 1971) that visual areas in marsupials may be interconnected, via the anterior commissure; the same may hold for echidnas. However, information could conceivably be transferred from one side of the brain to the other elsewhere than in the cortex—the massa intermedia of the thalamus, or even the commissure connecting the two superior colliculi may be serving the purpose; an experiment like that of Allison and Goff (1972)

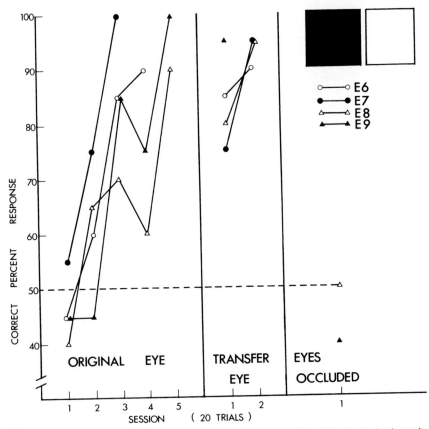

Figure 62. *Tachyglossus.* Transfer of training between eyes for a black-white discrimination task. The left-hand graph shows acquisition of a black-white discrimination to a criterion of 90% correct with one eye occluded. The middle graph shows acquisition of the same discrimination for the same animals with the trained eye occluded and the naive, transfer eye uncovered. The results on the right side of the figure show the performance of two animals with both eyes occluded. *(From Gates, 1973.)*

mentioned above to see whether or not short-latency homotopic interhemispheric responses follow stimulation of the visual cortex would help to determine the region of transfer.

Another possible route of transfer of visual information discussed by Gates is the uncrossed optic nerve fibers that run to the ipselateral geniculate body. Admittedly these account for as little as 1% of the number of optic fibers (p. 186) but Lashley (1939) has estimated that only $^1/_{50}$ of the number of neurones in the lateral geniculate nucleus is required to mediate discrimination of visual patterns in eutherians. Furthermore, Muntz and Sutherland (1963) have shown that rats, which also have a very small number of uncrossed optic fibers, can learn a visual discrimination using the uncrossed fibers only.

Position-Habit and Habit-Reversal Learning

Saunders *et al.* (1971b) found that *Tachyglossus* could learn to make choices up to 100% correct in a simple T-maze apparatus which offered a choice of left or right turn into a goal box. The results are summarized in Fig. 63; the speed referred to is the running time, from start box to goal box, which also increased as the habit was acquired. Extinction of the habit was quite rapid and by the fourth session all animals were responding at chance.

Similarly Saunders *et al.* (1971a) observed successful habit-reversal learning in echidnas. In this sort of experiment the animal is first trained to select one alternative of a two-choice task until a criterion performance level is reached. Then the animal is trained to choose the second alternative; this procedure is repeated for a number of times. The aim is to see if there is improved performance at later reversals compared with earlier reversals. It was found that echidnas are able to achieve rapid improvement during a series of position-habit

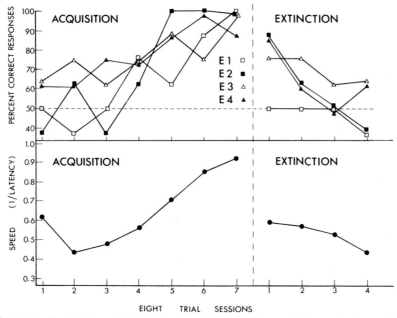

Figure 63. *Tachyglossus.* Choice performance (top) and running speed (bottom) during acquisition and extinction of the position habit in four echidnas. The top curves represent the performance in individual animals and the bottom curve the mean speed for all animals. *(From Saunders et al., 1971b; by permission of Dr. J. Saunders.)*

reversals so much so that during the later reversals they were able to make "one-trial" reversals which have often been observed in experiments with rats but never with nonmammalian (avian) species.

Maintenance and Social Behavior, Vocalization

By means of observation of a very large number of echidnas kept in the laboratory, outdoor cages or in outdoor enclosures, Brattstrom (1973) recognized or defined some 65 behavioral postures, most of which he illustrated. Examples of the sort of behavior observed are: scratching at various parts of the body with the grooming toe, various types of walking and investigative behavior, sniffing at objects and congenors, crawling positions adopted during sleep, rolling in dust, defensive postures, feeding behavior, etc. A lot of this sort of behavior can be seen in the CSIRO film *The Echidna* (1969). A matter of considerable interest arising from Brattstrom's study is the observation of dominance and its maintenance. This takes the form of a subordinate individual recognizing some dominance feature (apparently size) of another and endeavoring to avoid contact. These "avoids" as Brattstrom terms them are of various kinds: approach-avoid, two echidnas on the same path, the subordinate continuing on but to one side; approach-turn, two approaching head on, the subordinate turning away; and variations of this, including the approach-quick turn, and approach-turn-completely-around. Occasionally subordinate echidnas stand still as a dominant passes or the dominant may pass by walking over the approaching subordinate. However, as a rule, if subordinates do not move out of the way they are bumped by the dominant. The bumps may be to the front, side, or rear of the subordinate and are given by the shoulders of the dominant.

It is just conceivable that this sort of behavior could occur among a group of echidnas living in the bush. Most of the year echidnas, in my experience and that of others, are solitary, even at Kangaroo Island where they are relatively abundant. However, echidnas in small numbers have been found together in the bush but this has been observed only during the breeding season. There are at least five authentic records of this behavior:

3 echidnas taken together	2♂,1♀	Southern New South Wales	August 18, 1967
6 echidnas taken together	5♂,1♀	Heathcote Victoria (Augee, 1969)	August 3, 1968
5* echidnas taken together	4♂,1♀	Central Australia, Northern Territory	July 29, 1969
4 echidnas taken together	3♂,1♀	Southern Victoria	August 22, 1971
6* echidnas taken together	unknown	Homestead Ck., Northern Territory	August, 1971

*I am indebted to Mr. Lindsay Best, Canberra College of Advanced Education, and to Mr. William Bolton of Animal Industry Branch, N.T.A., for the Northern Territory records.

The echidnas at Central Australia were in a rocky outcrop walking in Indian file almost nose to tail. The Homestead Creek animals were all together in a rocky crevice and the Southern Victorians were found within a few feet of one another, all digging down into sandy soil in a stand of bracken fern, apparently alarmed by my approach. Since echidnas do come together to form groups, it is at least possible that dominant-subordinate behavior could occur but it has not been observed.

It was suggested earlier that olfaction might play a part in bringing echidnas together. It is also possible that vocalization may help. It has at last been established that echidnas can make sounds. Pridmore (1970) recorded on tape a sound emitted by a male specimen of *Tachyglossus aculeatus acanthion*; the sound was later analyzed on a sonograph machine, and the animal was also tested to see if it made ultrasonic noises, with an ultrasonic receiver but the result was negative. The call, repeated 4–12 times, was of about 20-msec duration broken by 20-msec intervals. The note emitted had frequency of 0.3 kHz. To Pridmore's ears it sounded like the coo of a dove and he found a close resemblance of the sonograph of the echidna noise to that of the ring-dove, *Streptopelia risaria*. It would be interesting to know how loud the call would have to be for another echidna to hear it—Johnstone (cited in Griffiths, 1968) did not determine CM below 0.5 kHz.

Behavior in Relation to Environmental Temperature

Augee *et al*. (1970) observed the movements of echidnas in and out of a pit of earth provided in their otherwise concrete covered living quarters. When in the pit they dug in and were partially covered by earth; movements in and out were monitored continuously by means of a telemetry system involving a transmitter placed surgically inside the animals.

It was found that, of several environmental factors examined, one only had a significant influence on movements in and out of the pits, this was ambient temperature. In winter both movements were correlated with the times at which air and ground temperatures crossed. In spring the best correlation with movement both in and out of the pits was temperature alone; the mean air temperature at the time of leaving was 18.1°C and 18.6° at the time of returning. However, during periods of torpor, rain, and air temperature over 33°C the animals remained in the pits.

Paradoxical Sleep

Allison and Van Twyver (1972) studied states of waking and sleep in five young but adult echidnas of both sexes by means of polygraphic recording from chronically implanted electrodes in olfactory bulbs, pyriform cortex, parieto-occipital cortex, hippocampus, neck muscles, and around the eyes. When the echidnas were asleep recordings described as typical of "mammalian slow wave sleep (SS)," i.e., large slow electrical changes indicating synchronous activity of many nerve cells, were obtained from parieto-occipital and pyriform cortex and from the hippocampus. This type of activity accounted for 35.8% of the recording time (24–120 hour continuous recordings). This was in marked contrast to the many small fluctuations recorded when the animals were awake but quiet. In both states eye movements were negligible. Allison and Van Twyver however, were unable to observe convincing evidence of rapid eye movement sleep or paradoxical sleep as it is called. During PS in a mammal like the cat, sporadic eye movements are very evident and the electrical activities of cortex and hippocampus are fast and rhythmic as in the waking state but the animal is obviously not awake—the musculature is limp and the animal is difficult to arouse. Hence the name paradoxical sleep to describe a state in which electrical activity of the brain approaches to that of wakefulness in a sleeping animal. These manifestations were not observed in sleeping adult echidnas.

Echidnas are not alone in this respect: rabbits exhibit PS only after a sojourn of several months in the laboratory, and donkeys not at all (Allison and Van Twyver, 1970). Very young mammals, however, exhibit more PS than adults do so it is possible that very young echidnas might also exhibit PS. The "dreaming" observed in George Bennett's young platypuses supports the notion since PS and dreaming are linked in man.

ZAGLOSSUS

To my knowledge no formal studies of behavior of *Zaglossus* have been carried out nor is there any information on cortical localization to hand. Kolmer (1925) however, has published some observations on the histology of the eye, inner ear, and of the tactile sense organs of the beak in one animal only. The preparations, however, were not good since the tissues were fixed an unknown period after death of the animal; despite this Kolmer seemed to be more concerned with details of dubious cytology than with gross anatomy and histology. Prior to its death the animal had lived 10 years in a Vienna menagerie.

Anatomy of the Eye

Size and shape unknown, the sclera is made up of connective tissue that encloses a cartilaginous cup far thicker than that in the eye of *Tachyglossus*—on the average it is 160 μ thick. Internal to the cartilage is a chorioid layer sparsely supplied with blood vessels; this is surprising since this layer with its blood vessels is the sole source of nutrients for the retina. The elements of the latter were unrecognizable but the whole retina was some 170 μ thick.

The ciliary body, as in *Tachyglossus*, consists of a series of processes or folds arranged around the circumference of the eye between iris and retina, interconnected anteriorly by the annular ciliary web. As far as ciliary musculature is concerned Kolmer takes a stand between that of Gresser and Noback who say ciliary muscles are well developed in the ciliary web of *Tachyglossus*, and that of O'Day, Walls and Prince who have it that there are none: "doch sind darinter wenig Muskeln vorhanden."

The iris is thin, has blood vessels on its anterior surface, and exhibits a strongly developed sphincter muscle of circularly arranged smooth muscle fibers with pigment in between the fibers. Kolmer says the lens is the flattest he has ever observed and indeed the flatness index of 3.00 is greater than that for the lens of *Tachyglossus*, 2.75. The lens is suspended by zonule fibers inserted at the extreme periphery of the lens as it is in *Tachyglossus*. In spite of the age of the animal, the lens was not as hard as one would expect; in fact, Kolmer says it was soft in the middle. If this proves to be so in all *Zaglossus* eyes and the ciliary body proves to have muscles, the eye may accommodate by alteration of lens shape by contraction of ciliary muscles as well as by deformation of the eyeball as suggested for *Tachyglossus* by Gates (1973).

The cornea apparently is keratinized since Kolmer found that the outer epithelium exhibits seven layers of squamous cells.

Anatomy of the Ear

From a few sketchy observations on this badly-fixed material Kolmer concluded that the inner ear is no different from that of other monotremes: the cochlea is curved, not coiled, exhibits an isthmus lagena, lagena, lagenar macula, and a true organ of Corti. It is apparent from Gervais' (1877–1878) illustration that the malleus has a long manubrium and gracile process and that the relations of that bone to the tympanic are the same as those in *Tachyglossus*. Kolmer made the interesting observation that the elastic cartilaginous ear trumpets are made of hyaline cartilage resembling that found in sauropsids.

Tactile Organs of the Beak

Kolmer (1925) described the skin of the beak as smooth, keratinized, and studded with the openings of sweat gland ducts. The sweat glands are located deep in the corium and associated with their ducts are rodlike structures (Hornzapfen) apparent at the surface of the skin and which pass down into the corium. Numerous Pacinian corpuscles are present in the corium and not infrequently they surround the bases of the rodlike structures. As well as Pacinian corpuscles numerous myelinated fibers and Merkel's cells can be seen in the corium. Kolmer gives no description of the structure of the Hornzapfen but says they are like the modified hairs described by Poulton in the epidermis of the muzzle of *Ornithorhynchus*. As Wilson and Martin and Bohringer have shown these "modified hairs" are multicellular rods, so at present we are completely in the dark as to the nature of Hornzapfen. Histological observations on well-fixed skin from the beak will doubtless settle the matter.

Organs of Taste

Circumvallate papillae are present posterior to the keratinized pad at the base of the tongue but unlike those of *Tachyglossus*, they occur at the surface of the tongue, not at the bottom of deep grooves (Kolmer, 1925). Foliate papillae are also located lateral and posterior to the keratinized pad.

8

Reproduction and Embryology

ORNITHORHYNCHUS

Female

OVIPARITY. Although much of the interest of the monotremes lies in their display of a curious mélange of reptilian, mammalian, and intrinsic monotreme anatomical and physiological characters, the one thing that really sets them apart from all other mammals is the fact that the females lay eggs and do not give birth to their young. It took 93 years from the time of their discovery to prove to the scientists that they were oviparous, but a lot of people living in Australia in the 19th century knew that both echidnas and platypuses were oviparous. The man, at first a skeptic himself, that convinced the scientists was a young Scot, W. H. Caldwell from Cambridge University, who had set up camp in April 1884 on the banks of the Burnett River in Queensland for the express purpose of collecting the eggs of our lungfish and a sequence of embryological stages from the uteri of echidnas and platypuses. Actually aborigines did the collecting but the story is best told in Caldwell's (1887) own words: "During part of June and July I spent many hours daily in the water, hunting everywhere for the eggs of *Ceratodus*. Towards the end of July the blacks began to collect *Echidna,* and very soon I had seg-menting ova from the uterus. In the second week of August I had similar stages in *Ornithorhynchus*, but it was not until the third week that I got the laid eggs from the pouch of *Echidna*. In the following week (August 24) I shot an *Ornithorhyn-chus* whose first egg had been laid; her second egg was in a partially dilated os uteri. This egg, of similar appearance to, though slightly larger than that of *Echidna*, was at a stage equal to a 36 hour chick. On 29th I sent in the telegram 'Monotremes oviparous, ovum meroblastic' to a neighbouring station, where it would meet the passing mail-man, addressed to my friend Professor Liversidge, of the Sydney University, asking him to forward it to the British Association at Montreal." The discovery was announced at Montreal Sept. 2 (Caldwell, 1884b). From July–August of the next year he took, with the help of 150 "blacks," between 1300 and 1400 echidnas of both sexes. Very little came of that fantastic slaughter—one paper labeled Part I (Caldwell, 1887) containing facts and misinterpretations concerning oogenesis and embryology in platypuses, echidnas, and koalas; the facts are as follow:

1. Ripe ovarian ova were found to be 2.5–3.0 mm in diameter. Actually maximal diameters are 4.3 mm for platypuses and 3.96 for *Tachyglossus*.

2. Two eggs were found in the dilated end of the Fallopian tube (the infundibulum). Both these had commenced to segment meroblastically and each had eight nuclei; usually segmentation commences when the fertilized ovum arrives in the uterus. Each of these eggs was enclosed by a thin vitelline membrane and surrounded by what he calls pro-albumen which he thought was laid down by the follicular epithelium during oogenesis; actually the "albumen" layer is deposited round the egg in the fallopian tube (C. J. Hill, 1933). Nevertheless the observation that the follicular epithelium does form a secretion is regarded by Flynn and Hill (1939) as a discovery of first rate importance since according to them, the secretion is the homologue of the liquor folliculi of the Graafian follicle of other mammals; thus the monotreme follicle is fundamentally different from the sauropsidan follicle which lacks follicular fluid.

3. An egg 6 mm in diameter was found in the uterus of a platypus; it was enclosed by a two-layered shell and meroblastic segmentation was taking place.

4. A laid egg measured 15 × 12 mm from which Caldwell concluded that the egg increases in size in the uterus by absorption of uterine secretions.

5. When dilute hydrochloric acid was placed on the shell of a laid egg of *Ornithorhynchus* a quantity of gas was evolved indicating the presence of a "calcic" salt, perhaps Caldwell meant carbonate. No calcic salts were detected in the echidna egg shell.

OOGENESIS AND FERTILIZATION.   Flynn and Hill (1939) have given a definitive account of oogenesis in *Ornithorhynchus* and *Tachyglossus* but they made no attempt to give separate accounts of what happens in the two genera; stages from either genus are used to give a sequential account of "monotreme" oogenesis and that is what will be given herein.

Flynn and Hill recognize three stages in the development of oocytes: at phase 1 the oocytes range in diameter from 0.06–0.15 mm and are located in the cortical zones of the ovary (Fig. 75). The nucleus is eccentric in position and lies close to what will later become the upper pole of the egg. The nucleus exhibits a reticulum in the strands of which many fine granules are dispersed, one large nucleolus, and numbers of smaller ones. A homogenous membrane, the zona pellucida, lies just below the follicular epithelium which at this stage consists of a monolayer of cuboidal cells; beneath the zona pellucida is a thin striate layer and below this is the egg membrane enclosing the cytoplasm which contains lipid droplets.

During the second phase of growth the oocytes achieve a diameter of ca. 0.5 mm. The fat droplets are now distributed so as to form a cortical zone and yolk sphere primordia can be detected among the droplets. Apparently the latter are incorporated into the yolk spheres since more and more yolk spheres are formed but less lipid droplets can be found as oogenesis proceeds.

The third stage is one of growth of oocytes from diameter of ca. 0.6 mm to the fully grown ovum of 4.0–4.5 mm diameter (3.96 mm in *Tachyglossus*) (Fig. 64).

During the early phases of this growth a layer of cytoplasm, more or less free of yolk, forms around the periphery. This layer is noticeably thicker above the eccentrically located nucleus; the latter and the thickened layer of the cytoplasm is the primordium of the germinal disc. Just beneath the peripheral layer of

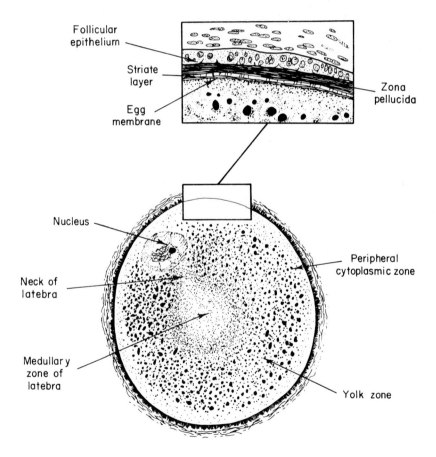

Figure 64. *Tachyglossus*. Section of oocyte 0.65 × 0.60 mm in diameter. *(From Flynn and Hill, 1939; with permission of The Zoological Society of London.)*

cytoplasm two zones are discernible—the outer containing yolk spheres and a central medullary zone with finely alveolar cytoplasm and yolk-sphere primordia. From the medullary zone a column of cytoplasm of the same structure ascends to the germinal disc. This is the latebra recognized by Semon (1894c) as having the same form and relations as the latebra of the sauropsidan egg. At the final phase of oogenesis the latebra is flask-shaped, the neck of the flask communicating with the germinal disc; the body exhibits a central core of vacuolated cytoplasm and eosinophil granules which are a source of fine-yolk granules. The latebra persists up to the short primitive streak stage and apparently it supplies fine-grained yolk to the developing embryo.

The follicular epithelium surrounding the fully-developed oocyte consists of an epithelium two cells deep, the cells exhibiting secretion granules which are the precursors of the liquor folliculi.

After full growth is achieved the first maturation spindle is formed followed by division of the nucleus and the formation of a polar body while the oocyte is still in the ovary. After that division a second spindle forms and the ovum is then shed into the infundibulum where it is fertilized. The infundibulum is lined by a single layer of columnar epithelium exhibiting ciliated nonsecretory and nonciliated secretory cells. The latter form a clear fluid that is shed into the infundibulum just before ovulation (C. J. Hill, 1933, 1941). The egg passes down into the fallopian tube and its passage is doubtless assisted by the ciliated cells. During its passage secretory cells in that region pass a product into the lumen which forms a two-layered coating around the egg. This has been called the "albumen" layer by Caldwell (1887) and by C. J. Hill. Recently, however, Hughes *et al.* (1975) state without evidence that it is a mucoid coat of acidic glycoprotein. Doubtless they have deduced this from the fact that marsupial eggs have an "albumen" coat which Hughes (1969) found, in the eggs of *Trichosurus vulpecula*, had the histochemical properties of a strongly acidic mucopolysaccharide.

Sometimes, in the platypus, this coat traps spermatozoa as it passes down the fallopian tube (Flynn and Hill, 1939); the mucoid coat around the eggs of marsupials also traps spermatozoa. During passage down the fallopian tube the second polar body has been formed and the male pronucleus (from the spermatozoan) has come to be near the female pronucleus in the germinal disc. Meanwhile the egg has descended to the tubal gland region of the fallopian tube; when this happens the glands achieve maximal secretory activity and at the same time a thin homogenous layer forms around the egg—the basal layer of the shell; C. J. Hill (1941) concluded that the tubal gland secretion is responsible for formation of that shell layer. Hughes *et al.* (1975) imply that it is composed of ovokeratin but no evidence is presented; in this connection it may be pointed out that Hughes (1969) found that the shell of the egg of *Trichosurus* has the histochemical properties of a keratin (see Appendix).

CONJUGATION, CLEAVAGE, AND FORMATION OF THE UNILAMINAR BLASTOCYST. The egg now passes into the top end of the uterus and here the two pronuclei conjugate and form a single spherical nucleus, but *en passant* to the body of the uterus a second shell membrane is laid down; it is probably formed from the secretions of the uterine glands of this region (C. J. Hill, 1941). This second layer is differentiated into a matrix substance in which rodlike bodies are embedded with their long axes normal to the surface of the egg. After formation of this layer, the egg, having now passed to the body of the uterus, is ready to undergo cleavage which, as Caldwell discovered, is meroblastic. He found that the first cleavage furrow divides the germinal disc into a larger and a smaller area and the second furrow is laid down at right angles to the first so that the four-celled stage consists of two large and two small blastomeres sitting on top of a sphere of yolk. He also found, he says, that cleavage in the marsupial *Phascolarctos cinereus* is meroblastic and that the four-blastomere stage is the same as in the monotremes except that the egg is much smaller and the yolk is white. He promised to tell us about further development of the koala egg in "Part II," but he never did—he became a successful business man in Scotland instead. If his observation is confirmed it should have some impact on arguments about relationships of Eutheria, Metatheria, and Prototheria.

From here on the descriptions of "monotreme" development are based on those of Wilson and Hill (1908, 1915) for *Ornithorhynchus* but mainly on those of Flynn and Hill (1947) for *Tachyglossus*. Division of blastomeres continues and at the 32-blastomere stage the blastodisc, as it is called, is differentiated into two sets of blastomeres—one of large marginal blastomeres open to the surrounding and underlying yolk, and an inner set of smaller cells delimited on all sides except underneath where they are open to the yolk. This arrangment is found in the eggs of Sauropsida at that stage; it would be of great interest to learn if this holds good for *Phascolarctos*. Cleavage proceeds and leads to the formation of a circular biconvex blastodisc 5 cells deep centrally and thinning out peripherally to one cell deep. The cells by now have limiting membranes and are marked off by a yolk membrane from the underlying yolk. Around the periphery of the biconvex blastodisc a large number of free cells appear—the vitellocytes of Flynn and Hill formed by division of the marginal cells of the disc. They are capable of migrating through the cytoplasm by means of pseudopodia. At this stage the diameter of the ovum is not very much more than at ovulation—4.7 mm (Fig. 65).

When the blastodisc is about six cells deep the marginal vitellocytes form a definite ring around the disc, the germinal ring. This commences to grow outwards over the surface of the ovum and at the same time the blastodisc does likewise, the thickness of the disc decreasing so that it ends up two to three cells deep. These processes give rise to a unilaminar blastoderm one cell thick and

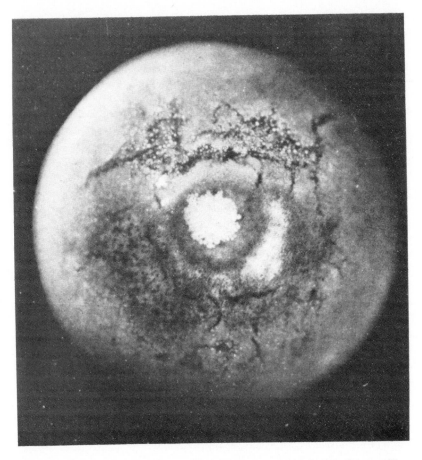

Figure 65. *Tachyglossus.* Surface view of blastodisc of an ovum 4.7 mm in diameter. The ragged edge is formed from vitellocytes. *(From Flynn and Hill, 1947; with permission of The Zoological Society of London.)*

almost completely enclosing the sphere of yolk except for a small area at the lower pole; the diameter of the blastocyst as it may now be called is ca. 7 mm. The increase in size is due to absorption of uterine secretions and is accompanied by growth, presumably by intussusception, of the two-layered shell; the rodlets also increase in length so that the shell is thicker.

FORMATION OF THE BILAMINAR BLASTOCYST AND EMBRYO. Long before completion of the unilaminar blastocyst stage the constituent cells of the blastoderm show signs of differentiation into prospective ectoderm and endoderm, the latter finally developing anastomotic connections and amoeboid pseudopodia. They sink below prospective ectoderm cells and form a network. The spaces in the network finally are filled in by mitotic division of its cells and by this means a complete layer of endoderm is formed beneath the ectoderm; thus the unilaminar blastocyst is converted into a bilaminar vesicle or blastocyst. The endoderm cells then commence to ingest yolk granules. At the lower pole of the blastocyst the hitherto small bare area of yolk is finally enveloped by ectoderm and endoderm cells that come together and form a thickening or scar—the yolk navel, a structure very like the yolk navels of the eggs of sauropsidan reptiles. At this stage in the monotreme the first sign of the embryo appears; the primitive streak develops in the blastodisc as a longitudinal thickening, about 0.6 mm long in the ectoderm.

The complete enclosure of the yolk enables the blastocyst to increase in size by cellular division and by absorption and storage of fluid secreted by the uterine glands. This fluid provides the nutrition for the growth of the embryo during the remainder of the gestation period and during the incubation period (a little over 10 days for *Tachyglossus,* p. 240). According to Flynn and Hill growth of the blastocyst to a diameter of ca. 12 mm is rapid, the two-layered shell swelling and coming into intimate association with the uterine wall; this no doubt facilitates passage of nutritive fluid into the blastocyst. During absorption of the fluid the yolk liquefies and mingles with the uterine fluid. The processes of secretion of the nutritive fluid are known for the platypus only (C. J. Hill, 1933, 1941). The secretion is formed in the superficial and middle regions of the uterine glands, which at this stage resemble the glands at the luteal phase in the uteri of the eutherians and metatherians (Fig. 77). The secretion takes the form of fine granules that pass to apical swellings in the cytoplasm projecting into the lumen. The granules are shed into the lumen where they liquefy and are absorbed by the bilaminar blastocyst. Nothing whatever is known of the composition of the uterine secretion but it is not unlikely it will prove to be similar to those of the marsupials *Macropus eugenii* and *Didelphis marsupialis* that have been studied to date (Renfree, 1972, 1973, 1975). In *Macropus eugenii* the embryo enlarges more than 200,000 times before attachment of the yolk sac membrane to the uterine mucosa; during that time the only contact with the mother is the uterine fluid that bathes the blastocyst. This fluid contains some proteins detectable in serum and lymph but uterine specific proteins are synthesized as well. The secretion from the opossum uterus has fewer uterine-specific proteins than that from *Macropus eugenii*.

It can be gleaned from the publications of Wilson and Hill (1908, 1915) that after the completion of the bilaminar blastocyst a medullary plate appears at one pole of the sphere, which acquires bilateral symmetry and a primitive streak. In an egg 9 mm in diameter a sheet of mesoderm derived from the primitive streak was interposed between ectoderm and endoderm. The mesoderm at this stage had extended beyond the limits of the medullary plate—the beginnings of trilaminar omphalopleure.

When the blastocyst has achieved a diameter of 11–12 mm, the mesoderm exhibits three to four pairs of somites and the third layer of the shell is apparent. The precursor of this layer appears in the cells of the uterine glands when the precursors of the nutritive fluid are being shed into the lumina of the glands. The shell precursor is coarsely granular and during formation of the third layer the uterine glands attain their peak of activity. The structure of the newly-laid egg shell has been described by Caldwell (1887), Semon (1894c), and J. P. Hill (1933). According to the latter the basal layer is 0.0016 mm thick, the rodlet layer 0.01–0.012 mm, but the third layer forms the great part of the thickness since the whole shell is 0.16–0.20 mm thick. There are two zones in this layer, an inner next to the rodlet layer consisting of coarse granules of irregular form and size between which open spaces occur; the outer zone exhibits larger and more irregular granules. Pore canals in the form of irregular clefts stretch from the inner zone and open to the exterior at the surface. The openings, or pores, are small and are not numerous. Hughes *et al.* (1975) have also discovered, with the aid of an electron microscope, that the shell of the full term egg in *Ornithorhynchus* is porous. This, they say, ''is obviously of primary importance as a pre-adaptation for viviparity.'' Others might think it useful for respiration (see Appendix).

Burrell (1927) has recorded the sizes of laid egss found in nests and on average single eggs measure 17.25 × 14 mm and the individuals of twin sets, 17.5 × 13.8 mm. He also speaks of twin eggs measuring 15 × 26 and 17 × 26 mm and that another set measured 15 × 25 and 16 × 25 mm. Flynn and Hill (1939) do not believe that platypus eggs can be as big as this and they say ''These latter measurements are far in excess of any that have come within our own experience and we suspect the numeral 2 in each case is a misprint for 1.'' Burrell also states that of 70 nests with tenants, 11 contained 1 egg or one young one, 54 contained two, and 5 contained three eggs or young.

As far as the composition of the shell of the egg is concerned, Neumeister (1894) applied many chemical and enzymatic tests to the shells of echidna pouch eggs and concluded that they were largely made up of a protein, containing 5% sulphur, allied to keratin.

At the time of laying the eggs of both genera exhibit 19–20 pairs of somites, optic vesicles at the anterior end of the neural tube, and a proamniotic head fold (Hill and Gatenby, 1926; Hughes *et al.*, 1975; Luckett, 1976). Evidently the major part of embryogenesis occurs during the incubation of the egg; very little is known about that in the platypus but something is known of development of the embryo and fetal membranes in *Tachyglossus* pouch eggs (p. 241).

CORPUS LUTEUM.   The duration of the gestation period in *Ornithorhynchus* is unknown but it is known that it is accompanied by the formation and growth of a corpus luteum (Hill and Gatenby, 1926). These authors distinguish five stages of growth of corpura lutea taken at different stages of the growth of the embryo.

Stage 1 corresponds to unsegmented eggs to one with 20 blastomeres. The corpus luteum exhibits an open hole at its distal end through which the ovum escaped. The opening leads into a large central cavity containing a mass of clotted extravasated blood and single cells or islands of luteal cells derived from the follicular epithelial cells. These form the folded inner lining of the cavity. External to this is the theca interna and the outermost layer is the fibrous theca externa. The latter sends actively ingrowing trabeculae to connective tissue carrying blood and lymph capillaries into the center of the developing gland. Hill and Gatenby found that the luteal cells are different from those of other mammals in that they lack lipid droplets.

Stage 2 is associated by an egg with a late cleavage stage of the blastodisc; the corpus luteum accompanying this ovum becomes solid throughout as a result of enlargement of the luteal cells and the opening is plugged up with the same cells. The luteal cells in the center of the gland are accompanied by theca interna cells and fibroblasts. The cytoplasm of the luteal cell is dense and very finely granular; very infrequently it exhibits a clear vacuolar space. These vacuoles according to Hill and Gatenby are not to be confused with the fat-containing vacuoles of the luteal cells in other mammals. The majority of the theca cells now lie along the septal ingrowths of the theca externa and apparently attain "their maximum functional activity."

Stage 3 is also found at a late cleavage stage of the blastocyst. The plug is now covered by a thin layer of cells continuous with the theca externa but the main difference from Stage 2 is the first sign of regression: polymorpho-nuclear leucocytes are detectable in the connective tissue and some have actually penetrated the cytoplasm of the luteal cells.

Stage 4 is a corpus luteum associated with a blastocyst exhibiting a primitive streak 6 mm long and a short head process. The corpus luteum measures 3.8 × 2.75 mm and the luteal cells within exhibit various states of degeneration: some are shrunken, many are spindle-shaped while others contain clear spherical vacuolar spaces, the largest occupying half the cell body; even the nuclei in some instances are vacuolate. Leucocytic invasion is slight and the theca interna cells appear to be unchanged.

Stage 5 is associated with two near-term eggs in the uterus; the corpus luteum measures 4 × 2.5 mm. The degenerative changes in the luteal cells are pronounced and throughout the tissue intensely stained homogeneous masses are encountered. These are completely degenerate luteal cells. Numerous polymorpho-nuclear leucocytes are present in both cytoplasm and nuclei of the luteal cells but theca interna elements are remarkably well preserved and show no signs of regression. Hill and Gatenby, therefore, are of the opinion that regression of the corpus luteum sets in before the egg is laid. This is not confirmed by a preliminary report (Hughes *et al.*, 1975) to the effect that corpora lutea, examined by electron microscopy, in a platypus bearing two near-term eggs did not exhibit necrotic tissue. It would be of great interest to learn what they did find there; it is possible they were looking at theca interna cells. In the peripheral plasma of that platypus Carrick *et al.* (1975) found a higher concentration of progesterone than in the plasmas of other platypuses (Table 28).

EGG LAYING, INCUBATION, AND HATCHING.   No one has observed any of the activities listed in the subsection for the platypus but a few published facts allow us to make informed guesses about what happens. The first fact is the female digs a serpentine burrow between 35 and 50 feet long (Bennett, 1835) from the entrance, which is in the side of the bank usually above water level, to a chamber containing a nest of grass, leaves, willow roots etc. According to Burrell (1927) males never enter these breeding burrows. When the nest is complete the female retires to it for some days (Fleay, 1944) (fact number two). The third fact is Burrell managed to expose a nesting chamber without disturbing the inmate who was curled-up asleep with her tail bent round and laid along her abdomen; between tail and abdomen two young, each 6.5 cm long, were located. The guesses are that the female lies on her back or side in the nest, curls up her tail over the abdomen and when the eggs are extruded from the cloaca they come to lie in the "incubatorium: formed between tail and abdomen. When laid the eggs are covered by a sticky secretion of unknown origin, so that they adhere to one another (Burrell, 1927) and presumably to hairs of the abdomen. My guess is that the incubation period is at least 10 days long since: (a) the incubation period for the egg of *Tachyglossus* in the pouch (temperature 32°C) is ca 10.5 days (p. 240); (b) the stage of development at the laying of the egg in both genera is an embryo with 20 pairs of somites; (c) the temperature of the platypus' putative incubatorium is 31.8°C at TA 20°C (p. 128). If the eggs are incubated in this way it seems inconceivable that once the process had started the female could drop the eggs in the nest before hatching, leave the burrow, return and start incubation again, yet according to Fleay's (1944) protocol his female platypus

TABLE 28

Progesterone and Estradiol-17β Concentrations in Platypus Peripheral Plasma[a]

Steroid (ng/ml)	Pregnant	Nonpregnant		Ovariectomized	Male
		(1)	(2)		
Progesterone	10.4	2.1	—	5.2	6.1
Estradiol-17β	0.16	0.01	0.09	0.07	0.04

[a] Data from Carrick *et al.*, 1975.

appeared 6 days after retiring to the nesting burrow, went into the water for a short time, and went back into the burrow. Only direct observations can solve that mystery. Perhaps the TB of the platypus in the nesting chamber is higher than we think, which could lead to a shorter incubation period.

To escape from the egg at the end of incubation the young probably tears the shell by the combined action of caruncle and egg tooth (see Fig. 79 for *Tachyglossus* emerging from egg). The egg tooth forms during incubation and Hill and de Beer (1949) have described its development. They confirmed the conclusions of Seydel (1899) and Green (1930) that the monotreme egg tooth takes its origin, like the transient toothlets of reptiles, as a mesodermal papilla which pushes out the buccal epithelium to form the external covering of the tooth. Within that epidermal covering is differentiated an enamel organ similar to that of the reptilian toothlet, consisting of enamel epithelium, stellate reticulum, stratum intermedium, and outer enamel epithelium. The superficial cells of the mesodermal papilla are differentiated to form a layer of odontoblasts which form a cap of dentine investing the papilla down to its base; the dental papilla becomes highly vascular. The egg tooth during its formation is attached to bone which is at first independent of the premaxillae, but it soon becomes attached to the latter. The finished product looks like that of *Tachyglossus* shown in Fig. 81A.

THE BREEDING SEASON AND SEASONAL CHANGES IN THE REPRODUCTIVE ORGANS. The platypus breeds only in the springtime (Burrell, 1927). Temple-Smith (1973) has recorded the size of ovarian follicles, the weights of ovaries, and of the uteri in 42 platypuses taken in all months of the year from November 1969 to December 1971. Marked changes were observed in the weights of the left ovaries and in the size of the follicles (diameters of the two largest were measured). Growth of uteri and ovarian follicles commenced in late June and attained maximal values in August; thereafter regression set in and minimal values were achieved by December. The average weights of the ovaries were 0.2 g/kg body weight in summer and 0.6 g/kg in August. M. Griffiths and M. A. Elliott (unpublished) found that the largest follicles measured less than 2 mm in summer but from June to October they ranged in size from 1.6–5.7 mm and that the left uteri were small in cross section in summer but large in August. Temple-Smith also found that the follicles measured less than 2 mm in summer and 3.8–4.9 mm in August.

No discernible changes were detectable in the right (nonfunctional) ovaries but the right uteri showed changes corresponding in time with those of the left, but in lesser degrees. Griffiths *et al*. (1973) noted briefly that the glands of the left uteri were in an inactive state until mid-July but from July to October the glands were numerous and some of their cells were ciliated. They also mention that they confirmed Bennett's (1835) observation that the right uterus hypertrophies along with the left, both uteri showing a luxuriant growth of glands containing both ciliated and secretory cells; the only difference being that the right uterus is much smaller in cross section than the left (4.0 mm versus 6.5 for the left).

Published data indicate that the breeding season of the platypus seems earlier in Queensland and northern New South Wales than in southern Australia: Caldwell on the Burnett River in Queensland got his segmenting ova between August 7th and 14th and first full-term egg August 24th. Wilson and Hill (1908) and C. J. Hill (1933) obtained uterine eggs at various stages of development from platypuses taken in undisclosed rivers between August 23rd and September 9th. Burrell (1927) working on northern New South Wales rivers records the earliest time of the year he got eggs from a nest was August 24th and the latest date was October 22nd. Kershaw (1912) obtained nest eggs of a platypus living on a Victorian river also on October 22nd. Carrick *et al*. (1976) got two full term eggs on October 5th from a platypus living on the Murrumbidgee River near Yass where George Bennett (1835) got uterine eggs between October 5th and 8th, three nestlings each ''in length 1⅞ inches,'' at the end of November and, as already related, two young each 10 inches long on December 28th.

Male

SPERMATOGENESIS AND SEASONAL CHANGES IN THE TESTES. Temple-Smith (1973) has studied seasonal changes in the testes of platypuses taken in all months of the year from November 1969 to December 1971. Testis weights and sizes were minimal from December to April; they increased rapidly in July and by mid-August maximal weight and volume were attained. From September to November testis size and weight declined and the summer and autumn resting level was reached in late December (Fig. 66).

M. Griffiths and M. A. Elliott (hitherto unpublished data) have also studied seasonal changes in the testes of adult platypuses trapped incidentally during a study of lactation. The animals were collected in all months of the year, except February and April (May 1967–November 1971), from the waters of the Murrumbidgee River and its tributaries in the Australian Capital Territory. Although our sample is smaller than Temple-Smith's (35 as against 65) and it was collected over a longer period of time our findings are almost identical to his (Fig. 66). Apparently we missed the peak of development since only one male was caught in August and its testes weighed 9.0 g/kg bodyweight; this is much less than the mean weight 14.5 g/kg recorded by Temple-Smith for that month. However expression of testis weight as a function of body weight may not necessarily indicate maximal development of the testis; one animal we took on July 26, 1967 had testes weighing 22.0 g the set, a figure almost identical to that recorded by Temple-Smith for the highest absolute weight of a set of testes—22.3 g; his animal was taken at practically the same time of the year, July 29, but 3 years later.

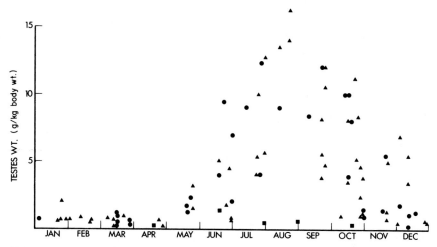

Figure 66. *Ornithorhynchus.* Seasonal changes in the weight of the testes (paired). ●, from Griffiths and Elliott (unpublished); ▲, from Temple-Smith (1973); ■, from Temple-Smith (1973), juvenile testes.

Concurrently with increase in weight, increase in size of the tubules of the testes takes place. Temple-Smith found that the mean diameter increased from 101 μm in March to 234 in July–September; Griffiths and Elliott found much the same thing (Table 29).

Along with those changes in weight and tubule diameter the histology of the testes changes. Griffiths and Elliott also studied seasonal variations in testicular

TABLE 29

Ornithorhynchus **Mean Diameter of Testis Tubules at Different Times of the Year**[a]

Month	Griffiths and Elliott		Temple-Smith	
	Number in sample	Tubule diameter (μm)	Number in sample	Tubule diameter (μm)
March	4	90	6	101
May	3	150	4	160
June	4	214	5	213
July–September	6	225	21	234
October	6	205	10	204
November–December	5	176	2	95
			(late December only)	

[a] Data from Temple-Smith (1973) and Griffiths and Elliott (unpublished).

histology in *Tachyglossus* and in *Zaglossus* (see pp. 248 and 249). The stages of growth and spermiogenesis in all three genera are so much alike that a preparation from the testis of any one of these can be used to illustrate that stage of development in the other two; this procedure has been adopted here. In cross section it can be seen that the tubules in the resting stages of summer and autumn have no lumina but consist of a peripheral ring of spermatogonia with large round nuclei; interspersed between the spermatogonia are the cell bodies of Sertoli cells whose elongated nuclei are radially directed towards the center of the tubule, giving a "cartwheel" appearance to the structure—the tubules of the juvenile testis also exbibit that appearance. In this condition the platypus testis is very like that of the resting tubule of the *Tachyglossus* testis (Fig. 67). Spermatogenesis begins in winter, the spermatogonia dividing to give rise to primary spermatocytes (Fig. 68); these in turn give rise by meiotic divisions to a generation of haploid secondary spermatocytes (Benda, 1906) accompanied by enlargement of the diameter of the tubule. The seminiferous tubules continue to enlarge and now consist of a peripheral layer of spermatogonia, primary and secondary spermatocytes, and undifferentiated spermatids at the center; these are loosely arranged indicating the beginning of the formation of the lumen (Fig. 69). The rounded spermatids now undergo the process of spermiogenesis, described in detail by Benda (1906): during attachment to and penetration into the cytoplasm of the Sertoli cell the spermatid metamorphoses into an elongated cell with a filiform nucleus; excess cytoplasm is sloughed off into the lumen of the tubule, but portion is retained as the cytoplasmic droplet. After maturation within the Sertoli cell is complete the spermatozoa are released into the lumen. They are like the spermatozoa of no other mammal except *Tachyglossus* and *Zaglossus*, but they are like those of reptiles and birds. The condition of the ripe tubule at the height of spermatogenesis and spermiogenesis is shown in Fig. 70. All testes collected in August and September exhibited this condition, but in October many of them were showing signs of regression. The first signs are a decrease in the number of spermatocytes and a general thinning out of the seminal epithelium giving an appearance of loosely packed cells (Fig. 71). This is followed by almost complete disintegration of the cells of the epithelium including some spermatogonia (Fig. 72). Then follows a marked decrease in the diameter of the tubules and in the lumina of some of these, masses of spermatozoa are trapped. These in turn are removed together with other cellular debris by phagocytes (Fig. 73). It will be noted that some of the tubules shown in this figure have returned almost to the solid resting condition. This condition is attained by very active mitotic division of those spermatogonia and Sertoli cells that have survived the disintegration and phagocytosis (Fig. 74).

The transition to the resting condition occurs in December and January but Griffiths and Elliott found spermateliosis still taking place in the tubules of a platypus taken in December (see p. 232, spermatozoa in epididymis).

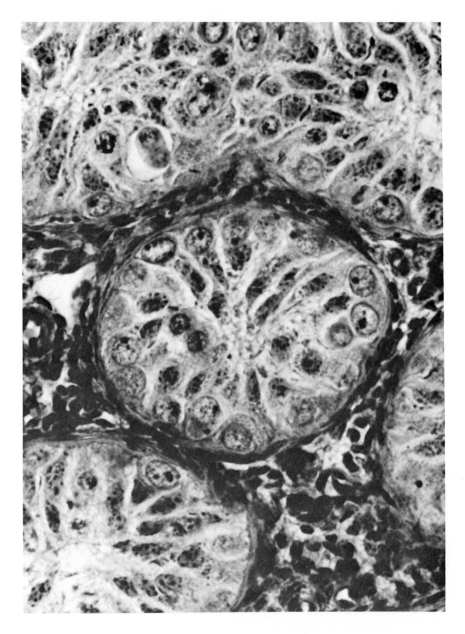

Figure 67. *Tachyglossus.* Tubules of testis in section; resting phase. The cells, with the large nuclei, at the periphery of the tubule are spermatogonia and the cells with elongated nuclei are Sertoli cells. × 700.

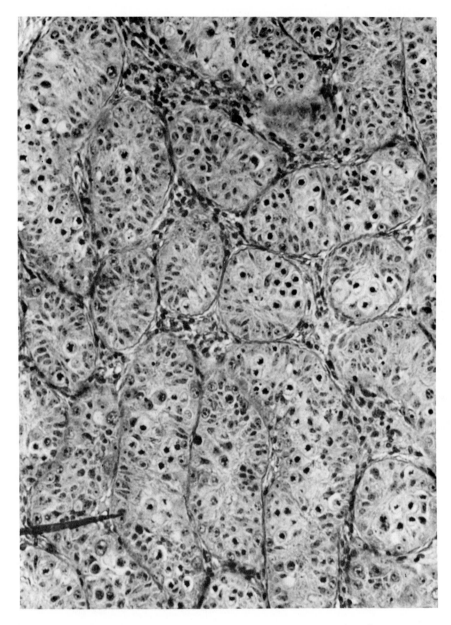

Figure 68. *Zaglossus.* Testis tubules in section; early spermatogenesis showing primary spermatocytes (dark staining nuclei). Heidenhain's iron hematoxylin. × 285.

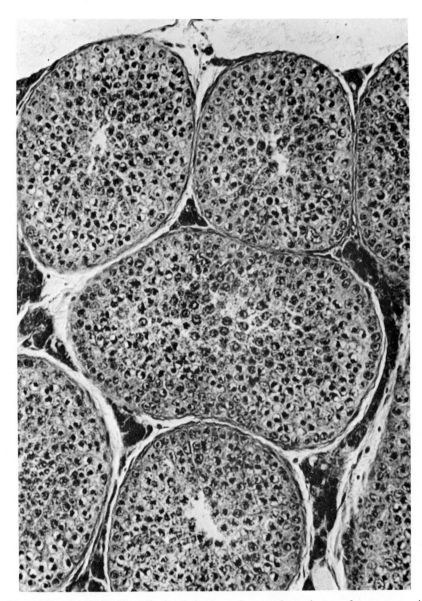

Figure 69. *Ornithorhynchus*. Testis tubules in section; at advanced stage of spermatogenesis. Spermatogonia at the periphery of tubule, primary and secondary spermatocytes towards the center; lumen just appearing with spermatids bordering on lumen. Heidenhain's iron hematoxylin. × 240.

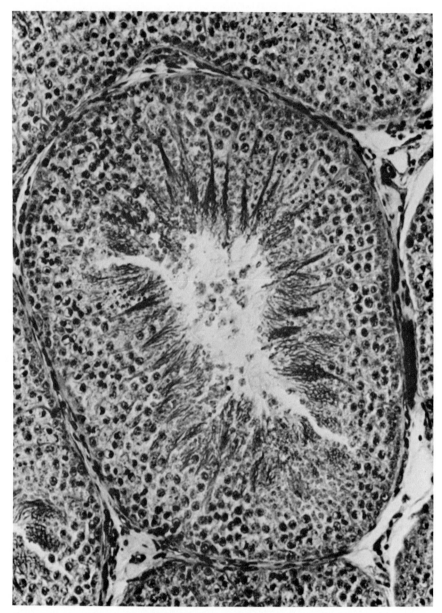

Figure 70. *Zaglossus.* Section of ripe tubule. Note spermatogonia. primary and secondary spermatocytes, spermatids, early and ripe spermatozoa with their coiled heads embedded in the elongated Sertoli cells. Heidenhain's iron hematoxylin. × 285.

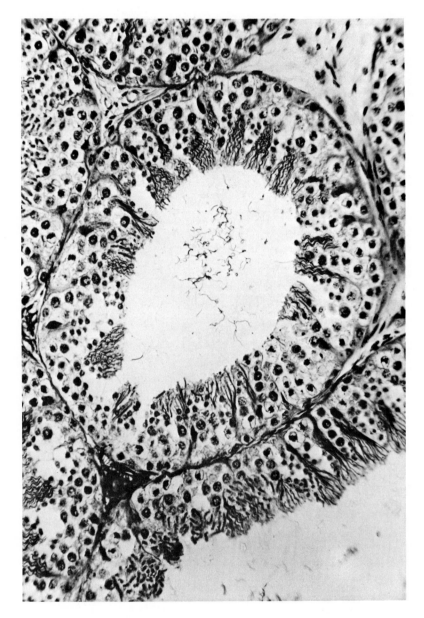

Figure 71.   *Tachyglossus*. Section of ripe tubule in testis showing signs of regression—decrease in the number of spermatocytes and thinning out of the seminal epithelium. Heidenhain's iron hematoxylin. × 282.

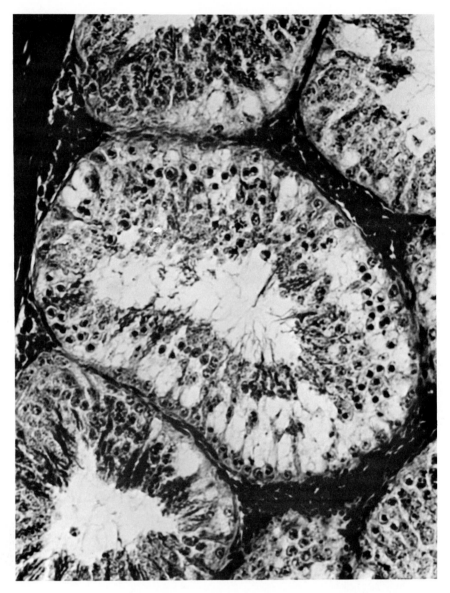

Figure 72. *Ornithorhynchus.* Sections of degenerating tubules of the testis showing general breakdown of the seminal epithelium including some spermatogonia. Heidenhain's iron hematoxylin. × 330.

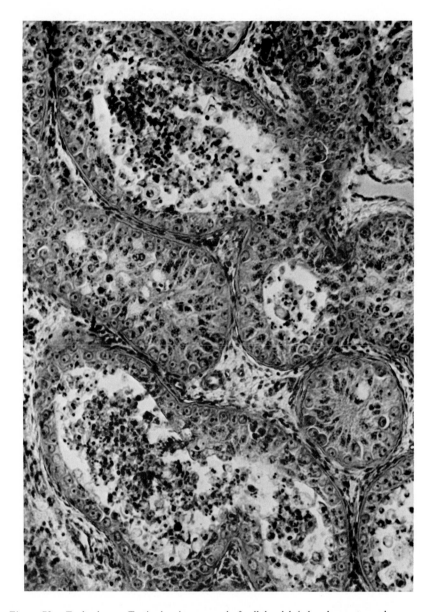

Figure 73. *Tachyglossus*. Testis showing removal of cellular debris by phagocytes and regeneration of the seminal epithelium. Some tubules have almost resumed the resting state. Heidenhain's iron hematoxylin. × 250.

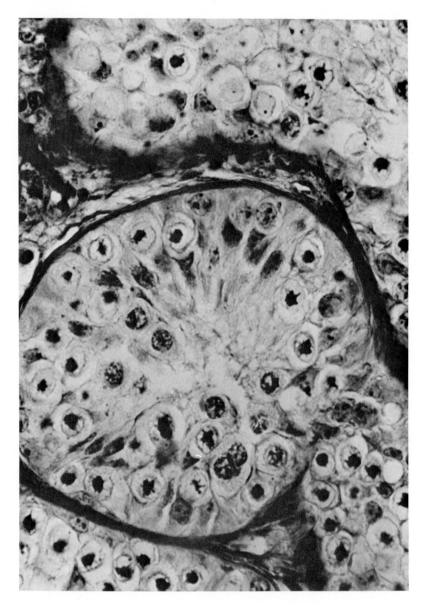

Figure 74. *Tachyglossus*. Testis showing transition to the resting state by mitotic division of surviving spermatogonia and Sertoli cells. × 600.

The testes of immature males consist of solid tubules containing a peripheral layer of spermatogonia and radially arranged Sertoli cells. Except for their small size they look very like the resting stage of the adult testes. Temple-Smith found the testes of the immatures showed no sign of spermatogenesis until August–September of their second year.

The interstitium shows seasonal changes closely associated with the seasonal cycle of the seminiferous tubules (Temple-Smith, 1973). At maximal tubule diameter the Leydig cells are restricted to the narrow spaces between contiguous tubules, and they exhibit large rounded nuclei along with hyperplasia of the cytoplasm. In regressing testes, however, the interstitium spreads gradually and finally encloses the tubules. In this condition the Leydig cells exhibit marked decline in the size of their nuclei and of the volume of their cytoplasm. At the same time small lipid droplets appear in the cytoplasm which coalesce into the larger droplets characteristic of the resting stage. These droplets persist throughout the nonbreeding season but are depleted during the early stages of testicular recrudescence coincident with increase in size of the Leydig cells. Temple-Smith pointed out that this has some resemblance to the events found in the Leydig cells of seasonally breeding eutherians, but there is, unlike in other mammals, some disintegration of Leydig cells straight after the breeding season accompanied by phagocytic activity and loss of lipid. This, however, is not on the massive scale exhibited by the testes of birds and reptiles at the end of their breeding seasons.

Testosterone (presumably secreted by the Leydig cells) has been identified in testicular venous plasma of a platypus taken during the breeding season. The secretion rate of testosterone into spermatic vein plasma has been found to be between 4 and 6 ng/min/g testis, a figure far higher than that found in some genera of marsupials and almost equal to that for *Megaleia rufa* (red kangaroo), 6.8 ng (Carrick and Cox, 1973).

Temple-Smith found that the epididymides undergo a seasonal waxing and waning synchronized with the testicular cycle. During testicular regression epididymides were minimal in weight and the epithelium lining the collapsed tubules consisted of cuboidal cells; spermatozoa were absent. About 1 month after the beginning of testicular recrudescence, the growth and development of the epididymides commenced. Weight and tubule diameter increased as did the height of the lining epithelium. Enlarged glandular epididymides were found in all adult males taken from August–October, the tubules of which were distended with large quantities of spermatozoa and seminal fluid.* Although testicular recrudescence preceded that of the epididymides, the maximum size of both

---

*It is interesting to note that in spermatozoa from the caput of the eutherian epididymis when treated with 1% lauryl sulphate at pH 9.0 the head swells up and the spermatozoa disintegrate. The spermatozoa from the tail, however, are resistant to this chemical. Spermatozoa from both caput and cauda in reptiles, birds, marsupials, and the platypus, disintegrate when treated with 1% lauryl sulfate solution (Allan Braden, CSIRO, personal communication).

organs was reached at the same time. Regression closely followed the regression of the testes. Those platypuses exhibiting early stages of regression in November and December still had large glandular epididymides with distended tubules lined by tall columnar epithelial cells; spermatozoa were still present in the seminiferous tubules of these platypuses and in each region of the epididymides. This suggests that such platypuses could breed late in the season.

ACCESSORY GLANDS. The paired Cowper's glands also exhibit seasonal waxing and waning but the other accessory glands, the disseminate urethral glands, do not. The latter are most interesting since anatomically they bear a remarkable resemblance to the disseminate prostatic tissue of marsupials. In fact the male reproductive system of the monotremes is very like that of didelphid marsupials [see Biggers (1966) and Setchell (1977) for reviews of anatomy and reproduction in male marsupials]. In both, the vas deferens leads to the anterior end of an elongated urogenital sinus which is surrounded by a pear-shaped mass, the disseminate bulbo-urethral (or disseminate prostatic) glands, and at the lower end of the urogenital sinus paired Cowper's glands open into it; in both the penis is bifid. In the platypus the tissue of the disseminate urethral glands consists of a series of tubules arranged radially around the urogenital sinus, lined by a single layer of tall columnar secretory cells with eosinophilic cytoplasm. The histology is uniform throughout the length of the gland and shows none of the zonation found in the disseminate urethal glands of marsupials. Temple-Smith found no evidence of regression of the secretory epithelium in nonbreeding males nor of glandular hypertrophy in breeding males.

TACHYGLOSSUS

Female

Oogenesis, cleavage, and early embryology have already been described in the previous section as "monotreme" processes. What follows largely concerns *Tachyglossus*; where the information may be applicable to *Ornithorhynchus* will be indicated.

GESTATION AND CORPUS LUTEUM. Recent observations on the gestation of the egg in the uterus have extended our information a little; the little confirms Broom's (1895) observation that it is of the order of weeks. The difficulty of getting exact data is due to the fact that echidnas will not breed to order in captivity, not thus far anyway.* There are accounts of *Tachyglossus* having bred

*T. J. Bergin of Taronga Zoo, Sydney, has recently embarked on a program of work designed to elucidate the conditions necessary to induce *Tachyglossus* and *Zaglossus* to breed in captivity under controlled conditions.

in captivity (Heck, 1908; Lang, 1967) in the zoos of Berlin and of Basel but these were not planned attempts to study reproduction; indeed the first intimation that they bred was the discovery by the keepers that their exhibits were carrying large pouch young. The facts so far collected are as follows:

1. Broom (1895) received a pair of echidnas taken in copulation on September 4. The pair were kept together in a box but the male refused to eat, took no interest in the female, and died on the 14th day of captivity. The female, however, developed a pouch, empty at the 26th day after copulation, but which contained an egg on the 28th day. From this it is surmised that the gestation period was about 27 days.

2. Griffiths *et al.* (1969) received a female echidna and two males taken together in August at Hoskinstown near Canberra. The female was kept apart from the males and her pouch area was examined daily to see if she had laid an egg. The pouch was empty on the 16th day at 0900 hours but at that time on the 17th it contained an egg. This was successfully incubated in the pouch and the egg hatched (see below).

3. After a similar series of observations on another female taken near Canberra in September and kept alone (Griffiths *et al.*, 1973), an egg was found in the pouch on the 9th day of observation, the pouch being empty on the 8th day. This egg also hatched (see below).

4. A female echidna taken in the company of four large males in arid country near Alice Springs on July 29, 1969 was segregated from the males and observed daily. After 34 days had passed the pouch was not very well developed so it was decided she was not pregnant. At autopsy, however, it was found that she had an egg 4.5 mm in diameter in one of her uteri and that the ovary on the same side had a corpus luteum (Fig. 75). The glands of the uteri, which were enlarged and hyperaemic, exhibited the luteal phase (Fig. 77); the egg itself contained a zygote at the 16-nuclei stage, and was furnished with a shell consisting of basal and rodlet layers. The degree of development of the shell, the number of nuclei in the zygote, and the condition of the uterine glands are all as Flynn and Hill (1939) C. J. Hill (1933), and J. P. Hill (1933) have described as normal for an egg of this size.

It would seem from the above evidence that the first three periods of time listed are geniune gestation times ranging from at least 9 days to a possible 27, but the period of 34 days to produce 16 blastomeres seems far too long to be in that category; marsupials for example, have putative gestation times ranging from 12 days 8 hours for *Isoodon macrourus* (Lyne, 1974) to 38 days for *Potorous tridactylus* (Hughes, 1962); even rabbits have a gestation period of only 28–30 days. Another interpretation of the matter is that echidna oviducts can store live sperm for a long time as can those of some reptiles and bats (see Racey and Potts, 1970, for recent account). The marsupials *Antechinus stuartii* and *Dasyurus viverrinus* are also known to store sperm (see Tyndale-Biscoe, 1973). In these animals the spermatozoa are retained alive in the fallopian tubes for days and are thus available to fertilize an egg when it is eventually shed into the oviduct. Consequently gestation periods as measured from time of observation of copulation have been found to be variable; we may be observing the same phenomenon in echidnas. Storage of live sperm in the oviduct may be a useful attribute in a cryptic and usually nongregarious species such as the echidna, the females of which might be caught, so to speak, before they are ready to ovulate.* Indirect evidence for sperm storage came from histological examination of the corpus luteum in the echidna with an apparently very long gestation period; the corpus luteum had attained a degree of differentiation corresponding, in the platypus, to intermediate between Stage I and II of Hill and Gatenby (1926). That is to say it was a very young corpus luteum, the luteal cells of which exhibited a finely granular cytoplasm occasionally containing a clear vacuole (Fig. 76); furthermore the central cavity of the structure was filled with a coagulum (Fig. 75), a sign of a very young corpus luteum and evidence that ovulation had taken place recently and that the ovum had been recently fertilized presumably by a stored spermatozoon. It is interesting to note in connection with all the above that Flynn and Hill (1939) found sperms in the lumina of the uterus and the uterine glands of a platypus with oocytes not full grown.

As far as structure of the corpus luteum in *Tachyglossus* is concerned Hill and Gatenby (1926) described one (stage of development of the egg unknown) very like that shown in Fig. 75. They found it contained a large central cavity filled by a dense homogeneous coagulate that they likened to the colloid in a thyroid gland. The luteal cells were large and well formed but some vacuolation was apparent. The interstitial tissue was in direct continuity with the theca externa and some of it formed capsular investments around individual luteal cells. Along the strands of connective tissue small theca interna cells were located. Another corpus luteum described by Hill and Gatenby and designated stage 6 in their series of "monotreme" corpora lutea, came from an echidna that had laid an egg 1–2 days previously. It was "in a fairly advanced stage of regression: the luteal cells exhibiting colloidal and granular degeneration of the cytoplasm and vacuolation of the nuclei. However, around the periphery of the gland many luteal cells were encountered devoid of vacuoles and with large nuclei apparently normal. The theca interna cells had a shrunken and ill-defined cytoplasm and apparently were quite inactive. There was an abundance of polymorpho-nuclear leucocytes throughout the gland.

In two echidnas sampled 11 and 19 days after laying eggs the corpora lutea were found to be degenerate (Griffiths *et al.*, 1969). In the corpus luteum sampled 11 days after, the luteal cells were for the most part intact but in some the nuclei were vacuolated and their cytoplasms exhibited both vacuoles and colloi-

---

*Nothing is known of the occurrence and duration of estrous cycles in monotremes.

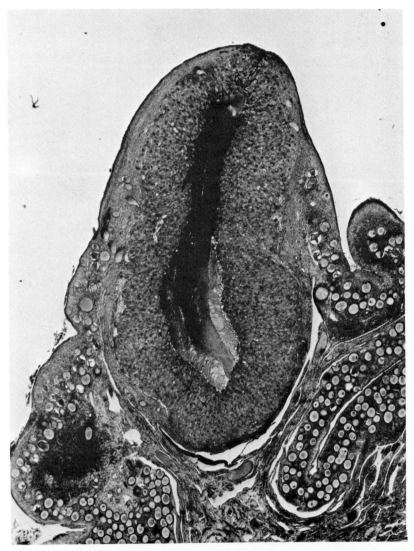

Figure 75. *Tachyglossus*. Sagittal section of a young corpus luteum showing central cavity filled with a coagulum; distal opening to exterior now plugged with luteal cells. Note immature oocytes in cortex of the ovary. Heidenhain's iron hematoxylin. × 30.

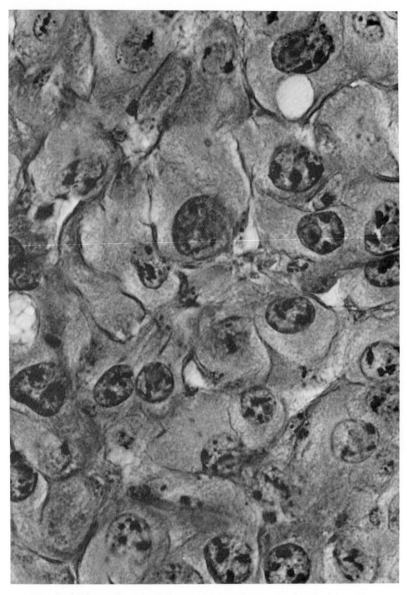

Figure 76. *Tachyglossus*. Luteal cells in young corpus luteum showing finely granular cytoplasm and incipient vacuolization. Heidenhain's iron hematoxylin. × 630.

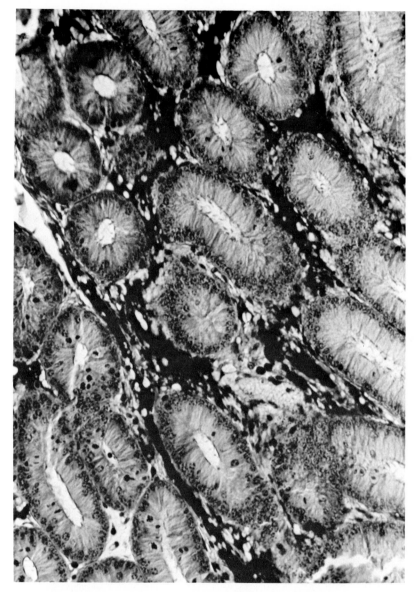

Figure 77. *Tachyglossus*. Uterine glands at luteal phase; this uterus contained an egg exhibiting an embryo at the 16-blastomere stage. × 260.

dal degeneration. That sampled 19 days after was in a ruinous condition: the luteal cells were hardly recognizable as such, some were highly vacuolated and others cytolysed giving rise to masses of granular unorganized cytoplasm and colloid.

INCUBATION AND EMBRYOGENESIS.   The events leading to the laying of eggs and the incubation period are as follows: a female taken August 18, 1967 had a pouch that exhibited thick tumescent lips (Fig. 78). She laid an egg on September 4, 1967; for some time afterwards when she was picked up it was noted that the cloaca practically reached into the posterior end of the pouch (see also CSIRO, 1969). If the animal was allowed to curl up it entered the pouch which suggests that the egg could have been deposited therein in this way. The egg diameters were 16.5 × 13.0 mm; the temperature of the pouch was 32.5°C, ambient being ca. 20°C. During the 10th day of incubation the egg hatched and a young one 1.47 cm long and 378 mg in weight was found in the pouch clinging to

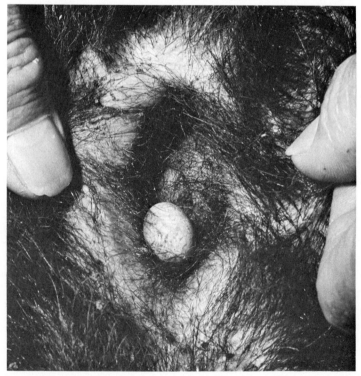

Figure 78.   An egg in the pouch at 8th day of incubation; note presence of tumescent lips of pouch, which before egg-laying, is a sign of pregnancy. *(From Griffiths, 1972.)*

Figure 79. *Tachyglossus.* Young emerging from egg after ca. 10 days, 6 hours incubation. *(From Griffiths, 1972.)*

hairs over one of the two mammary areolae (Griffiths, 1968; Griffiths *et al.*, 1969). The newly hatched had remnants of the fetal membranes adherent to the abdomen (Fig. 80) and an egg tooth at the rostral end of the upper jaw. Four hours later the rests of the fetal membranes had disappeared leaving a smooth abdomen but the egg tooth was still present (Fig. 81A). It was concluded the incubation period was 10–10.5 days.

The other incubation observed enabled us to define period a little more accurately: on September 16, 1969 the pouch of this echidna was empty but at 1215 hours on September 17 the animal was found to be lying on her side with the back arched and the lips of the pouch in close apposition. When the pouch was examined an egg was found there and from the animal's unusual position it was concluded it had just been laid. Thereafter the pouch was inspected daily. On the 8th day the egg, which was a creamy yellow color and had a rubbery feel to it, was measured and found to have diameters 15 × 15 mm. The temperature of the pouch was 32°C. At 0800 hours on the 10th day the egg was as described above but at 1600 it exhibited a dimple on one side and on removal from the pouch, movements could be detected under the indentation. The egg was replaced in the pouch and 1 hour later it was found to be stretched into a pear shape. By 1810 hours the hatchling had burst open or cut open the shell and it could be seen

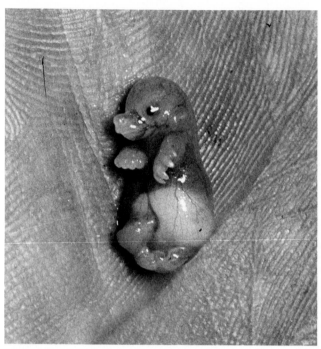

Figure 80. *Tachyglossus*. Young shortly after hatching showing relatively enormous development of forelimbs and its resemblance to the marsupial neonatus (Fig. 82). Remnants of fetal membranes are still adherent to abdomen. Milk has been ingested and is visible through the transparent wall of the abdomen. Note caruncle. *(From Griffiths et al., 1969; with permission of The Zoological Society of London.)*

struggling, with a continuous writhing motion of the head and forelimbs, to emerge from the shell (Fig. 79, and CSIRO, 1974). The head and visible part of the body were rosy pink in color and the hands, snout, and caruncle were a translucent white. Twenty minutes later the hatchling was almost clear of the shell and when handled it slipped out into the pouch bearing at the abdomen a transparent vesicle almost as big as itself; this was the allantois filled with fluid. At 1835 hours the allantois was still intact but by 2250 hours it had ruptured and the rests were adherent to the ventral surface. If it is taken that incubation ended at the time of rupture of the shell the incubation period was the interval between 1215 hours September 17 and 1810 hours September 27, i.e., 10 days 6 hours which is close to the figure of 10–10½ days given for the previous incubation. It is also close to an approximate figure of 12 days for an incubation observed by Ann Miller of "Bilbah," Tara, Queensland who kindly sent Griffiths *et al.* (1973) her protocol of observations for comparison with their data. The time taken to hatch another echidna egg also shows that the incubation period is of the order of 10 days or more: an echidna caught by W. Braithwaite of CSIRO in a burrow in a mallee fowl's incubation mound at 0900 hours August 23, 1973, was found to have a pouch and an egg therein. She was brought to the laboratory and induced to eat regularly. At the 9th day of observation the egg was intact but at 0815 hours September 2 a young one was found to be struggling to clear itself of the egg shell, i.e., 9 days 23.25 hours after the egg was found in the pouch. This young survived and was raised by the mother; the rate of growth, the first ever to be determined in a monotreme from hatching onwards, is discussed in Chapter 9.

The greater part of embryo-genesis takes place during the 10–11 days of incubation since in an egg that had just been laid the embryo had only 19 pairs of somites (Hill and Gatenby, 1926). It will be recalled in this connection that the near-term platypus egg studied by Hughes *et al.* (1975) also contained an embryo at practically the same stage—20 pairs of somites. This reminds one of the situation in the wallaby, *Macropus eugenii*, which takes ca. 17 days to achieve a 15–20 somite stage but thereafter embryogenesis is rapid and the full term fetus stage is achieved during the next 11 days of gestation (Renfree, 1972).

The first studies of embryogenesis in *Tachyglossus* were carried out by Richard Semon (1894b,c) and his colleagues. Semon came here in 1891 under the auspices of the University of Jena and, doubtless profiting by Caldwell's experience, collected at the Burnett River with the help of aborigines, a magnificent series of uterine and pouch eggs of echidnas. The results were published in the form of beautifully illustrated monographs known collectively as Zoologische Forschungsreisen in Australien und dem Malayischen Archipel. Semon himself was concerned mainly with external features of the embryos and embryonic membranes; the stages of development were numbered 40–53, numbers 46–53 being embryos from pouch eggs. The earliest, stage 40, is an embryo 7 mm long with blunt lobate forelimbs, buds for hindlimbs, very prominent pharyngeal pouches, ca. 39 pairs of somites, and a well-developed tail. However, Luckett (1976) who had access to 38 laid eggs from both *Ornithorhynchus* and *Tachyglossus* examined earlier stages and found in agreement with Hill and Gatenby (1926) and Hughes *et al.* (1975) that the embryo at hatching has 19–20 pairs of somites, a rudimentary proamniotic head fold, no allantois, but has a yolk sac with trilaminar and bilaminar omphalopleure. This has been achieved as follows: The extraembryonic mesoderm previously mentioned splits to form two layers, splanchnopleure and somatopleure. The endoderm of the yolk sac is covered by splanchnopleure and where ectoderm, endoderm, and mesoderm come into contact trilaminar omphalopleure is formed. During incubation the yolk sac becomes completely vascularized and comes to occupy half the inner surface of the chorion.

By the 27-somite stage a vascular allantois is present which during its growth pushes splanchnic mesoderm outwards until contact is made with the serosa; it fuses with the latter at the 40-somite stage (Semon's stage 40), thus forming with the yolk sac omphalopleure a practically complete trilaminar investment around the vesicle within the shell. Glomeruli are present in the mesonephros at the 40-somite stage (Luckett, 1976).

At Semon's stage 44 the forelimbs are pentadactyl, the tail has shortened, and the pharyngeal pouches have been incorporated into the neck. Just before this, however, at stage 43, the anlagen of the egg tooth appears (Seydel, 1899); its further development is similar to that of *Ornithorhynchus*. By stage 45 it has a

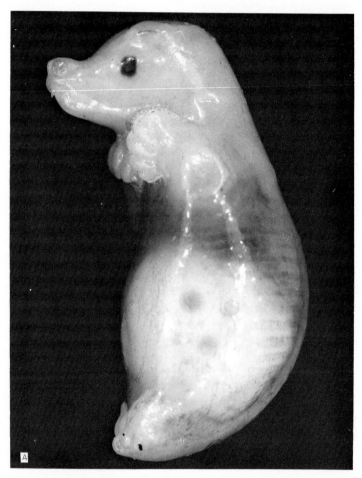

Figure 81.   A, *Tachyglossus*. Same specimen as in Fig. 80, 4 hours older. The rests of the fetal membranes have disappeared but the egg tooth is still present. B, *Tachyglossus*. Pouch young 3 days old; hindlimbs are now growing at rapid rate relative to the forelimbs.

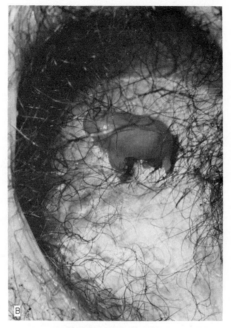

Fig. 81 *(continued)*

sharp point and is invested with solid enamel while the highly vascular dental pulp is enclosed by bone attached to the symphysis of the premaxillae. This is the condition just prior to hatching; Fig. 81A shows the egg tooth a few hours after hatching.

BREEDING SEASON. Recent observations on the time of the year eggs were laid and of the size of pouch young taken at different times of the year in the bush, confirms previous indications (see Griffiths, 1968, for review) that the breeding season of *Tachyglossus* all over Australia, including Kangaroo Island off the coast of South Australia, occurs in the months of July and August. The results are summarized in Table 30.

It would appear, however, that the breeding season is longer than hitherto thought since an egg was laid as late as October 22. Flynn and Hill (1939) state "The latest date in the year of which we have any record of the discovery of an intra-uterine egg is 4 September 1929." Semon (1894a) found "Ende August hatten fast alle ausgewachsenen Weibchen Eier im Uterus oder Beutel oder Junge im Beutel."

Figure 82. *Megaleia rufa*. Newborn red kangaroo just attached to a teat. Note strongly developed forelimbs, weakly developed hindlimbs, and the caruncle. *(Photo by Ederic Slater, CSIRO.)*

TABLE 30

Time of Year Eggs Were Laid, Found in Uteri or Pouches, and Time of Year Young Were Found in Pouches of Echidnas Caught in the Bush in Various Parts of Australia 1963–1973

	Eggs	Pouch young		
	Date laid found in pouch or in uterus[a]	Date found in pouch	Body weight (g)	Body length (cm)
Gladstone, Tasmania	Aug. 15, 1969 (L, two)			
Rankins Springs, New South Wales	Aug. 23, 1973 (P)			
Tara, Queensland	Sept. 3, 1969 (L)			
Hoskinstown, New South Wales	Sept. 4, 1969 (L)			
Alice Springs, Northern Territory	Sept. 5, 1969 (U)			
Canberra, Australian Capital Territory	Sept. 17, 1969 (L)			
Orange, New South Wales	Oct. 22, 1969 (L)			
Casino, New South Wales		Sept. 7, 1969	60, 69 (Twins)	11.0, 12.0
Kangaroo Island, South Australia		Sept. 15, 1970	153	14.5
Kangaroo Island, South Australia		Sept. 19, 1965	100	11.5
Kangaroo Island, South Australia		Sept. 19, 1969	36.5	9.2
Kangaroo Island, South Australia		Sept. 19, 1969	134	13.0
Kangaroo Island, South Australia		Sept. 19, 1969	151	13.5
Kangaroo Island, South Australia		Sept. 19, 1969	175	14.5
Kangaroo Island, South Australia		Sept. 21, 1965	100	11.5

(continued)

TABLE 30 (continued)

	Eggs	Pouch young		
	Date laid found in pouch or in uterus[a]	Date found in pouch	Body weight (g)	Body length (cm)
Kangaroo Island, South Australia		Sept. 21, 1965	—	11.0
Kangaroo Island, South Australia		Sept. 22, 1970	72	10.5
Inverell, New South Wales		Sept. 25, 1969	87	10.5
Kangaroo Island, South Australia		Oct. 2, 1969	120	13.5
Kangaroo Island, South Australia		Oct. 2, 1969	175	14.5
Kangaroo Island, South Australia		Oct. 3, 1969	210	15.5
Rankins Springs, New South Wales		Oct. 12, 1971	90	12.0
Kangaroo Island, South Australia		Nov. 1, 1967	266	18.0
Rankins Springs, New South Wales		Nov. 18, 1963	241	16.0

[a] L, date egg was laid; P, found in pouch; U, found in uterus.

Male

SEASONAL CHANGES IN THE TESTES. Semon (1894a) was aware that the testes regress at the end of the breeding season: "Die Hoden Schrumpfen auf Bohnengrösse zusammen." I have recorded seasonal changes in weight, tubule diameter, and the histology of testes of adult echidnas taken in southeast Australia in all months of the year except January, February, and June from April 1966 to August 1973. The lack of data for January, February, and June was not due to choice—I simply have not been able to find any echidnas in those months. Testes weights as a function of body weight were minimal from October to March (Fig. 83) but larger testes were encountered in early April, in fact one was very large and exhibited spermiogenesis. Testes taken in May also exhibited spermiogenesis and an average tubule diameter of 340 μm, the second largest encountered in the series. In July, August, and September spermiogenesis was encountered in all testes but one, and all tubule diameters were large (160–360 μm) compared with those encountered October–March (70–150 μm). From these data it would appear that echidnas are capable of breeding earlier in the year than platypuses—it will be recalled that Flynn and Hill (1939) took an echidna with an egg in its uterus on June 29. It is also apparent that regression sets in sharply after mid-September and that average tubule diameter is a little higher in tachyglossid testes (see p. 249 for those of *Zaglossus*) than in those of *Ornithorhynchus*, both in the regressed and in the fully developed conditions.

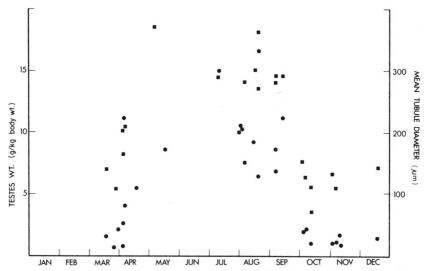

Figure 83. *Tachyglossus.* Seasonal changes in weight of the testes (2) and mean tubule diameters. ●, Weight as function of body weight; ■, mean tubule diameter. *(From M. Griffiths and M. A. Elliot, unpublished.)*

The extremes of testis weight as a function of body weight, however, are remarkably similar in all three genera despite the disparity in body weights of the adults:
1.6–2.4 kg for *Ornithorhynchus*, 3–5 kg for *Tachyglossus*, and 6–8 kg for
Zaglossus.

The histological changes encountered with change of season are exactly the
same as those found in *Ornithorhynchus* testes and it would only be repetitive if
they were described. As already mentioned some of the illustrations of the stages
of monotreme testis development and regression (p. 222) are of those of *Tachyglossus* testes.

The testes of only one juvenile echidna were examined during a breeding
season. The animal weighed 2.3 kg and was 36 cm long, and was sampled
September 21, 1971. Histologically the tubules (average diameter 100 μm)
looked exactly like those of juvenile *Ornithorhynchus* consisting of radially
arranged Sertoli cells and a peripheral ring of spermatogonia. From comparison
with a known-age juvenile it is considered to be 2 years old. This suggests that
male *Tachyglossus* takes longer to achieve sexual maturity than *Ornithorhynchus*
does.

ZAGLOSSUS

Male

BREEDING SEASON.   Whether or not *Zaglossus bruijnii* has a breeding season
is still a matter of conjecture due simply to lack of data but the picture is
becoming clearer. During the period July 5–14, 1972 I sampled testes from
adults caught near the villages of Mondo and Umboli on Mount Tafa (2500 m) in
the mountains of Central Papua; in addition Carolyn Murtagh of Macquarie
University, New South Wales kindly gave me a fixed testis of an adult caught in
the same general area October 27, 1973. The testes weights and the mean tubule
diameters are given in Table 31.

From this it is apparent there is a great range in weight and mean tubule
diameter in testis taken at the same time of the year just as there is in those of
Tachyglossus in April in Southeast Australia.

Of the five males taken in July 1972 four had testes filled with mature tubules
in which spermatogonia, primary and secondary spermatocytes, spermatids,
spermatids in Sertoli cells, and mature spermatozoa in Sertoli cells were present
(Fig. 70). The average tubule diameter ranged from 240–350 μm. In some of the
larger tubules there was incipient regression equivalent to that shown in Fig. 71.
In the fifth male taken at that time, an adult of body weight 7.6 kg and apparently
in good health, the testes were minute and the average tubule diameter was 84
μm. Histologically the solid tubules consisted of a peripheral ring of sper-

TABLE 31

Weight and Mean Tubule Diameters of Testes of Adult *Zaglossus* from Central Papua

Date sampled	Body wt. (kg)	Testis wt. (2) (g)	Testis wt./ body wt. (g/kg)	Mean tubule diameter (μm)
July 5, 1972	8.0	100	15.0	350
July 7, 1972	6.0	86	14.3	300
July 7, 1972	6.6	66	10.0	240
July 10, 1972	7.6	4	0.5	84
July 14, 1972	7.0	88	12.6	300
October 27, 1973	7.0	65 (32.5 × 2, one only weighed)	9.3	240

matogonia some of which had divided mitotically to produce primary spermatocytes which in turn were in process of dividing, i.e., the testis was at an early stage of spermatogenesis (Fig. 68). The various tubules of a sixth male taken October 27, 1973 exhibited a mixture of full spermatogenesis and partial regression similar to that illustrated in Fig. 72, and dense accumulation of trapped spermatozoa characteristic of the advanced regression of testes in *Ornithorhynchus* and *Tachyglossus*.

Female

BREEDING SEASON. Of two females taken in July 1972 along with the males mentioned above, only one was an adult. The other 4.5 kg in weight had ovaries which, when examined after serial sectioning, exhibited small oocytes (maximum diameter to 0.23 mm), and no corpora albicantia, i.e., it had probably never ovulated. The adult, 7.4 kg in weight, exhibited ovaries with large oocytes (maximum diameter 2.1 mm) but not as large as the fully grown oocytes of *Tachyglossus*, 3.96 mm (Flynn and Hill, 1939). The left uterus of this animal contained an egg 18 mm long; it proved to have absolutely no contents; apparently it had become trapped due to premature regression of the uterus. It would seem this had happened some time ago since no corpus luteum was detectable but a large (2.5 mm diameter) spherical body with a thick wall of connective tissue and a central body of clear colloid, which may be the remains of a corpus albicans or of a very large atretic follicle (see Hill and Gatenby, 1926), was found projecting from the surface of the left ovary. A number of recognizable atretic follicles were present also. The glands of the endometrium of the left uterus were sparsely packed and were at the proliferative stage. Those of the right uterus however were numerous and consisted of ciliated and nonciliated cells. The mammary glands were comparable to those of *Tachyglossus* at the start of the breeding season (p. 268).

These data and those from the males suggest that there is a breeding season: one male and the adult female caught in July were at a reproductive stage comparable to that of *Tachyglossus* at recrudescence after a resting condition; the other males taken in July were at the peak of breeding condition; the testes of another male taken 3½ months later in the season exhibited stages of regression. On the other hand the Fuyughé villagers living in the area, for whose opinions on wildlife I have respect, say the animal breeds all the time. They know what a breeding season is because they volunteered the information without questioning, that some birds breed only for a short time once a year, but Saangi (*Zaglossus*) breeds all the time. However, a prolonged period of lactation with the young associated with the mother for months, as quite likely happens since it is so with *Tachyglossus* and *Ornithorhynchus,* could give the impression of continuous breeding.* It would seem the best way to determine the matter is to make another collection of *Zaglossus* in January or February.

OVERVIEW OF MONOTREME AND MARSUPIAL DEVELOPMENTAL PROCESSES

The eggs of marsupials, although not as large as those of the monotremes, contain a considerable amount of yolk. After ovulation the egg, invested by a zona pellucida, passes rapidly down the fallopian tube to the uterus arriving there about 24 hours after ovulation (Tyndale-Biscoe, 1973) but en route it is fertilized and acquires a mucoid coat of acid mucopolysaccaride (Hughes, 1969). Mucoid coats, however, are not confined to prototherian and metatherian eggs; those of the mare, bitch (Blandau, 1961), and rabbit (Braden, 1952) are also coated with acid mucopolysaccharides secreted by glands in the fallopian tube. However, the marsupial zygote is invested with another layer—the shell membrane thought to be homologous with the basal layer of the monotreme egg shell by C. J. Hill (1933). In *Didelphis* it is formed by a shell gland situated at the lower end of the fallopian tube (McCrady, 1938) but according to Hughes (1969, 1974) it is secreted by glands in the uterus in *Trichosurus vulpecula* and is keratinous in nature.

After fertilization and formation of the egg shell the yolk is extruded as a single discrete yolk body within the shell in *Dasyurus* (Hill, 1910) and as several

---

*The observation of Van Deusen (1971) that a very young *Zaglossus*, apparently furred and with spines, was taken on July 15, 1964 along with an adult is hard to interpret in this connection since the sex of the adult was not mentioned.

bodies in the instance of the egg of *Didelphis* (McCrady, 1938). Lyne and Hollis (1976), using the electron microscope, found in *Perameles nasuta* that the "yolk material was dissipated into the space which had formed between the cells." After extrusion of the yolk cleavage takes place, holoblastic in *Dasyurus, Didelphis,* and *Perameles,* meroblastic according to Caldwell (1887), in *Phascolarctos* but nothing is known of the events following the early divisions in that animal. The blastomeres formed as a result of holoblastic segmentation lose contact with one another, migrate outwards, and flatten themselves against the zona pellucida. There they divide mitotically and form a complete hollow vesicle or blastocyst containing yolk and fragments of protoderm cells, the whole being surrounded by zona pellucida, mucoid coat (containing spermatozoa), and shell membrane. This resembles closely the monotreme unilaminar blastocyst containing yolk, the vesicle also being surrounded by zona pellucida, "albumen" (presumably mucoid coat), and basal shell layer, with the difference that it is surrounded by another shell membrane, the rodlet layer.

Having achieved the unilaminar stage the marsupial blastocyst is converted to a bilaminar condition by processes practically identical to those of the monotreme blastocyst. In *Dasyurus* and the bandicoots (Hollis and Lyne, 1977) some of the cells of the unilaminar blastocyst acquire pseudopodia and migrate inwards from the outer layer to form a cell network that is filled in by mitotic division to form a complete new vesicle inside the outer layer, which now may be called the ectoderm. In *Didelphis* much the same happens except that migratory cells are large and rounded at first, but they too acquire pseudopodia, migrate to the outer layer, and there flatten themselves to form a complete vesicle of endoderm as in *Dasyurus* and the bandicoots. By this time the zona pellucida and the mucoid layer have disappeared leaving a bilaminar blastocyst containing the remains of yolk and surrounded by a keratinous egg shell, but as yet there is no sign of an embryo.* Soon, however a medullary plate in the ectoderm forms at one pole of the blastocyst followed by development of bilateral symmetry and a primitive groove from which mesoderm proliferates laterally, as it does in all amniotes. Somites in the mesoderm appear at the 8th day of gestation in *Didelphis* (full term is at 12.75 days), and at the 12th and 16th day in *Trichosurus* and *Setonix*, respectively (see Tyndale-Biscoe, 1973). The arrangement of the fetal membranes of the *Tachyglossus* embryo described on p. 241 is very like that in the blastocyst of *Phascolarctos*; Caldwell (1884a) and Semon (1894b), in greater detail, showed that the allantois in the latter genus grows out so that its

---

*Müller (1969) and Lillegraven (1969) seem to be unaware that complete unilaminar and bilaminar blastocysts are formed during both monotreme and marsupial ontogeny before an embryo is discernible—the latter's Fig. 48 stated to be a summary of monotreme early development is errone-ous. Müller seems to be under the impression that there is no uterine gestation period in monotremes during which uterine secretions are absorbed by the vesicle.

splanchnic mesoderm layer fuses (verwachsen) with the serosa. There is, however, an important difference from the condition in the *Tachyglossus* egg in that the extra-embryonic mesoderm penetrates only half way round the blastocyst between ectoderm and endoderm so that the yolk sac is invested, in great part, with nonvascular bilaminar omphalopleure as in the earlier stages of the monotreme blastocyst. This limited extent of vascular trilaminar omphalopleure is characteristic of all marsupial blastocysts (Semon, 1894b; McCrady, 1938; Sharman, 1961; Hughes *et al.*, 1965) but the allantois does not fuse with the serosa in all instances. Tyndale-Biscoe (1973) has admirably summarized the variations in allantois morphology, including the allantoic placentas of bandicoots, and has compared them with the morphology of the allantois in *Tachyglossus*, in his Fig. 2.8 on p. 61. In *Didelphis* the egg shell persists for 10 days of the 13-day gestation period (Renfree, 1975) but in most marsupials the shell breaks down during the last third of the gestation period; Tyndale-Biscoe (1973) and Hughes (1974) give accounts of the varying degrees of attachment of the yolk sac placentas to the uterus.

A matter of interest is that during ontogeny the embryos of *Trichosurus vulpecula* and *Phascolarctos* exhibit rudiments of an egg tooth (Hill and de Beer, 1949). The structure of the tooth is similar to that of *Tachyglossus* at stage 43. The pouch young of these two marsupials also exhibit anlagen of a caruncle as do those of *Megaleia rufa Perameles nasuta, Didelphys aurita*, and *Caluromys philander* but the last four mentioned have no vestige of the egg tooth (Fig. 82).

The collection of facts mentioned in the foregoing has led to the notion that marsupial developmental processes are slightly modified monotreme developmental processes (Hill, 1910; Flynn and Hill, 1947; Hill and de Beer, 1949; Luckett, 1976). The notion is reinforced when one compares the marsupial neonatus with the monotreme newly hatched. Both look amazingly alike with their well-developed forelimbs and rudimentary hindlimbs (Figs. 80, 82). The big forelimbs are, apparently, associated with coming into the world in a near-fetal condition and having to move about to find, unaided, the outlets of mammary glands. Weights and sizes are likewise comparable: one of the *Tachyglossus* hatchlings mentioned above weighed 378 mg and was 1.47 cm in length (crown-rump) whereas the large wallaby *Petrogale xanthopus* gives birth to a single young as small as 1.60 cm long and 460 mg in weight (J. C. Merchant, personal communication). The weights of neonatuses of small wallabies giving birth to single young are as follow: *Potorous tridactylus*, 333 mg; *Bettongia lesueur*, 320 mg; *Setonix brachyurus*, 340 mg (Waring *et al.*, 1966). However, some marsupials bearing many young at once have a much smaller neonatus—*Dasyurus*, for example, weighs, on the average, only 12.5 mg at birth (Hill and Hill, 1955). Along with this tremendous difference in size, it should be kept in mind that marsupial neonates among themselves exhibit profound variation in internal organization. For example *Dasyurus* at birth lacks cranial nerves II, III,

IV, VI, and VIII (Hill and Hill, 1955) whereas all cranial nerves are present in the *Trichosurus* neonate (Müller, 1969).

The eyes and aural apertures are covered over by a layer of epitrichium but the external nares are widely open and the olfactory organs appear to be fully functional in *Dasyurus* at least (Hill and Hill, 1955), and in *Tachyglossus* (Fig. 100 and p. 280).

Monotremes lack the peculiar oral shield found on the muzzle in *Didelphis* and *Dasyurus* neonates but not in those of other marsupials (Hill and Hill, 1955).

The marsupial neonatus has no gape, the cheeks being covered with epitrichium, and the hatchling has only a small gape. Müller (1969) considers this condition to be necessary to prevent undue movement of the jaws while the dentary squamosal articulation is being established. Doubtless the small gape of the monotreme hatchling, which also, as yet, lacks the dentary squamosal articularion, limits movements of Meckel's cartilage with its ventrolateral investment of dentary bone. This articulation forms during development in the uterus in eutherians and according to Müller a transitory closing of the cheeks occurs in utero; this may have a similar function in limiting jaw movements during ingestion of amniotic fluid. The cheek closures in marsupials persist for a sizeable part of pouch life, a condition no doubt related to the necessity for marsupials to be attached firmly to a teat, since soon after birth many lose the ability to reattach themselves should they fall off, or be removed from, the teat [Merchant and Sharman (1966) and my own experience].

Since the incus remains joined to Meckel's cartilage for some time after hatching and after birth the two kinds of young have no sense of hearing but the membranous labyrinth with its semicircular canals is well developed in the newly-hatched *Tachyglossus* (Alexander, 1904) and in newborn *Didelphis* (Larsell *et al.*, 1935); in *Dasyurus*, however, the semicircular canals are present only as rudiments.

The newly hatched and the neonatus exhibit some similarities in internal organization. In the lungs for example true alveoli are not present in the neonatus nor in the hatchling. In both, the bronchi end blindly in modified bronchioles which function as respiratory chambers (the "geräumiger Luft-kammern" of Narath, 1896), for some weeks after hatching and birth (Selenka, 1887; Bremer, 1904; Hill and Hill, 1955; Krause and Leeson, 1973). These chambers are lined with respiratory epithelium from which later in pouch life alveoli arise. In *Tachyglossus* the ductus arteriosus is obliterated at hatching and all blood from the right ventricle passes to the lungs via the pulmonary artery. Hill and Hill say of *Dasyurus* that the aortic arch and the pulmonary artery are two distinct vessels at birth and that there is no ductus arteriosus.

Concerning the excretory system of the marsupial neonatus (Hill and Hill, 1955) state "As is well known, the mesonephroi in the later uterine embryo and the new-born young of the marsupials reach a relatively large size and are functionally active excretory organs both before and for some little time after birth (Buchanan and Fraser, 1918, pp. 45–6, and for experimental evidence, see Gersh, 1937).'' However, they found that the glomeruli present in the mesonephros of *Dasyurus* differ in detail from those in the mesonephroi of other marsupials neonates. In *Tachyglossus* well before hatching (stage 44) the pronephros (Vorniere) is a shrunken remnant but the mesonephros is well developed exhibiting glomeruli (Keibel, 1904a) but an anlagen of the metanephros is present also. This is the condition at hatching, stage 46, and the metanephros now exhibits a few definitive glomeruli. The mesonephros is still present at stage 51 but shows signs of atrophy.

The skeletons of neonatus and hatchling also present common features. The chondocranium of *Dasyurus* is even a more sketchy affair than that of *Ornithorhynchus* (p. 13). Seen in median longitudinal section it is shallow and saucer-shaped but the chief constituent parts ''have already become joined up to form a continuous, though as yet very incomplete, case for the support of the brain and the olfactory organs, the olfactory capsules being the only parts which can be regarded as relatively well formed'' (Hill and Hill, 1955). The beginnings of premaxillae, maxillae, palatines, and dentaries are the only ossifications present. The chondocranium of the new-born *Trichosurus*, while still very incomplete, approaches to that of *Ornithorhynchus*, and the anlagen of the following membrane bones are present; premaxillae, maxillae, palatines, squamosals, pterygoids, and dentaries (Broom, 1909); pilae antoticae, of course, do not appear. In the monotreme at hatching the anlagen of the same membrane bones are present along with four more: nasals frontals, septo-maxillae, and parietals.

The shoulder girdle of the neonatus is like that of the monotreme hatchling in that coracoid cartilages connect the scapulae with the sternum and the anlagen of an interclavicle is present (Broom, 1897; Watson, 1918; Hill and Hill, 1955).

It would appear from all these comparisons of ontogeny in the monotremes and marsupials that ancient marsupials were oviparous, and that their embryos were surrounded by egg shells that had to be torn open by the actions of egg teeth and caruncles; what emerged from the egg apparently resembled the newly hatched of the present day monotreme. This raises the following interesting questions: did ancient marsupials have teats and if so how did the mothers feel when their newly hatched latched onto their teats with a sharp egg tooth?

Oviparity implies incubation; perhaps the fact that marsupial embryos like those of *Macropus eugenii* take ca. 17 days to achieve a 15–20 somite stage of development is an echo of times when marsupials had to store enough uterine fluid in their large blastocysts to tide them over an incubation period outside the uterus. It would seem that extrusion of yolk from the ovum and blastomeres, and its subsequent liquefaction and dilution by uterine fluids, is an arrangement entailed by replacement of meroblastic cleavage in the eggs of ancient marsupials, with holoblastic cleavage in their descendants.

9

Lactation, Composition of the Milk, Suckling, and Growth of the Young

ORNITHORHYNCHUS

Seasonal Changes in Nonlactating Mammary Glands

Along with the reproductive system, the mammary glands exhibit marked seasonal changes in their anatomy and histology (Griffiths *et al.*, 1973); in most platypuses they are quiescent from May until about the middle of July. In this condition the lobules, which are only about 1 cm long, consist of solid cords or closed ducts invested with myoepithelium and occasionally of ducts with small but patent lumina (Fig. 84). In a few platypuses the glands are not detectable to the naked eye—a fact noted by Bennett (1835); it seems likely that these are the glands of juveniles and had the criteria for determining age of young platypuses (p. 48) been available to Griffiths *et al.* the matter might have been determined.

Towards the end of July, or earlier in some cases, signs of growth and differentiation are apparent in the lobules: the connective tissue surrounding the lobule becomes hypercellular and strands of mesenchyme derived from this hypercellular tissue invade the adipose tissue that often surrounds the lobules. The mesenchyme is followed by streams of ectodermal cells from the lobules themselves and at the same time the duct system becomes less tightly packed. At the periphery the lobule develops outgrowths that penetrate the newly-formed connective tissue matrix. Finally the new solid duct system acquires lumina. These processes result in a gland exhibiting lobules ca. 3.5 cm in length. If the platypus fails to become pregnant the gland can remain in this condition until January after which time it reverts to the quiescent state. If the platypus lays eggs and suckles young the tubules expand, branch, and subdivide into a quasi-alveolar condition. One such gland was taken early in the month of October but

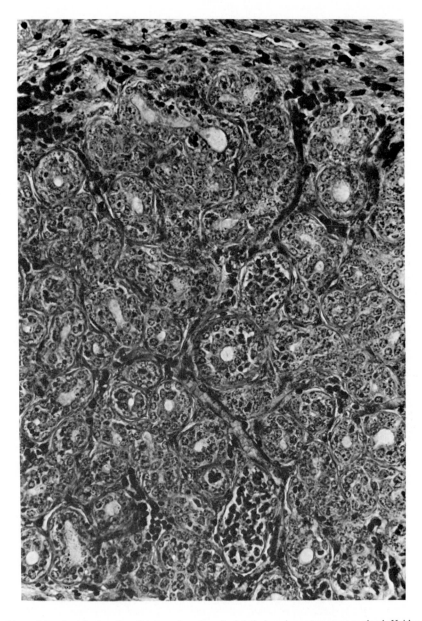

Figure 84. *Ornithorhynchus.* Section of portion of a lobule in quiescent mammary gland. Heidenhain's iron haematoxylin. × 386. *(From Griffiths et al., 1973; with permission of The Zoological Society of London.)*

the stage of development of the young was unknown. The mammary glands of lactating platypuses taken later in October through to March are fully alveolar.

Structure and Ultrastructure of the Lactating Gland and the Effect of Oxytocin Thereon

Lactating platypuses can be taken any time between mid-October and late March in the waters of the Murrumbidgee River and its tributaries. An example of a fully lactating gland with its club-shaped lobules was shown in Fig. 2. Internally the lobules consist of alveoli (Fig. 85) lined by a secretory epithelium as they do in all mammals, metatherian and eutherian and, as in those mammals, the shapes of the epithelial cells change with the amount of milk secreted into the lumen. Those alveoli filled with milk are lined by a thin flattened epithelium while in other areas of the gland alveoli can be found devoid of milk in which case they are lined by tall columnar cells. Griffiths *et al.* (1973) found that the secretory cells exhibit microvilli on the luminal surface and that their bases abut onto myoepithelial cells and a limiting basal membrane. The cell nuclei are large and rounded, the endoplasmic reticulum is well developed and the Golgi apparatus particularly so. Casein granules (the characteristic protein of milk) are present in the Golgi complex and in apical vacuoles which appear to empty into the lumen (Fig. 86). Fat droplets, however, are present in the cytoplasmic matrix, especially in the apical region of the cell where fat droplets could be found ballooning out the apical plasma membrane (Fig. 87) (see Wooding, 1977).

The myoepithelial cells, as in the mammary glands of all mammals are filamentous and form a kind of network over the alveoli. In electron microscope preparations it can be seen that their cytoplasm contains myofilaments running parallel to the long axis of the cell (Fig. 86). There is good evidence (see Richardson, 1949; Linzell, 1955; Folley, 1969; Linzell and Peaker, 1971) that contraction of these cells in the mammary glands of the Eutheria raises intra-alveolar pressure leading to expulsion of milk into the ducts. There is also good evidence that myoepithelium is induced to contract by the action of the pituitary posterior lobe hormone, oxytocin (see p. 151). This hormone is released in response to many stimuli such as sucking at the teat or, as in the instance of cows, visual and auditory stimuli such as the sight of the milking shed or rattling of milk buckets. In the lactating eutherian injected oxytocin will induce contraction of myoepithelium and ejection of milk. Injection of oxytocin also induces milk ejection in the lactating platypus (Griffiths *et al.*, 1973) and, as in *Tachyglossus* (Griffiths, 1968) and in the red kangaroo (CSIRO, 1974) it also induces marked distortion of the lobules of the mammary gland. Examples of this are shown in Fig. 88A–D. At injection the lobules have a flat flaccid appearance, the superficial blood vessels lying loosely on the surface of the lobule. Five minutes after intramuscular injection of oxytocin the lobules exhibit a turgid mottled

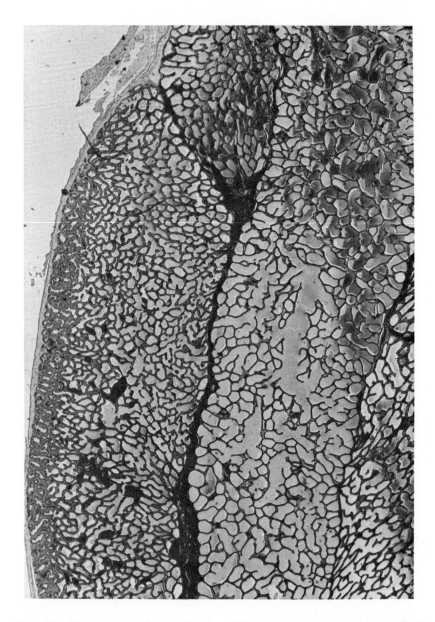

Figure 85.   *Ornithorhynchus*. Section of lobules of fully lactating mammary gland showing alveolar structure. Heidenhain's iron hematoxylin. × 25. *(From Griffiths et al., 1973; with permission of The Zoological Society of London.)*

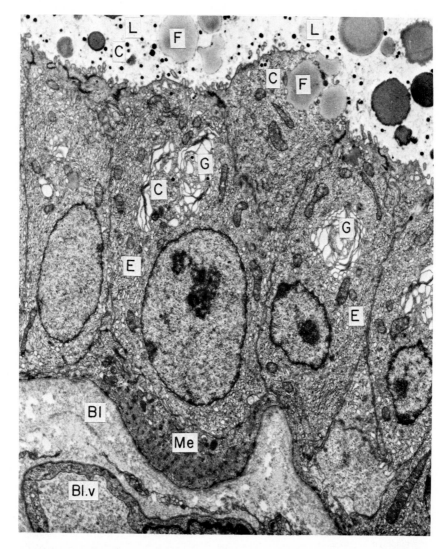

Figure 86. *Ornithorhynchus*. Electron micrograph of the alveolar lining of the lactating gland. E, secreting epithelial cell; Me, myoepithelial cell; Bl, basal lamina; L, lumen of alveolus; C, casein particle; F, fat globule; G, Golgi apparatus; Bl.v, blood vessel. × 5600. *(From Griffiths et al., 1973; with permission of The Zoological Society of London.)*

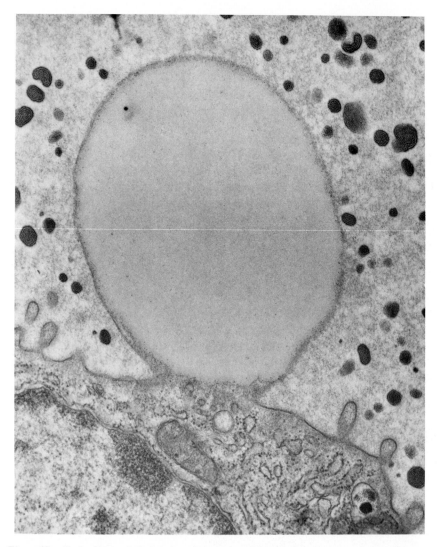

Figure 87.   *Tachyglossus*. Apical region of secretory cell in alveolus of mammary gland showing fat globule ballooning out the apical plasma membrane just before being shed into the lumen. The electron dense bodies are casein granules. × 32,000. Preparation by Dr. G. I. Schoefl.

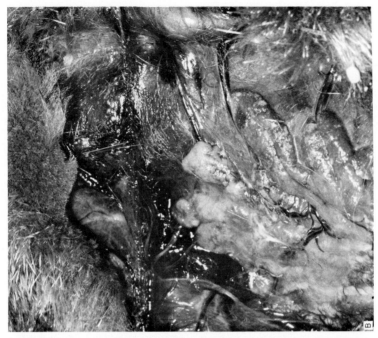

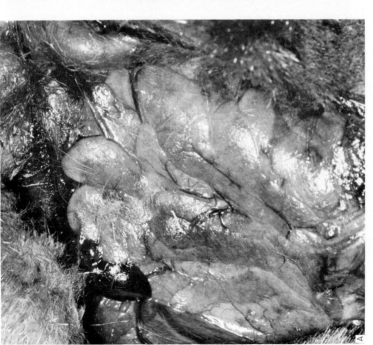

Figure 88. A, *Ornithorhynchus*. Portion of mammary gland exposed to show flaccid, flat lobules. × 2.2. (*From Griffiths et al., 1973; with permission of The Zoological Society of London.* B, Same portion of mammary gland 10 min after injection of oxytocin. × 2.2.

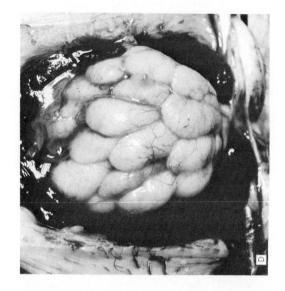

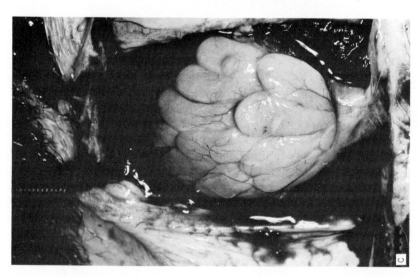

Figure 88. C, *Tachyglossus*. Portion of mammary gland exposed. × 2.2. D, Same 7.5 min after injection of oxytocin. (*From Griffiths, 1968.*)

TABLE 32

Fatty Acid Complement (g/100 g methyl esters) of the Triglycerides of the Milk of Platypuses Taken at Various Times of the Lactation Season[a]

Date sampled	Saturated											
	C_{12}	C_{13}	$C_{13}Br$	C_{14}	C_{15}	C_{16}	$C_{16}Br$	$C_{16}iso$	C_{17}	C_{18}	C_{19}	C_{20}
October 30	8.4	0.1	—	4.4	0.5	25.2	—	0.2	0.8	5.5	—	1.0
November 6–March 22 (avg. 11 samples)	1.1	0.2	0.5	2.7	0.8	19.5	0.2	2.1	0.8	3.5	0.5	0.8
November 4, P33	1.1	0.2	0.1	2.8	0.9	20.1	0.6	3.7	0.5	3.4	0.9	0.7
March 4, P33	1.2	0.3	0.2	3.6	2.0	16.4	1.0	5.1	0.6	1.9	1.0	0.5

Unsaturated														
$C_{14:1}$	$C_{15:1}$	$C_{16:1}$	$C_{17:1}$	$C_{18:1}$	$C_{18:2}$	$C_{18:3}$	$C_{20:1}$	$C_{20:2}$	$C_{20:3}$	$C_{20:4}$	$C_{20:5}$	$C_{22:4}$	$C_{22:5}$	$C_{22:6}$
2.4	1	15.0	0.8	25.9	2.5	2.1	4.4	0.2	—	0.3	0.1	—	0.1	—
1.2	0.1	10.6	1.6	25.9	7.5	7.4	1.7	1.1	0.4	3.1	1.9	0.4	2.4	1.1
2.2	0.3	7.2	1.9	23.3	8.9	9.0	1.6	2.0	0.5	3.8	1.3	0.3	2.6	0.7
1.1	0.4	11.1	2.7	23.3	8.6	6.6	1.6	0.4	0.6	3.6	1.1	0.5	1.9	0.6

[a] Data from Griffiths et al. (1973); reproduced by permission of The Zoological Society of London.

appearance and at 10 minutes the blood vessels appear to be sunk into the substance of the lobules due to distortion. The lobules by now also have a nodular appearance consistent with the notion that myoepithelium is contracting around the alveoli. In the intact lactating platypus injection of oxytocin facilitates milking; in fact it is almost impossible to get samples adequate in amount for analysis without it.

Duration of Lactation

Apparently lactation can continue for as long as 3½ months: a platypus taken November 20, 1970 was found to be lactating so she was milked, marked, and released. She was caught again March 4, 1971 and proved to be lactating still. The fatty acid composition of the milk was the same on both occasions (Table 32). The glands of other platypuses taken late February and early and late March showed signs of regression: the alveoli were invaded by long strands of epithelial cells and leucocytes in the process of removing milk. The glands of one platypus taken March 5 consisted largely of flesh-colored lobules returning to the solid cord condition but some lobules still contained milk. The latest time of the year at which a gland containing milk was observed was April 18; in that gland many alveoli were filled with leucocytes. All this evidence indicates that platypuses suckle their young for months on end.

In the completely regressed gland the myoepithelium forms a spirally, or longitudinally, arranged investment over the tubules and solid cords. Gegenbaur (1886) was the first to describe this arrangement of myoepithelium but he states that the mammary gland he found it in was that of a male (see p. 9).

Composition of the Milk

FATTY ACID COMPLEMENT OF THE TRIGLYCERIDES.   The fatty acids in the milk fat have been studied by Griffiths *et al.* (1973). Samples of milk were taken soon after platypuses were caught by the netting technique (p. 45). Upon squeezing the mammary glands, after intramuscular injection of oxytocin, the milk wells up through the hair over the areolae from where it can be sucked up by pipette— about 10 ml can be taken with ease at a milking (see CSIRO, 1974).

The samples were taken at different times of the lactation season (Table 32) but since the platypuses were caught in the water the sizes and ages of the sucklings were unknown.* The samples were collected over a long period of the lactation season to see if any changes in fatty acid content were detectable that

---

*It may prove possible to catch lactating platypuses in nets, milk them, and attach a radio-transmitter to each in order to locate them, after release, in their underground nesting chambers. Provided these are not under large rocks and trees, as usually seems to be the case, it may be possible to dig down, expose the chambers, and make observations on the young.

might reflect changes associated with the growth of the young; such changes occur in the milk of the red kangaroo (Griffiths *et al.*, 1972).

All samples were creamy white in color and showed none of the pink flush seen in *Tachyglossus* milk (p. 289). Total solids ranged from 19.8 to 14.0% w/w and the fatty acid complements of the triglycerides showed little difference in samples taken in the period November 6 to March 22 (Table 32). However, the sample taken October 30 contained more lauric acid (C_{12}) and more palmitic acid (C_{16}) than the other samples did suggesting that milk supplied to very small young may prove different from that for older young. Obviously what is needed to determine the matter is analysis of samples taken from the same females over a lactation period ranging from hatching to weaning.

A matter of interest, already mentioned, that can be found in Table 32 is that a platypus, P33, was milked, marked, and released November 20, 1970; when caught again 3½ months later, March 4, 1971, she proved to be lactating still and the fatty acid composition of the milk was found to be virtually the same as that taken in November. The similarity in composition supports the notion that platypuses suckle their young for months on end.

The fatty acid complement of the triglycerides of these mature milks calls for no special comment; it is like that of some eutherians and it is quite like that of woman. However the milks of other eutherian species exhibit fantastic differences among themselves. This will be discussed in the section on composition of *Tachyglossus* milks.

CARBOHYDRATES.   Hopper (1971), incidental to a study of whey proteins in platypus milk, found in two samples, one taken midseason (December) and another late in the season (March), that they contained very little free lactose but they did contain a lot of carbohydrate with chromatographic properties identical to those of fucose. Subsequently Messer and Kerry (1973) confirmed that a sample of milk taken late in the season (February) contained very little free lactose and in fact total free carbohydrate was in low concentration—1.7 g/100 g milk; of this they say 0.91 g is the monosaccharide fucose. However, as it turns out most of the fucose is in the form of a tetrasaccharide, difucosyl-lactose, which accounts for 56% of the carbohydrate; free lactose was found to account for only 1% of the total carbohydrate. Obviously more samples will be needed to establish the identity of the other 43% of the carbohydrate. Messer and Kerry also found that there is practically no sialic acid in platypus milk. These findings will be compared with those for *Tachyglossus* and other milks (see Appendix).

PROTEINS.   It has already been shown that platypus milk contains casein (Fig. 86). Hopper (1971) found that the milks were highly viscous so that separation of the casein and fat required prolonged centrifugation at 48,000 × G. The whey so obtained is a pale yellow color; this could mean that it contains less iron-binding protein or less iron than *Tachyglossus* milk which is pink in color due to the presence of a high concentration of iron and of iron-binding protein (see p. 297). Hopper and McKenzie (1974) found in platypus whey subjected to gel filtration on Sephadex G75 in Tris-H 1 buffer at pH 7.9 an elution profile exhibiting three peaks only. No lactose synthetase activity was found in these fractions; however, whole whey had activity (Table 38). Hopper and McKenzie also remark that the number of milk specific whey proteins revealed by starch-gel electrophoresis was less than in the wheys of other mammals including *Tachyglossus*.

Suckling

Practically nothing is known of how platypuses suckle their newly-hatched young*, nor any other stage for that matter but one observation indicates how they might do it: Kershaw (1912) dug out a female from a nesting chamber and found a newly hatched attached firmly to her skin and that a second young fell off. This suggests that the female remains at rest in the burrow after hatching and that the hatchlings emerging from the eggs (held against the abdomen by the tail) snuggle down through the fur to the areolae and stay there imbibing milk until they have acquired the size, strength, and tough skin, enough to withstand scrabbling about in the rough materials of the nest. Alternatively it is just conceivable that the mother leaves the burrow to feed after her fast over the putative 11-day incubation period, taking the newly hatched with her, clinging with their relatively enormous forelimbs to the areolae, buried in dry fur, and breathing air trapped between the hairs.

TACHYGLOSSUS

Development of the Mammary Glands

The anlagen of the mammary glands in all mammals are identical and consist of a lenticular thickening of the skin on each side of the ventrolateral aspect of the embryo located towards the posterior end of the organism; these are the primary primordia of Bresslau (1920). Unfortunately, whether elongated or not, these have been termed milk streaks by embryologists working on eutherians, because in some species the anlagen appear as longitudinal ridges along the trunk; however in those eutherians in which the nipple rows do not extend beyond the level

---

*Burrell (1927) asserts his belief that the newly hatched do not imbibe milk until they are a week old! How he imagines these delicate organisms survive and grow during that week without food is not explained.

of the navel, the primary primordia are identical to those in echidnas and metatherians. The primary primordia in *Tachyglossus* are visible in the stage 42 embryo (5-mm crown rump length) and take the form of two ectodermal thickenings on either flank located dorsal to where the hindlimb buds will form. The tissue beneath the thickenings exhibits an accumulation of nuclei and blood vessels. Soon after hatching the skin heals over the navel opening but the muscle layer in the surrounding area does not thicken as it does elsewhere on the abdomen; an oval area of very thin muscle is left in the middle of the panniculus carnosus of the belly in which the pouch will form later in life. The primary primordia, meanwhile, have moved toward the midline and are not located at the anterolateral regions of the pouch area. Both the pouch area and the rest of the skin exhibit sparsely distributed hairs but the primary primordia are devoid of hair at first. However, when the pouch young is 4–5 cm in length follicles within the primary primordia exhibit a vigorous growth of hair, more so than in the surrounding skin. Each follicle exhibits a downgrowth of ectoderm and depending on the position of the follicle in the primary primordium form two kinds of gland, those in the center of the primordium become mammary gland and those at the periphery give rise to sweat glands otherwise known as the Knäueldrüsen of Gegenbaur (p. 7). Further development involves elongation and subdivision of the mammary anlagen into lobules in which tubules form, while sebaceous glands are also formed in association with the hair follicles. The distal tubules form the ducts that come to open to the exterior at the base of a special hair—the mammary hair. These have nothing to do with other hairs in the pouch area; they are special hairs associated only with the ducts of the mammary glands (Bresslau, 1920).

The mammary glands of the Metatheria exhibit a development very like those of *Tachyglossus*, somewhat modified, however, by the formation of teats, but these before eversion exhibit true mammary hairs that are shed following eversion; the teat of the koala is an exception—it exhibits mammary hairs for a long time after eversion. Mammary hairs do not develop in the primary primordia of eutherian mammary glands.

Seasonal Changes in Nonlactating and Structure of Lactating Mammary Glands

Like those of the platypus the mammary glands of *Tachyglossus* exhibit a seasonal waxing and waning Griffiths *et al.*, 1969). Outside the breeding season the lobules of the gland are thin, flat, and microscopically they exhibit a tightly packed convoluted solid duct system in which lumina are rarely seen. The lobules are separated one from the other by thin connective tissue sheaths which exhibit relatively few nuclei (Fig. 89) and adipose tissue, often found as a fat pad at the periphery of the gland, exhibits very little mesenchymatous tissue.

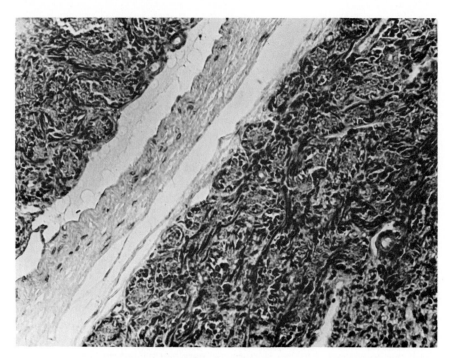

Figure 89. *Tachyglossus*. Quiescent mammary gland showing lobules made up of convoluted solid cords. Interlobular connective tissue sheaths thin, exhibiting few nuclei. Heidenhain's iron hematoxylin. × 225.

During the breeding season mammary glands of some nonpregnant females can show a tubular grade of organization, the tubules elaborating a secretion that appears as a coagulum in the lumina. Very often, and particularly in glands with a solid-cord duct system, the myoepithelium, stained with Heidenhain's iron hematoxylin is prominent.

The lobules of the glands supplying milk for a large suckling are fully alveolar; the ultrastructure of the secretory epithelium of these alveoli and of the myoepithelium are similar to those of *Ornithorhynchus* with the difference that casein granules are particularly prominent and have a whorled structure (Figs. 87 and 90) apparently unique since it has not been observed in other mammary glands, metatherian or eutherian, nor in those of the platypus. The significance of this configuration is unknown.

The processes of regression from the fully alveolar to the solid-cord state have been described by Griffiths *et al.* (1969) in two echidnas taken in the bush December 2, 1967 and January 18, 1968, respectively. Both were lactating but neither had pouches. Pouches regress after the young leave the pouch following

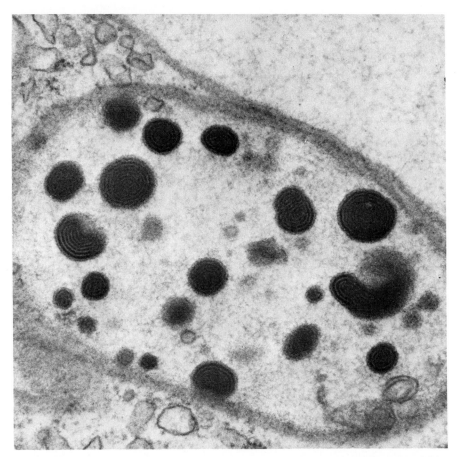

Figure 90. *Tachyglossus.* Casein granules in lumen of an alveolus of a lactating mammary gland. Granules exhibit concentric layering. × 6420. *(From Griffiths et al., 1973; with permission of The Zoological Society of London.)*

development of their spines, but suckling and lactation continue so in all probability these two echidnas were suckling young living in burrows (see p. 299). The milk contained ca. 50% solids, a figure consistent with that for milk being supplied to young of 400–800 g weight.

Biopsy samples taken soon after capture revealed that the alveoli were thin-walled and distended with milk. The mammary gland of No. 1 echidna (taken December) was examined 71 days, and that of No. 2, 29 days later. No. 1 was lactating at one gland only and No. 2 at both. A biopsy sample from No. 2 presented an histological appearance quite unlike that found 29 days earlier: the connective tissue around some of the alveoli had hypertrophied, while the epithelium and lumina of other alveoli still distended with milk had been invaded by large populations of lympocytes. Ninety days after capture of this echidna the gland at autopsy exhibited a solid-cord organization and regression was complete.

The glands of No. 1 echidna were examined at autopsy 125 days after capture. Both glands exhibited to the naked eye a well-defined area about 0.5 cm in diameter which still contained milk; the rest of the glands consisted of shrunken flattened lobules. Some of these had thin-walled alveoli, others had alveoli stuffed full of pleomorphic macrophages, others had reverted to tubules and in some instances to solid-cords (Fig. 91). Thus the sequences of metamorphosis from distended alveolus to solid-cord duct system are as follow: the alveolus is invaded by lymphocytes, large lymphocytes, and macrophages. When these have removed nearly all the milk the macrophages become pleomorphic filling the lumen of the regressing alveolus. The macrophages finally vacate the alveolus

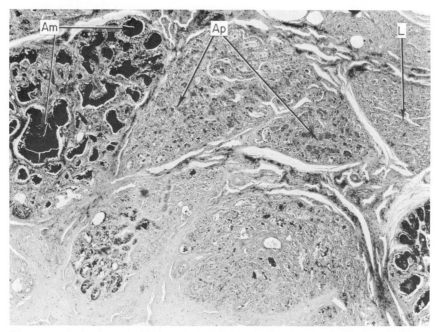

Figure 91. *Tachyglossus.* Lactating mammary gland regressing to the quiescent state: some lobules still alveolar and contain milk, in others the lumina of the alveoli are packed with pleomorphic macrophages, while other lobules have reverted to the solid-cord condition. × 44. Am, alveoli still full of milk; Ap, alveoli full of pleomorphic macrophages; L, lobule reverted to solid-cord state. *(From Griffiths et al., 1969; with permission of The Zoological Society of London.)*

leaving an empty space lined by a pavement epithelium—this becomes columnar and fills in the lumen until a solid-cord condition is achieved.

Effects of Steroid Hormones on Growth of Mammary Glands and Lactation

Cowie and Tindal (1971) state "We are not aware of any studies on the hormonal control of mammary growth in either Prototheria or Metatheria." Since that time Hearn (1972) has studied, inter alia, the effects of hypohysectomy in lactating tammar wallabies, *Macropus eugenii*, and a start has been made on a study, by myself, of the effects of the hormones estradiol and progesterone on the mammary glands of adult female and male echidnas. The experiments were carried out on both sexes in the nonbreeding season. The glands of the females at that time have already been described (Fig. 89). In the males the duct system of the minute lobules at all times of the year is of a tubular grade and secretion is often present in the lumina (Fig. 92A). The adipose tissue surrounding the lobules is largely free of mesenchymal elements and the connective tissue exhibits very few nuclei.

The hormones were administered in the form of pellets implanted subcutaneously under anesthesia; dosages were 1×20 mg or 2×10 mg pellets of estradiol proprionate allowed to act for 20–52 days, progesterone 1×100 mg implanted 16–22 days after implantation of estradiol. In two instances estradiol and progesterone were implanted simultaneously and allowed to act for 26 days in an intact echidna and for 52 days in an ovariectomized animal. The results from the females are summarized in Table 33. In all instances treatment with either estradiol alone or with progesterone as well gave rise to glands that were enlarged and could be palpated, but it was not a normal growth: a pouch was not formed and the glands took the form of protrubent round bodies lying immediately dorsal to the areolae, showing no sign of the fan-shape of the normal gland. When palpated the glands felt hard and nodular. The main increase in size was due to increase in thickness: the average thickness of glands from eight controls was 0.4 cm (range, 0.3–0.5) whereas in ten treated females in which the glands were measured it was 1.0 cm (range, 0.7–1.5). Internally the treated glands exhibited to an exaggerated degree the structure found during the seasonal recrudescence: the cords or ducts were loosely arranged and at the periphery the lobules appeared to merge without clearly defined boundaries into the surrounding masses of mesenchyme migrating from the hypertrophied connective tissue between the lobules (Fig. 93A). These mesenchymal elements also invade the adipose tissue (Figs. 93A and B) replacing it to form the substratum of new lobules. Formation of new tubules is achieved by further invasion of the mesenchyme and adipose tissue by long files of ectodermal cells streaming out of the old lobules into the new (Fig. 94).

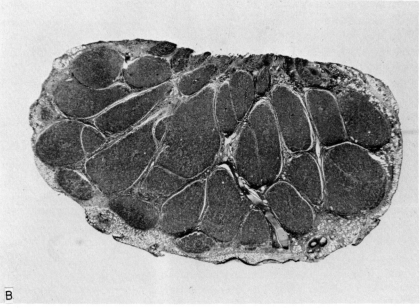

Figure 92. A, *Tachyglossus*. Sagittal section of mammary gland in an adult male. Heidenhain's iron hematoxylin. × 8.9. B, *Tachyglossus*. Sagittal section of mammary gland in an adult male treated with estradiol and progesterone. × 8.9.

TABLE 33

Effect of Estradiol with or without Progesterone on the Mammary Glands of Adult Female Echidnas in the Nonbreeding Season

Number of animals	Avg. body wt. (g)	Treatment	Condition of glands
12	4477 (3500–5900)	Nil	Lobules small, tightly packed solid duct system, connective tissue with relatively few nuclei, adipose tissue free of mesenchyme
2	3700 3900	Estradiol implanted	Lobules enlarged, loosely packed solid duct to tubular system, mesenchymal and ectodermal cells streaming into adipose tissue forming new lobules
5	4445 (3600–5260)	Estradiol plus progesterone implanted	Lobules enlarged, loosely packed solid duct to tubular system, mesenchymal and ectodermal cells streaming into adipose tissue forming new lobules
2	3900 4430	Ovariectomized. Estradiol in one, estradiol plus progesterone in other	Lobules enlarged, loosely packed solid duct to tubular system, mesenchymal and ectodermal cells streaming into adipose tissue forming new lobules
2	4745 4900	Estradiol implanted 7 × 5 mg daily injections of reserpine	Lobules enlarged, loosely packed solid duct to tubular system, mesenchymal and ectodermal cells streaming into adipose tissue forming new lobules but milk found in tubules of one animal

The glands in females treated with progesterone plus estradiol could not be distinguished from those of females with estradiol alone.* The changes described above were also elicited in two echidnas that had been ovariectomized.

Two intact females carrying estradiol pellets were given during the last 7 days of the implantation period daily intramuscular injections of 5 mg reserpine (Ser-

*The uteri in these treated females also exhibited marked differences from those of the controls. In the untreated females (six) the average diameter of cross sections of the uterus was 3.8 mm (range, 2.5–5.0). The mucosae of these shrunken uteri consisted, for the most part, of sparsely distributed small tubular glands lined by a relatively undifferentiated columnar epithelium. There was one exception however: in an animal taken early in January the mucosa consisted of large degenerating glands filled with secretion and smaller ones which exhibited an epithelium containing ciliated cells. It

pasil, Ciba). This compound is known to induce a reflex secretion of prolactin from the pituitary gland in rats (Arai *et al.*, 1970) so it was considered likely that it might have a lactogenic effect in echidnas. The injections had no effect as far as external signs of milk secretion were concerned but histological examination showed that in one animal the tubules were greatly enlarged and contained milk.

Mammary growth was also induced in the males by implantation of the ovarian hormones* but surprisingly the effects were far more marked than those seen in the females. It is well known, of course, that estrogens can induce mammary growth in male eutherians and as Jacobsohn (1961) says "In many species the male mammary gland is equipotential with the female gland in its response to ovarian hormones." However, in the echidna the effects of ovarian hormones on the mammary glands surpassed those in the females. As in these there was overlap of the effects of estradiol alone and of estradiol plus progesterone (Table 34) and the same thickness of gland was achieved, 1.0 cm (range, 0.8–1.2). A comparison of Figs. 92A and B will indicate the degree of difference from the untreated and the treated male gland. Moreover in four of the animals treated with estradiol plus progesterone an alveolar condition accompanied by secretion of milk was achieved (Fig. 95). After intramuscular injection of oxytocin it was found that these animals could be milked. One of them also given 5 mg reserpine daily for the last 7 days of the treatment gave enough milk for the determination of the fatty acid complement of the triglyceride; samples from the other three were pooled for analysis; the results are given below in the section on composition of the milk.

It is possible that the greater response of the male gland to ovarian hormones is due to the fact that, although male glands are much smaller than those of females in the nonbreeding season, they have a tubular grade of organization capable of greater response to the hormones than the solid duct system of the female gland. At all events the mammary glands of one kind of prototherian, at least, respond to ovarian hormone treatment like those of eutherians. The response of intact eutherians to estradiol alone or to estradiol plus progesterone varies with the species: in some, treatment with estrogen alone will lead to growth culminating

---

would seem that this was the uterus of a female that had bred the previous spring, that had not regressed to the resting state.

In the treated females the mean diameter of cross section (9 animals) was 7.5 mm (range, 6.0–10.0); the mean diameters of cross section were the same in the estradiol and estradiol plus progesterone groups, i.e., 7.6 and 7.5 mm, respectively. The uterine mucosae were quite unlike those of the controls; in spite of the greater diameter of the organ the tubular glands were sparsely distributed in a stroma that contained an enormous population of leucocytes not evident in the controls. The cells of the tubular glands were both of the secretory and ciliated types and the lumina of the glands contained a secretion.

*As far as size and histology were concerned the testes of these treated males were similar to the resting phase of normal males sampled during the nonbreeding season (see Fig. 67).

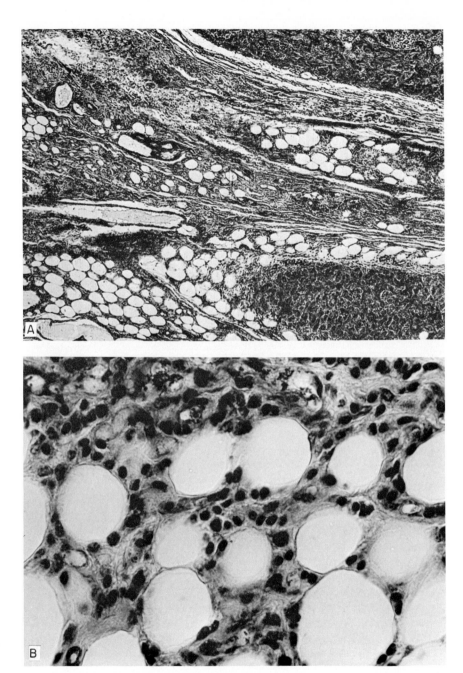

Figure 93. A, *Tachyglossus*. Mammary gland of adult female treated with estradiol and progesterone; section showing hypertrophied connective tissue sheaths between two lobules, and mesenchymal elements invading adipose tissue. × 50. B, *Tachyglossus*. Mesenchyme invading and replacing adipose tissue, forming substratum of new lobule. Heidenhain's iron hematoxylin. × 500.

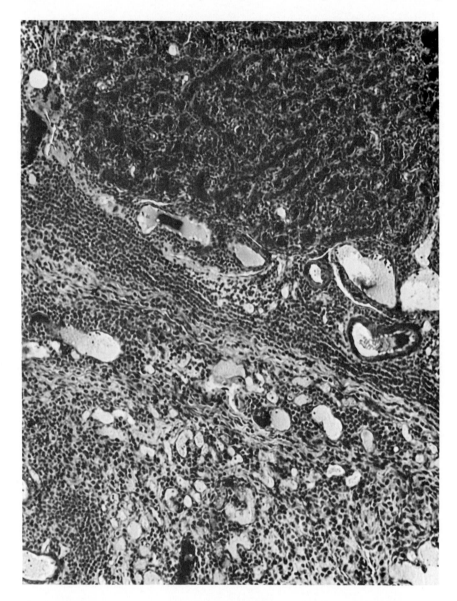

Figure 94. *Tachyglossus*. Mammary gland of adult female treated with estradiol and progesterone. Section showing migration of ectoderm cells from preexisting quiescent lobule (lower part of photo) into the mesenchymal substratum of a new lobule. Heidenhain's iron hematoxylin. × 65.

TABLE 34

Effect of Estradiol with or without Progesterone on the Mammary Glands of Adult Male Echidnas

Number of animals	Avg. body wt. (g)	Treatment	Condition of glands
11	4380 (3250–5500)	Nil	Lobules minute, duct system tubular, secretion often present in lumina, connective tissue with relatively few nuclei, adipose tissue free of mesenchyme
2	4160 5644	Estradiol implanted	Lobules enlarged, duct system in one animal loosely arranged tubules, mesenchyme invading adipose tissue. Other animal— also alveolar in parts, milk in alveoli
2	4268 5150	Estradiol plus progesterone implanted	As for previous two animals
3	3905 3164 3926	Estradiol plus progesterone implanted	Lobules greatly enlarged, alveolar, milk could be collected after injection of oxytocin
1	4694	Estradiol plus progesterone 7 × 5 mg daily injections of reserpine	As for previous three animals but larger volumes of milk could be collected

in the formation of alveoli, while in others treatment with estrogen plus progesterone is necessary to induce lobulo-alveolar growth (see Cowie, 1972, for review). The results with the male echidnas suggest that estrogen plus progesterone is necessary for lobulo-alveolar growth in prototherians. The fact that the "best" effect was observed in a male treated with reserpine as well suggests that prolactin is implicated in the lactation of echidnas. Experiments on ovariectomized and on hypophysectomized echidnas will be necessary to determine these matters. The effects of hormones on mammary glands cultured *in vitro* would also be of great assistance in unraveling mechanisms of control of mammary growth and lactation. For example Wood *et al.* (1975) found that insulin, prolactin, growth hormone, progesterone, estradiol, cortisol, and aldosterone were implicated in maintenance of lobulo-alveolar growth and secretory activity in explanted mammary glands of mice cultured *in vitro*. This gives some idea of the complexity of the hormonal control of mammary growth, differentiation and secretion; it is interesting to note that the mice were pretreated for 9 days with

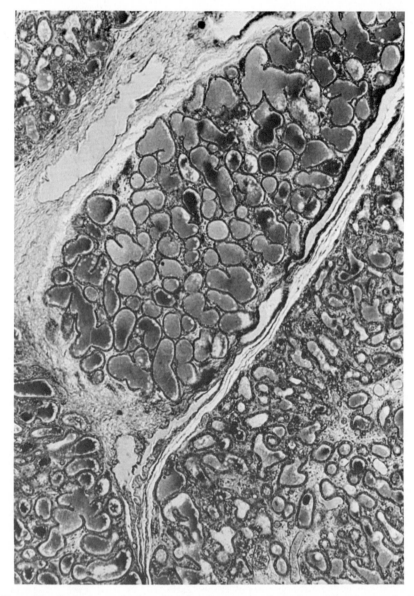

Figure 95. *Tachyglossus.* Section of lobules of mammary gland of an adult male treated with estradiol and progesterone. Note quasi-alveoli filled with milk. Heidenhain's iron hematoxylin. × 63.

subcutaneous injections of estradiol and progesterone. At the end of this pre-treatment and therefore at the start of the *in vitro* experiments "the morphology of the mammary gland was essentially ductal with some end buds." End buds presumably being the beginnings of alveolar structure.

Condition of the Mammary Gland and Areola at Hatching; Structures of the Hatchling Related to Ingestion and Digestion of Milk

At hatching the gland is not very far advanced from the tubular condition (Griffiths *et al.*, 1969). The tubules or quasi-alveoli are thick-walled (Fig. 96), intra-alveolar connective tissue is prominent, and the secretory epithelium varies from cuboidal to flat; in some instances it is organized into two rows giving the impression of a tall columnar epithelium. The lumina in many instances are distended with milk. The tubules communicate with ductules, these in turn lead to ducts which convey the milk to the bases of hair follicles in the areola. The ducts at this stage of lactation are of narrow bore and are quite unlike the swollen cisterns in the mature areola (Fig. 97).

The first thing the newly hatched has to do after slipping out of the egg is to find the areola in the relatively enormous pouch* (Fig. 98). It can do this even before the fetal membranes dry off since even at this early stage the stomach of one hatchling was found to be engorged with milk (Fig. 80) which was visible through the transparent abdomen. This hatchling was fixed and serially sectioned; the milk in the stomach was found to have the same appearance when stained as a smear of milk, taken from the mother, did when treated in the same way.

Just how the newly hatched finds its way to the areola appears to be a problem similar to the one of how the newborn marsupial finds its way, unaided by the mother, from urogenital opening to a teat inside a pouch. One hundred and seventy-two years ago Barton (1806) described the Herculean task facing the tiny larvae in these elegant terms: "The young opossums, unformed and perfectly

---

*Owen (1865) had it that the young lived in one of two slitlike pockets on the ventral surface. The existence of these mammary or marsupial pockets (Mammartaschen, p. 129) and their putative function were accepted without question for the next 40 years; their acceptance gave rise to various theories of the phylogeny of the metatherian marsupial and mammary glands. The fact was (see Bresslau, 1920) that an incision 5 inches long had been made in the belly of Owen's echidna and when it was preserved the consequent shrinkage distorted the edges of the cut, removed all trace of the pouch, and formed two small pockets over the areolae. When Owen received this specimen he had his artist draw it not as it was but as he thought it ought to be, so marsupial pockets came into existence and persisted until Bresslau pronounced them extinct with the comment that the episode "forms, as it seems to me, an interesting chapter in the history of scientific blunders." Unfortunately Van Deusen in 1971, apparently ignorant of Bresslau's book, reproduced the illustration and presented it as fact.

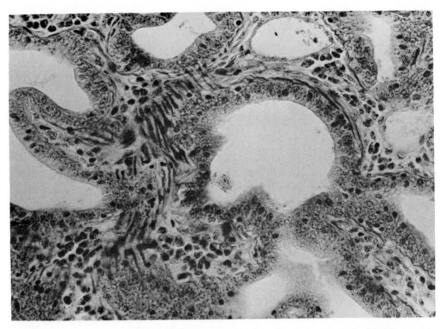

Figure 96.   *Tachyglossus.* Section showing condition of mammary gland just after hatching; quasi-alveoli are thick walled and lined by a cuboidal epithelium; the elongated bipolar elements are myoepithelial cells. Heidenhain's iron hematoxylin. × 285.

sightless as they are at this period, *find* their way to the teats by the power of an invariable, a *determinate* instinct, which may, surely, be considered as one of the most wonderful that is furnished to us by the science of natural history.'' In a footnote he adds ''it is not true, as has often been asserted, that the mother *puts* the young ones into the pouch.'' Once in the pouch it is likely that olfaction plays a part in locating a teat since this is the one special sense organ differentiated, and apparently functional, that the newborn has (Hill and Hill, 1955). Likewise the olfactory organs of the newly hatched are well developed (Griffiths *et al.*, 1969). Seydel (1899) found that cavum nasum of the newly hatched (Semon's stage 46) was very large (Fig. 99) but he gives no indication of the state of differentiation of the olfactory epithelium nor of olfactory nerves and bulbs. All three kinds of tissue are well developed and apparently functional since olfactory sense cells are present in the epithelium (Fig. 100) and fibers from them can be seen entering the olfactory bulbs (Fig. 101) which exhibit differentiated neurones unlike the relatively undifferentiated cells found in many other parts of the nervous system. Whether or not sense of smell plays a part in location of the areola is not known but the presence of anatomically differentiated olfactory organs suggests that it does.

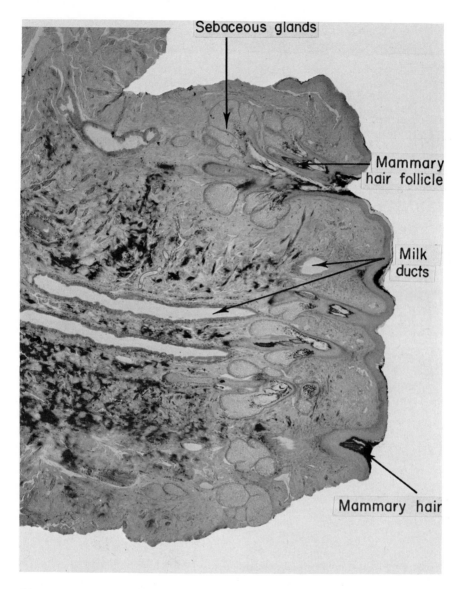

Figure 97. *Tachyglossus.* Section of portion of areola taken from a female just after hatching of her egg. Heidenhain's iron hematoxylin. × 30. *(From Griffiths et al., 1969; with permission of The Zoological Society of London.)*

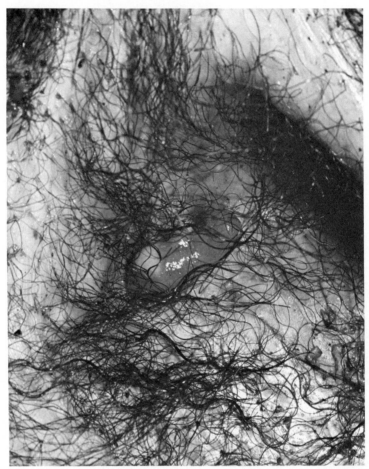

Figure 98. *Tachyglossus.* Newly hatched in pouch. Note milk visible in stomach through the transparent abdominal wall. *(From Griffiths et al., 1969; with permission of The Zoological Society of London.)*

The tongue of the marsupial neonatus bears a wide longitudinal groove on its dorsal surface for the reception of the long teat and the tongue itself is very deep dorsoventrally due to the presence of vertically arranged longitudinal fibers of striated muscle. By the action of these muscles they suck milk from the teat; they can even suck fluid uphill from a glass pipette (M. Griffiths, J. C. Merchant, and E. Slater, hitherto unpublished). However, the tongue of the newly hatched proved to be quite unlike this and, indeed, quite unlike that of the adult echidna. Overall length was 2.34 mm; it was 0.96 mm wide at the base, 0.40 mm wide three-

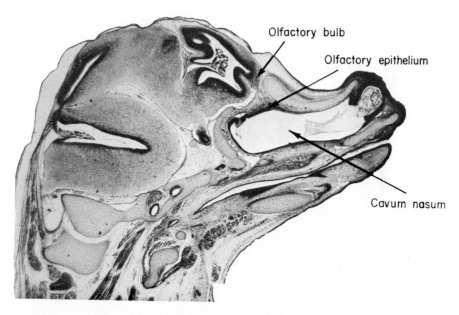

Olfactory bulb

Olfactory epithelium

Cavum nasum

Figure 99. *Tachyglossus*. Newly hatched; parasagittal section of head. Heidenhain's iron hematoxylin. × 20. *(From Griffiths et al., 1969; with permission of The Zoological Society of London.)*

quarters of the way along the length, and 0.34 mm thick, i.e., in this region it was rounded in cross section but flat at the posterior end. Faint indications of the lingual grinding pad were present on the posterodorsal surface. Circular muscles were not detected but the core of the tongue consisted of a peculiar median sheet of transversely arranged striated muscle fibers. This core can be seen cut in longitudinal section in Figs. 102 and 103. The anlagen of paired sternoglossal muscles lie externally along the central core and, opposed in action to the sternoglossi, are paired genioglossal muscles. Altogether the tongue gives the impression that it cannot be extruded and that therefore it is not used for licking up milk from the areola. There are, however, at least two ways that one could conceive of this tongue being involved in sucking up milk from the areola: it can be made thicker and thinner alternately by contraction and relaxation of the transversely arranged fibers of the median core, i.e., the dorsal surface of the tongue could be raised and lowered in the buccal cavity much as it is in the marsupial neonatus by the contraction of the vertical muscles. Griffiths *et al.*

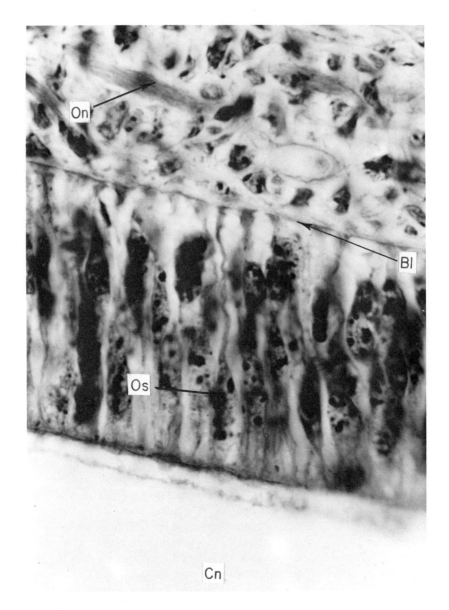

Figure 100. *Tachyglossus.* Newly hatched; portion of the olfactory epithelium in section. × 1480. On, fibers of olfactory nerve; Bl, basal layer; Cn, cavum nasum; Os, olfactory sense cell with neurofibril. *(From Griffiths et al., 1969; with permission of The Zoological Society of London.)*

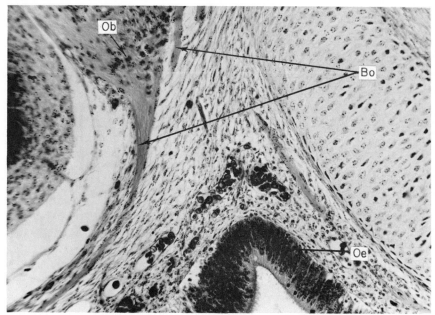

Figure 101. *Tachyglossus.* Newly hatched; longitudinal section through the olfactory bulb and branches of the olfactory nerves. × 225. Ob, olfactory bulb; Bo, branches of olfactory nerve; Oe, olfactory epithelium. *(From Griffiths et al., 1969; with permission of The Zoological Society of London.)*

(1969) suggest another mechanism of deformation of the tongue involving the combined contraction of median core musculature and of the sternoglossi to form a short rounded tongue at the posterior half of the buccal cavity; the sudden relaxation of this musculature could give rise to a flattened tongue and to a partial vacuum in the buccal cavity. Older pouch young certainly suck (p. 298) and do not lick milk from the areolae as is frequently stated in the literature.

The imbibed milk passes into a stomach lined not with squamous stratified epithelium as in the adult but with a columnar epithelium (Oppel, 1896) which persists until the suckling is at least 850 g in weight (Griffiths, 1965b). No signs of secretory activity are detectable in the cells of the epithelium in the hatchling nor at later stages of growth. Krause (1972) has described the ultrastructure of this epithelium in two pouch young, ca. 60 g in weight, imbibing mature milk. The esophageal stratified squamous epithelium changes abruptly to the tall columnar epithelium of the stomach. The cells of the latter exhibit short microvilli on the apical plasma membrane while the basal cell surface rests on a delicate basement membrane. Most of the cytoplasm contains numerous large mitochondria, indicating an active metabolic role, which often lie close to lipid droplets; a

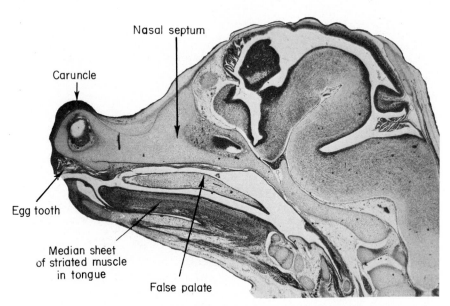

Figure 102. *Tachyglossus.* Newly hatched, median longitudinal section through the head. × 20. *(From Griffiths et al., 1969; with permission of The Zoological Society of London.)*

few Golgi profiles and only scant amounts of endoplasmic reticulum are present. This simple nonsecretory cell is thought by Krause to be involved in absorption of lipid.

   In contrast to the stomach the small and large intestines are furnished with villi (Griffiths *et al.*, 1969) and at the bases of these villi are rudimentary crypts of Lieberkühn equipped with granular Paneth cells that are quite unlike the vacuolated cells clothing the distal parts of the villus (Fig. 104). Krause (1972) by analogy with what he found in the intestines of 60-g pouch young concludes that the vacuoles contain glycoprotein. These vacuoles are characteristic of some suckling mammals and are thought to be of importance in intracellular breakdown of absorbed materials. Griffiths *et al.* (1969) found that the pancreas of the hatchling was differentiated into acini and Keibel (1904b) found that the pancreas communicates via the ductus pancreaticus et choledochus with the small intestine at Semon's stage 45a, i.e., just before hatching. It might be inferred from the above that the gut of the newly hatched is capable of digestion of substrates and of absorption of the products of digestion. It remains to be seen whether or not

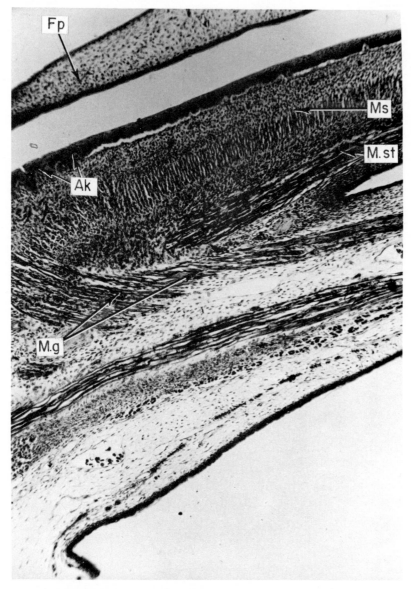

Figure 103. *Tachyglossus*. Newly hatched; parasagittal section through the base of the tongue. ×
115. Fp, false palate; Ms, median sheet of striated muscle; M.st, M. sternoglossus; M.g, M.
genioglossus; Ak, anlage of keratinized grinding pad. *(From Griffiths et al., 1969; with permission
of The Zoological Society of London.)*

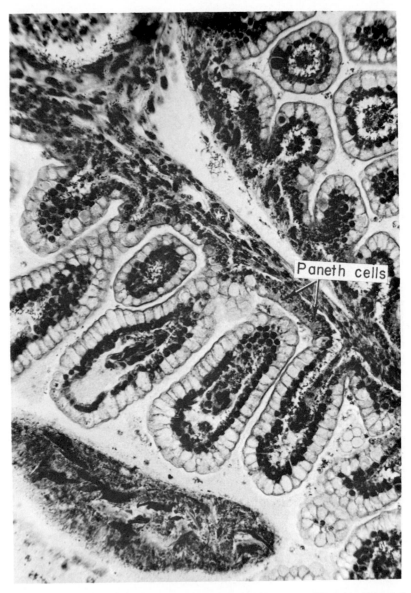

Figure 104. *Tachyglossus.* Newly hatched; section through portion of small intestine. Heidenhain's iron hematoxylin. × 282. *(From Griffiths et al., 1969; with permission of The Zoological Society of London.)*

immune globulins pass the gut epithelium unchanged, but it is possible that monotremes get their passive immunity from the mother via the egg.

Composition of the Milk

FATTY ACID COMPLEMENT OF THE TRIGLYCERIDES OF MILK TAKEN AT HATCHING AND AT LATER STAGES OF GROWTH OF THE YOUNG. The fatty acids of the milk fat have been studied by Griffiths *et al.* (1969, 1973). Samples of milk were taken by squeezing the glands after injection of oxytocin and sucking up the milk as it pours out at the areolae in the pouch (Fig. 105). Very little milk is obtainable without injection of oxytocin and only small amounts can be collected from glands supporting very small pouch young even with the help of oxytocin.

At hatching the milk is yellowish in color and in the one sample ever obtained (Griffiths *et al.*, 1969) total solids were in a concentration of 11.95% w/w, of which crude lipid accounted for 1.25%. Samples of mature milk are pink in color due to the presence of large amounts of the iron-binding protein lactotransferrin (p. 296). The figures for total solids in mature milk range from 44–53% w/w and crude lipid ranges from 14.8–35% w/w (Griffiths, 1968; Griffiths *et al.*, 1969). The data in Table 35 indicate that the fatty acid complement of that lipid varies with the age of the pouch young: early milks have much less oleic acid ($C_{18:1}$) and far more arachidonic acid ($C_{20:4}$) than the mature milks do. These fatty acid compositions shown in Table 35 cannot be said to be unusual in any way since those of different species of eutherians show fantastic differences (Glass *et al.*, 1967; Glass and Jenness, 1971) ranging from fat with no short chain acids (C_6, C_8, C_{10}, C_{12}) to fat in which a very large proportion is made up of those fatty acids. However, as in eutherians and woman (Insull *et al.*, 1959), the fatty acid complement of the milk fat in *Tachyglossus* can be changed by changing the fatty acids in the diet. Samples of milk taken from females immediately after capture in the bush exhibit an average of over 63% oleic acid and only 18% palmitic acid (C_{16}) (Griffiths *et al.*, 1973), whereas "captive" milks have 45% oleic acid and 25% palmitic. This difference in all probability is due to differences in diet since it was found that the fatty acid composition of the milk fat could be changed in captive echidnas by changing the composition of the lipid portion of the diet, especially if termites supplemented with oleic acid were fed (Table 36). It is concluded that the fat of the ants and termites ingested by wild echidnas contains a large amount of oleic acid leading to a milk fat rich in oleic acid.

The fatty acid complements of the triglycerides of the milks obtained from the male echidnas (which were fed the same diet as the captive females) mentioned above were of two kinds: in the pooled sample, which was yellow in color, the fatty acid complement was remarkably like that of the early milks from normal captive females including even a high proportion of arachidonic acid. The milk

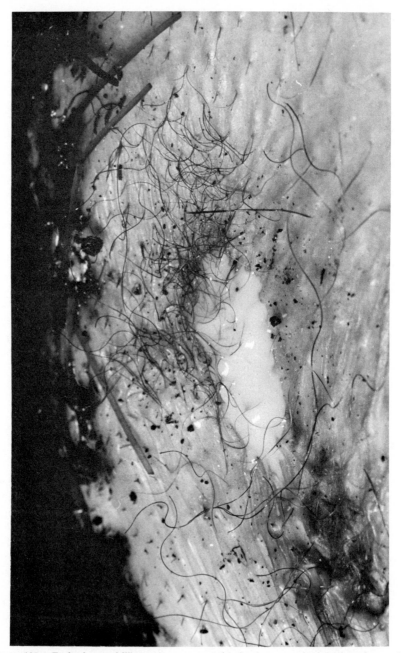

Figure 105. *Tachyglossus*. Milk pouring out at areola after intramuscular injection of oxytocin. Note mammary hairs in areola. × 4.1. *(From Griffiths, 1968.)*

TABLE 35

Fatty Acid Complement (g/100 g methyl esters) of the Triglycerides of the Milk of Echidnas Suckling Young of Different Ages

No. of days after hatching milk sampled and no. of animals	Diet of the mother offered daily, water ad lib.	Ratio of $C_{16}/C_{18:1}$ in diet	Saturated							Unsaturated								
			C_{12}	C_{14}	C_{15}	C_{16}	C_{17}	C_{18}	C_{20}	$C_{14:1}$	$C_{16:1}$	$C_{17:1}$	$C_{18:1}$	$C_{18:2}$	$C_{18:3}$	$C_{20:1}$	$C_{20:2}$	$C_{20:4}$
1 day, 1			—	0.8	—	28.9	0.5	7.2	0.9	0.2	6.5	0.5	32.1	7.3	0.1	3.5	2.7	5.0
12 days, 2			2.3	2.4	T	26.6	T	12.1	T	1.7	4.6	—	25.8	2.4	3.6	2.2	0.9	11.1
19 days 3	Custard of 1 egg, 100 ml cow's milk		—	0.8	0.2	26.0	1.9	7.1	—	0.4	8.6	1.0	38.0	9.0	1.8	1.4	1.9	1.2
31 days	5 g glucose, 25 g termites	24:41	—	1.5	0.5	24.0	1.7	7.1	—	1.4	7.8	1.2	42.6	11.3	0.3	0.2	—	T
Unknown but females suckling young 39–630 g in weight avg. of 9 samples from 8 females			0.3	2.9	1.1	25.2	0.7	8.0	0.4	0.1	4.1	—	45.8	7.9	0.6	0.1	T	T

TABLE 36

Effect of Change in Oleic Acid Content of the Diet on Oleic Acid Content of the Triglycerides of the Milk in Two Lactating *Tachyglossus*

Animal and place of origin	Weight of pouch young (g)	Diet offered daily	Oleic acid in lipid fraction of diet (%)	Oleic acid in triglyceride fraction of the milk (%)
(1) Rankins' Springs, New South Wales, October 12, 1971	150	Custard (1 egg, 100 ml cow's milk, 5 g glucose, 25 g termites)	40.4	43.9
	317	300 g termites	49.9	52.8
	700	250–300 g termites plus 12.5–15.0 g oleic acid	65.3	66.4
(2) Rankins' Springs, New South Wales August 23, 1973	93	Custard (1 egg, 100 ml cow's milk, 5 g glucose, 50 g termites)	42.7	42.6
	250	Custard (1 egg, 100 ml cow's milk, 15 g oleic acid, 100 g termites)	58.7	55.7
	594	Custard (1 egg, 100 ml cow's milk, 5 g glucose, 5 g palmitic acid)	33.7	37.8

from the male treated witb reserpine as well as ovarian hormones was pink in color like the mature milks of normal females and had a fatty acid complement like those milks (Table 37).

It is a matter of interest that the milk fat of *Tachyglossus* living in the bush is quite unlike that of its near relative *Ornithorhynchus*: the latter has no great preponderance of oleic acid over palmitic acid and it exhibits seven unsaturated and polyunsaturated acids not found in *Tachyglossus* milk fat (Griffiths *et al.*, 1973). Possibly some of this difference is the result of difference in diet but this is not the whole answer; *Tachyglossus* milk can change during lactation although diet remains the same. Echidnas are not unique in this respect, similar but far more marked changes occur during the lactation of the red kangaroo, *Megaleia rufa*: milk for the neonatus contains 51% palmitic acid and 15.6% oleic but these proportions gradually change with growth of the pouch young until mature milks exhibit ca. 25% palmitic and 53% oleic acid (Griffiths *et al.*, 1972). Not only does this happen but kangaroos can make two kinds of milk simultaneously: milk for a neonatus in one gland and mature milk in another for a young at heel— obviously diet has nothing to do with those differences. This suggests that the difference of platypus milk fat from that of *Tachyglossus* could be genetic rather than the result of difference in diet. Whatever the mechanisms involved the fact remains that the anatomically similar young of the monotremes are raised on very different milks. The reason why this is so, and indeed the reasons why eutherian and marsupial young of different species are raised on vastly different milks, are not known.

CARBOHYDRATES. It has been reported (Marston, 1926) that a sample of echidna milk had the equivalent of 2.81 g lactose %, and that a lactosazone was prepared from the sample. Subsequently Kerry and Messer (cited in Griffiths, 1968) analyzed another sample and found a reducing value equivalent to 2.8 g lactose per cent—a figure remarkably close to Marston's, however, they could not detect lactose using a chromatographic method for its identification. Subsequently Messer and Kerry (1973) and Messer (1974) reported that the principal free carbohydrate in three samples of *Tachyglossus* milk (one sample from a female suckling, a young one 300 g in weight; two samples from another when its young one was 150 g and later, 312 g) is sialyl-lactose* and it accounts for 50% of the total carbohydrate. Other components are fucosyl-lactose (29%), difucosyl-lactose (13%), and lactose (8%). However the total carbohydrate content (average of the three samples) was very low—1 g/100 g milk. It would appear from this figure and that for the platypus that carbohydrate is of little importance as a source of energy to monotreme sucklings, at least during the later stages of lactation; it may be present in larger quantities in early milks which contain less fat. Messer and Kerry suggest that the high content of fucosyl

*Identified as *N*-acetyl-4-0-acetylneuraminyl-lactose.

TABLE 37

Main Component Fatty Acids (g/100 g methyl esters) of Milk Triglycerides from Male Echidnas[a] Treated with Estradiol plus Progesterone or with Estradiol plus Progesterone plus Reserpine

Treatment	Saturated								Unsaturated					
	C_{14}	C_{15}	C_{16}	C_{17}	C_{18}	C_{19}	C_{20}	$C_{16:1}$	$C_{18:1}$	$C_{18:2}$	$C_{20:1}$	$C_{20:2}$	$C_{20:4}$	$C_{22:1}$
Estradiol plus progesterone implanted	0.8	0.6	30.2	0.2	9.3	—	4.2	5.5	30.0	5.4	2.8	2.2	8.1	—
Estradiol plus progesterone implanted followed by 7 × 5 mg daily injections reserpine	0.5	—	28.2	1.0	4.5	0.9	1.5	6.2	44.8	7.9	1.8	—	—	1.1

[a] These animals were fed the diet listed in Table 35.

oligosaccharides, which is a characteristic of monotreme milk, may be essential for the growth and development of the sucklings. They also suggest that the lactosazone obtained by Marston from the milk was probably formed by hydrolysis of the relatively labile fucosyl-lactose and sialyl-lactose bonds during preparation of the lactosazone.†

Proteins

The source of the lactose for synthesis of the fucosyl-lactoses and sialyl-lactose is of considerable interest since the lactose synthetase system in echidna milk appears to be different from that of eutherians and metatherians (Hopper and McKenzie, 1974). Watkins and Hassid (1962) showed that particulate fractions of mammary gland tissue from cows and from guinea pigs catalyzed the reaction:

$$\text{UDP-galactose + glucose} \xrightarrow{\text{Mn(II)}} \text{lactose + UDP}$$

Ebner *et al.* (1966) later showed that the enzyme system consists of two proteins, one of which is α-lactalbumin, inactive by itself, and the other is a galactosyl transferase known as protein A. Both those proteins occur in the wheys of eutherian and marsupial milks from which they can be isolated (Hopper and McKenzie, 1974). Closely related to α-lactalbumin is another protein in milk, lysozyme, which catalyzes the cleavage of β (1–4) glycosidic linkages but lysozymes from eutherian milks have no lactase synthetase activity and α-lactalbumins have no lysozyme activity.

The wheys of echidna and platypus milks exhibit lactose synthetase activity but it is low relative to those of cow, woman, and red kangaroo (Table 38); the whey of echidna milk, however, has very high lysozyme activity (Table 39). The lysozyme was separated from other whey proteins that were tested for lactose synthetase activity and were found to have none. This apparent lack of α-lactalbumin prompted Hopper and McKenzie to examine the lysozyme fractions, along with protein A, for its ability to synthesize lactose; it proved to be active, i.e., the lysozyme behaved like an α-lactalbumin.

This matter, however, is far from resolved since the lysozyme prepared from the whey of milk sample A1 (Table 39) although present in large concentration, had no synthetase activity; nevertheless some lactose synthetase activity was detected during fractionation of the whey; Hopper and McKenzie were unable to identify the protein fraction responsible for that activity but doubtless when further samples of milk become available the problem will be settled.

Griffiths (1968) noted that the whey of echidna milk had the color of a strong solution of oxidized cytochrome C, and found that it had an absorption maximum

---

†Marston's sample was not a good one since it was obtained from a recently-killed female whose pouch young had been removed 14 days previously. It is quite possible that that milk contained free lactose formed by hydrolysis of fucosyl-lactose in what must have been a regressing gland.

TABLE 38

Lactose Synthetase Activity in the Milks of Different Species[a]

Species	Sample vol. (ml)	Time of milking days postpartum or stage of lactation	Lactose synthetase activity (cpm)
Cow	0.10	Late	2262
Woman	0.10	ca. 150	2285
Red kangaroo	0.05	200	2700
Tachyglossus	0.05	70–80	266
Platypus	0.05	Late	977

[a] From Hopper and McKenzie (1974). Reproduced by permission of *Molecular and Cellular Biochemistry*.

at wavelength 460 mμ identical to that of human serum transferrin (Roberts *et al.*, 1966). Jordan and Morgan (1969) noted that the color is intensified by addition of iron to the whey. The total iron binding capacity of whey was found to be within the range 24.1–36.2 μg/ml and that it was much higher than that of blood serum, 3.3–5.0 μg/ml. Three iron-binding components were found in the whey but only one in the serum as shown by starch gel electrophoresis and

TABLE 39

Lysozyme Activity of the Milks of Different Species[a]

Species		Sample vol. ml	Time of milking days postpartum or stage of lactation	$\dfrac{\Delta T^b}{\Delta t}$	Approx. conc. lysozyme in milk (mg/dl)
Cow		0.10	Late	0.38	2
Pig		0.10	Mid	0.03	$\ll 1$
Horse		0.10	Mid	4.20	24
Megaleia rufa		0.10	200	0.30	<2
Tachyglossus	M4	0.05	30	2.80	15
	M2	0.05	40	1.00	6
	M3	0.05	60	5.60	33
	M1 + M5	0.10	70–80	14.60	83
	A1	0.01	70–80	73	ca. 450
Platypus	1	0.10	Late	0.40	<2
	2	0.10	Mid	0.40	<2

[a] From Hopper and McKenzie (1974). Reproduced by permission of *Molecular and Cellular Biochemistry*.

[b] Transmittance (T) versus time (t). The activity is expressed in terms of the initial slope ($\Delta T/\Delta t$) of the plot of T versus t where T is the decrease in turbidity of *Micrococcus lysodeikticus*.

autoradiography from which Jordan and Morgan conclude that the lactotransferrin is synthesized in the mammary gland. The iron content of the milk is discussed below.

Gel filtration and immunoelectrophoresis of serum and whey showed that the proteins in the whey were of serum type and also milk specific. At least three proteins in the latter group exhibited the electrophoretic mobility of α globulin and to these must be added the milk-specific lysozyme described by Hopper and McKenzie. The serum-type proteins in the whey corresponded to albumin, β_1, and γ-globulin fraction of serum (see p. 110).

It has been reported (McKenzie and Murphy, cited in Griffiths, 1968) that the total protein content of a sample of milk from a captive female suckling, 386 g young, was 7.3 g/100 ml and that 60% of the protein is casein; the samples examined by Jordan and Morgan were richer in protein, 11.3 to 13.0 g/100 ml.

IRON CONTENT OF THE MILK. The milks of eutherians have very little iron (those of women and cows, for example, have only about 0.5 μg/ml). However newborn eutherians have enough iron in their livers to tide them over until they can ingest iron in their definitive diets. The neonatus of the marsupial and the newly hatched echidna, however, are minute and their tiny livers cannot possibly store iron enough to tide them over a prolonged period of life in a pouch where they have to live on milk alone. Consequently their milks are rich in iron: Kaldor and Ezekiel (1962) found that the iron content of the milk of *Setonix brachyurus* ranges from 8.0–32.0 μg/ml during the lactation period; a sample of echidna milk taken at hatching contained 8.6 μg iron/ml and nine samples of mature milk from five echidnas had an average content of 37.1 μg/ml (range, 23.7–47.8).*

H. A. McKenzie and B. Treacey (personal communication) found in one of the above samples that iron content of the micelles was 9.5 μg/g and of the whey 32.8 μg/g, adding up to 42.3; the total iron found by analysis was 41.5 μg/g. The blood serum iron of the female at the time of milking was 12.5 μg/ml.

It will be of great interest to compare these figures with those for platypus milk when they become available. It is quite possible they will be relatively low since young platypuses do not live in a pouch so it is quite possible they could get iron from the dirt of their burrow much as young piglets get their iron from the filth of their sty; if young piglets are raised hygienically they get anemia.

Suckling, Growth of the Young, and Duration of Lactation

Contrary to the widespread belief entrenched in the literature, *Tachyglossus* does not lick up its milk from the areola, but it sucks it in like any other mammal

*My best thanks go to Barbara Treacey of the Australian National University and to Imre Kaldor of the University of Western Australia for the iron analyses.

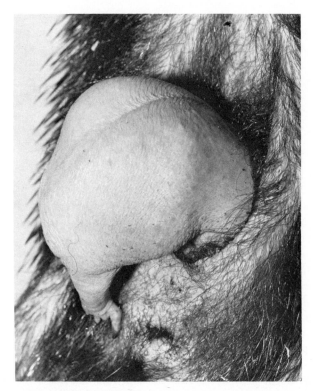

Figure 106. *Tachyglossus*. Position adopted for suckling when the young gets too big to enter the pouch. *(From Griffiths, 1965b.)*

does. The sucking action is vigorous and involves movement of practically the whole body which contributes to a sort of butting action against the areola by the snout; the whole performance reminds one of the vigorous sucking action of piglets (see CSIRO, 1974). Young are carried in the pouch until they are ca. 55 days of age (Table 40) but after they are ejected or dropped from the pouch the mothers continue to suckle them in the following way: the female approaches her usually sleeping young one and wakes it by gently nudging it under her body with snout and forelimb until it lies under her belly. The young then rolls onto its back, hangs onto the belly hair of the mother with its powerful forelimbs, and thrusts its snout into the pouch.* Let-down follows and the young sucks its milk

*When the young is dropped permanently from the pouch the latter gradually regresses but suckling continues, as described above, the young hanging onto the mother's belly hair.

from the areolae. Figure 106 shows how the young hangs on upside down with its head in the pouch; this young, however, was not being suckled at that particular time.

The noise of sucking is audible and sonagram sound spectra show that it reaches a pitch of 6 kHz (Griffiths, 1965b). The snout of the suckling (Fig. 107) is like that of the adult, i.e., it is elongated and flattened on the undersurface at the anterior end where the mouth is situated. This flattened intake surface is well suited for sucking up milk from the flattened but protrubent nipplelike areola. The data in Fig. 108 show how amazingly effective these arrangements are for ingesting milk; the intakes shown amount to ca. 10% of the body weight, enough to last the suckling for a day or two. These amounts are ingested in a matter of 20–30 minutes but the regularity shown in the graph is an artifact; we induced the female to feed the young (no longer in the pouch) at 24-hour intervals for another purpose. The data simply show how much an echidna can deliver and how much a young one can imbibe in a matter of minutes; they also show that echidnas apparently have a very efficient let-down mechanism which is in agreement with the fact that the mammary glands have myoepithelium responsive to exogenous oxytocin (p. 257). Using changes in body weight as a guide it is apparent that echidna young when left to their own devices may not take in milk for 3 or 4 days at a time (Fig. 109).

The echidna young is carried in the pouch until it starts to grow spines when, understandably, the mother has to eject her progeny. Well-authenticated accounts (Semon, 1894a; Barrett, 1942; Hodge and Wakefield, 1959; Frauca, 1969) show that the young is dropped in a burrow but it is not known whether of her own digging or of another animal; as we have seen (p. 136) the young

Figure 107. *Tachyglossus.* Suckling 300 g in weight ca. 55 days old and still a pouch young but not for long; note spines poking through the skin. *(From Griffiths, 1972.)*

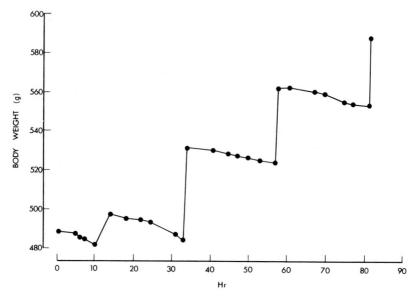

Figure 108. *Tachyglossus*. Weights of milk ingested at four consecutive sucklings. *(From Griffiths, 1965a.)*

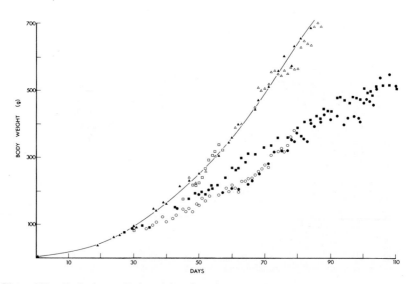

Figure 109. *Tachyglossus*. Body weights of pouch young and sucklings as a function of time. The continuous line is the one curve obtained from hatching onwards. See text for origins of other curves.

TABLE 40

Tachyglossus **Body Weights and Ages of Pouch Young at Time of Ejection from the Pouch**

| | | Young raised in captivity | | Young found in pouches of females in the bush |
| | BW of mother (g) | BW of pouch young at time of ejection from pouch (g) | Duration of pouch life estimated (E) by extrapolation of growth curves or known (K) (days) | BW of pouch young (g) |
Subspecies				
Aculeatus	4300	400	62 (E)	
	3950	230	50 (E)	
	3500	232	63 (E)	
	4800	410	63 (K)	
Multiaculeatus	3000	170	52 (E)	
	3500	180	48 (E)	
	2762			266
	3500			173
	3400			191
	3460			173
	3200			215
	Not recorded			175

itself is quite capable of burrowing and of extending a burrow. It is assumed that the female visits the burrow periodically to suckle her young but no one has kept records of this other than to note that the young remain fat and well for weeks on end. Just how long the female looks after the young in this way is not known but observations on the laboratory give an idea of what happens: Fig. 109 shows growth rates of six pouch young suckled by females held in captivity. Only one of the curves starts at hatching; the other five are growth rates of pouch young whose mothers were brought in from the bush, and the weight of the pouch young at capture has been superimposed on the one curve starting at zero time.

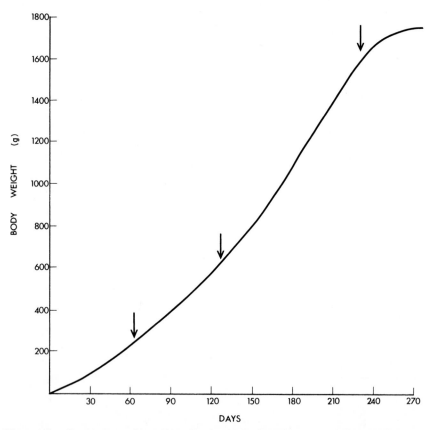

Figure 110.   Growth curve of an echidna 90 g in weight when first observed; the curve has been extrapolated back to zero to give an approximate age. First arrow indicates time of leaving pouch, second indicates when attempts were first made to induce the young to eat a diet containing termites, the third indicates when the mother was finally removed and the young was completely weaned onto a diet of termites alone.

From this it has been possible to get a rough idea of the age of the five pouch young by extrapolation of their growth curves back to zero time. From the data in Table 40 and Fig. 109 it is apparent that young are dropped from the pouch at vastly different sizes but their ages do not differ so very much, 48–63 days, a result of different growth rates. These differences are particularly marked from 50 days onward so that attempts to age young weighing more than 300 g in weight could be grossly inaccurate. Also in Table 40 it can be seen that females in the bush carry young in their pouches as large as some of those dropped permanently by their mothers living in the laboratory, but the one weighing 266 g fell out of the pouch when the mother was released into the bush. The young then crawled off for about 3 m, the female followed, went to the young whereupon it

TABLE 41

Ratio of Body Weight to Body Length in Echidna Pouch Young Taken in the Bush and in Young of Females Living in Captivity

Bush (14 pouch young)			Captivity (10 pouch young)		
Subspecies	BW (g)/BL (cm)		Subspecies	BW (g)/BL (cm)	
Multiaculeatus	40/ 9.1	4.4	Aculeatus	0.38/ 1.47	0.26
	72/10.5	6.8		34/ 8.4	4.0
Aculeatus	87/10.5	8.3	Multiaculeatus	59/10.5	5.6
	90/12.0	7.5	Aculeatus	64/10.5	6.1
Multiaculeatus	100/11.5	8.7	Multiaculeatus	97/12.5	7.8
	100/11.5	8.7	Aculeatus	99/11.8	8.4
	120/13.5	9.0		147/13.2	11.1
	135/13.0	10.4	Multiaculeatus	150/12.0	12.5
	152/13.5	11.2	Aculeatus	150/14.5	10.2
	153/10.5		Multiaculeatus	154/14.5	10.6
	175/14.5	12.0		174/13.5	12.9
	175/14.5	12.0	Aculeatus	180/14.0	12.8
	210/15.5	13.6	Multiaculeatus	194/16.5	11.7
	266/18.0	15.0	Aculeatus	205/16.0	12.8
Aculeatus	710/27.0	26.0	Multiaculeatus	213/17.0	12.5
Multiaculeatus	810/31.5	25.7	Aculeatus	214/16.0	13.4
Aculeatus	900/29.0	30.0	Multiaculeatus	215/15.5	14.0
Various, avg. of	1690/33.0	51.0		225/15.0	15.0
14 specimens			Aculeatus	241/16.0	15.0
(range BW			Multiaculeatus	250/16.0	15.6
1100–2300 g			Aculeatus	308/19.3	16.0
BL 32.0–38.0)				448/21.5	20.8
			Multiaculeatus	525/23.0	22.0
			Aculeatus	700/25.5	27.0
			Aculeatus	1700/34.0	51.0

turned over on its back, the female stood over its young one who entered the pouch (see Fig. 106 for illustration of this activity). She then carried it off to a nearby pile of logs. It would seem that this young was close to leaving the pouch for good.

Suckling continues for a long time after the young is dropped (Fig. 109); in fact the one baby echidna we managed to raise was still being suckled at ca. 7 months of age (Fig. 110). We finally taught this echidna to eat termites and was released into the bush at 9 months of age. This long period of suckling is possibly an artifact of laboratory life but one lonely observation, made early in the month of March, of a very small echidna in the company of a very large one (both evaded capture) is a faint indication that a young stays with its mother ca. 6 months.

It might also be suggested that the rates of increase in body weight shown in Fig. 109 are also artifacts of laboratory life since the young are nurtured on milk from sedentary and possibly overfed females. However, if one compares the ratios of body weight to body length in pouch young caught in the bush with those of young being raised in captivity (Table 41) it can be seen that they are practically the same in both groups; the progeny of the captive females, therefore, are not obese and are apparently normal.

Comparison of the Mammary Glands and Lactation of the Prototheria, Metatheria, and Eutheria

In the lactating platypus and echidna the lobular structure is quite apparent beneath the thin connective tissue investment; exactly the same structure is found in the mammary glands of marsupials (Eggeling, 1905; Griffiths et al., 1973). Furthermore the latter authors have shown that the elongated lobules in the mammary glands of *Megaleia rufa* (Fig. 111) and of *Trichosurus vulpecula* are filled with thin-walled alveoli ultrastructurally similar to those of monotremes and eutherians. One difference of monotreme alveoli from those of metatherians is a tendency of the former to be larger than the latter: mean diameters of alveoli in various areas of platypus glands ranged from 0.073–0.081 mm, 0.071–0.118 in echidnas, 0.053–0.083 in a gland of *Trichosurus vulpecula* suckling, a 135-day old young, and in two mammary glands of *Megaleia rufa* suckling young 190 and 320 days old mean diameters of alveoli ranged from 0.053–0.083 mm, respectively. Thus the alveoli of monotremes may be considered large but average diameter of alveoli in some areas of marsupial glands exceeds the average diameter of monotreme alveoli.

It has already been pointed out that the monotreme areola, particularly that of *Tachyglossus*, is very like the human nipple, but just before eversion the marsupial teat closely resembles the tachyglossid areola. The teat anlage passes through a phase consisting of a plaque of epidermis and dermis exhibiting ducts, Knäueldrüsen (which persist in the adult), sebaceous glands, and mammary hairs

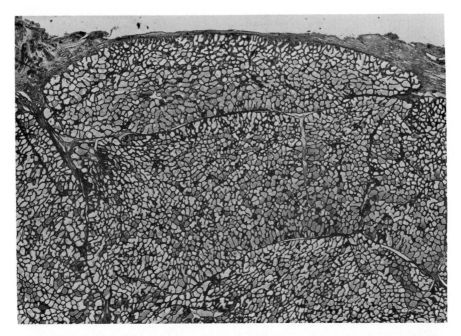

Figure 111. Mammary gland of *Megaleia rufa* supporting 190-day-old pouch young. Note lobules separated from each other by connective tissue sheaths as in the monotreme mammary gland. Heidenhain's iron hematoxylin. × 19. *(From Griffiths et al., 1973; with permission of The Zoological Society of London.)*

(Bresslau, 1920; Griffiths *et al.*, 1972); all very much as in the echidna areola.

The marked seasonal waxing and waning of the mammary glands is not unique to monotremes; such fluctuations occur, at the same time of the year, in the mammary glands of the European fox in Australia. The regressed glands in this species (in January) consist of ducts and small lobules embedded in fat and connective tissue. Within the lobules are solid cords or ducts which occasionally display a small lumen. Histologically these cords are very like those in the quiescent lobules in the platypus gland. In July the gland in the nonpregnant vixen exhibits an open duct system and swathes of mesenchymal elements invading the adipose tissue, forming new lobules. This process of formation of new mammary tissue in monotremes, fox, and marsupials is identical to that described for woman by Dabelow (1957).

The rapid and massive let-down of milk in the suckling echidna is paralleled by those of the European rabbit and the tree shrew *Tupaia belangeri*. Wild rabbits (*Oryctolagus cuniculus*) in Australia feed their young once a day and the suckling takes only ca. 10 minutes. Tree shrews suckle every 2 days and manage to give their young enough in 4–10 minutes to last until the next suckling in 2 days time (Martin, 1968). The milk of this species contains 40% w/w total solids (25.6% fat, 10.4% protein, 1.5% carbohydrate, 2.9% unidentified), a composition very like that of echidna milk. It would be of great interest to determine the fatty acid complement of *Tupaia* milk for comparison with that of *Tachyglossus*.

Finally it may be mentioned that the processes of regression of the monotreme glands are the same as those reported for the glands of domestic mammals (Morrill, 1938; Turner, 1952) and woman (Berka, 1911). In those glands some alveoli become distended with milk and thin walled, while others have thick walls and small lumina; various types of cells appear in the epithelium and lumina (lymphocytes, plasmacytes, and mast cells) and the infiltrations of these may persist after milk and many of the secretory cells bave been removed by their activity. When involution is complete the glands consist of a skeletal system of ducts embedded in a pad of fat, which is very like the condition of the mammary glands of monotremes in the nonbreeding season. From Berka's account it is apparent that the regression process in woman is heterogeneous as it is in

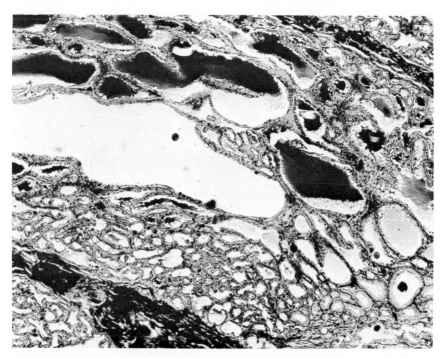

Figure 112. *Zaglossus.* Section of portion of the mammary gland of a male. Note tubular, almost alveolar grade of organization; the coagulation present in the ducts appears to be milk. Heidenhain's iron hematoxylin. × 55.

monotremes, i.e., not all lobules exhibit the same stage of regression at any one time.

ZAGLOSSUS

A normal lactating mammary gland has not yet been observed in this genus. The lobules in the gland of the adult female mentioned on p. 42 exhibited a duct system solid for the most part, but loosely arranged and some tubules had lumina; their myoepithelial investments were particularly prominent. Mesenchymal cells were migrating out of the connective tissue sheaths into the adipose tissue at the periphery of the gland. All this indicates that the gland was growing and new lobules were in process of formation as in the gland of *Tachyglossus*. There is reason, therefore, to believe that when a fully lactating gland comes to hand it will prove to be identical to that of *Tachyglossus*. The Fuyughé villagers mentioned previously told me that the young one is carried in a pouch.

There is reason to believe, however, that the fatty acid complement of the mature milk will prove to be different from that of *Tachyglossus*. We have seen that feeding lactating *Tachyglossus* diets low in oleic acid leads to a low concen-

TABLE 42

Total Fatty Acid Content (g/100 g methyl esters) of the Lipid Fraction of Earthworms from Kosipe Pass, Wharton Ranges, Papua and Mt. Gingera, Brindabella Range, Australian Capital Territory

			Saturated						
			Anteiso			Anteiso			
Place	C_{12}	C_{13}	C_{14}	C_{14}	C_{15}	C_{16}	C_{16}	C_{17}	C_{18}
Kosipe Pass (2300 m)	2.4	—	1.1	4.3	3.3	3.6	8.2	5.7	6.8
Mt. Gingera (1857 m)	1.6	1.9	1.2	4.2	4.1	3.9	7.0	5.8	7.4

				Unsaturated						
$C_{12:1}$	$C_{14:1}$	$C_{16:1}$	$C_{17:1}$	$C_{18:1}$	$C_{18:2}$	$C_{18:3}$	$C_{20:1}$	$C_{20:3}$	$C_{20:4}$	$C_{22:1}$
6.5	6.3	6.8	3.8	15.0	9.9	2.0	2.9	2.7	6.7	1.4
1.8	4.2	5.2	1.9	14.2	11.0	2.6	6.7	2.4	4.4	7.1

tration of oleic acid in the milk. It would appear that the diet of *Zaglossus* is relatively low in oleic acid since analysis of the fatty acid complement of earthworms collected at random in montane forests known to harbor *Zaglossus* revealed a low percentage of oleic acid (Table 42). The fatty acid complement of Australian montane earthworms, given in that table, was determined under standard laboratory conditions as a control for possible formation of artifacts since extraction of the Papuan earthworms in the chloroform-methanol mixture could not be carried out quickly nor could the product be stored cold continuously. The data in Table 42 show that the fatty acids in the two different collections of worms were reasonably similar so that one can place confidence in the analysis of Papuan earthworm fat. On the basis of the low oleic content of those earthworms relative to that of termites (Table 36) it is predicted that *Zaglossus* milk will prove to have a relatively low oleic acid content.

The lobules of the male mammary glands had a tubular grade of organization and one set exhibited quasi-alveoli (Fig. 112). This animal had very large testes (50 g each) in which spermatogenesis and spermiogenesis were in full swing.

Apparently nonpregnant female *Zaglossus* can also exhibit an alveolar mammary gland since Kolmer (1925) found this condition in the glands of his *Zaglossus* specimen that had been living an uneventful life on her own in a menagerie in Vienna for 10 years. The glands appeared to be abnormal, however, in that there was no pouch and the glands themselves were hard rounded bodies like those in the estrogen-treated *Tachyglossus* females described above.

10

The Affinities of the Monotremes

The relationship of the monotremes to the rest of the mammals continues to be a matter for debate and four widely differing views have been propounded recently.

VIEW 1

The most remarkable of these is that of Giles MacIntyre (1967). His views can be best described in his own words: "Most of the oddities of living monotremes are easily understandable, for example, if one thinks of them as living therapsid reptiles, nearly as remote from mammals as alligators are from birds." He proposes the term quasi-mammal to cover ictidosaurs, tritylodonts, eozostrodonts, triconodonts, multituberculates, and monotremes, and proposes a definition of the class mammalia which would exclude the "quasi-mammals." Even if one agreed with his classification the name is not felicitous; if one studies quasi-mammals one might be called a quasi-mammalogist.

In another part of his paper he poses the question "What is a mammal?" In the light of the facts presented by Griffiths (1968) and in this present work this question can easily be answered especially if one admits that a marsupial is a mammal. A mammal is an amniote with the following structures and physiological properties:

(a). Middle ear equipped with three ossicles: malleus, incus, and stapes. Organ of Corti in cochlea; cochlear microphonics smooth functions of sound pressure level.

(b). Dentary-squamosal articulation of jaw.

(c). Homeothermic; covered with hair, each hair consisting of a keratinized cuticle, cortex, and central medulla; the keratin has the α configuration; sweat glands in skin.

(d). Mammary glands exhibiting alveoli surrounded by a network of myoepithelium responsive to oxytocin. Milk has fat globules, casein granules and oligosaccharides containing hexoses. Composition of milk can be changed by change in diet. Growth and differentiation of mammary glands influenced by ovarian hormones. Milk imbibed by sucking.

(e). Posterior lobe of the pituitary releases oxytocin.

(f). Completely separate right and left ventricles in heart, the left ventricle pumping aerated blood to the body via the left aortic arch.

(g). Red blood cells non-nucleated discs, hemogloblin polymorphic.

(h). Kidneys discrete or compound reniform, not diffusely lobulate as in Sauropsida. Renal portal system not present, blood being supplied by a renal artery and drained away by a renal vein.

(i). Ureotelic, uricase in tissues.

(j). Alveolar lungs and diaphragm. Bronchial tree asymmetrical.

(k). Brunner's glands in gut.

(l). Circumvallate papillae in tongues.

(m). Medial wall of cavum epitericum membranous, portion of side wall and floor formed by alisphenoid (see below). Nasal capsules with maxillo- and ethmoturbinals. Stylohyals attached to cristae paroticae.

(n). Seven cervical vertebrae, first two modified to form atlas and axis, latter bears a dens.

(o). Pelvis with elongated anteriorly directed iliac blade, reduced pubis, large obturator foramen.

(p). The young, for the most part, are small, naked, and have to be raised on milk until able to fend for themselves. Young indulge in play.

(q). Brains exhibit enormous hypertrophy of the neopallium, the left and right halves of which are linked by anterior and hippocampal commissures. Caudal to anterior body parts represented progressively medially to ventrally on the cortex. Pons Varolii; roof of mesencephalon with corpora quadrigemina; large cerebellum with lateral lobes; cortico-spinal tracts.

The above is a list of some of the attributes of eutherians and metatherians; it also happens to be a list of some of the attributes of monotremes so let us retain them in the Class Mammalia.

VIEW 2

The second version of the affinities of the monotremes has been put forward by Kühne (1973, 1977). The evidence he brings forward confirms, he states, Gregory's (1947) conclusion that the monotremes have a close phylogenetic relationship to the marsupials and that this relationship is best expressed by grouping them in a taxon, the Marsupionta, the Class Mammalia being classified as follows:

Class Mammalia
Subclass Marsupionta

Order Monotremata
Order Marsupialia
Subclass Placentalia

In essence Gregory's thesis is that monotremes were derived from ancient marsupials and that they have retained many early marsupial characters but have become specialized for an amphibious or an anteating mode of life. There is a lot to be said for this concept and notwithstanding Kühne's statement that Gregory's paper has hardly been discussed recently, the characters common to marsupials and monotremes put forward by Gregory and many others were dealt with in some detail by Griffiths (1968). To the list of those characters Kühne has added another which he claims with remarkable logic, irrefutably establishes close phylogenetic relationship of marsupials to monotremes—the character in question is the milk dentition in *Ornithorhynchus* and marsupials. Of these he says "Reduction of tooth-replacement in marsupials is highly apomorphic. The tenor of the evolution of the mammalian dentition has been—among other features— the reduction of tooth replacement: the Marsupialia have exceeded the Placentalia, only one tooth in the upper and lower jaw being replaced. The picture given by Green of tooth replacement of *Ornithorhynchus* matches exactly the condition in marsupials. This is synapomorphy proven for the two taxa concerned and the end of the argument." However, the superstructure of his logic is based on tenuous anatomical evidence: the dental lamina in the suckling platypus is divided into an anterior portion in which rudiments of incisors and canines form, and a posterior giving rise to premolars and molars, some of which erupt. In the posterior part the teeth anlagen have been designated v, w, x, y, and z by Wilson and Hill (1907) and Green (1937). Wilson and Hill considered that a series of five epithelial nodules in the lamina represented a milk dentition. Simpson (1929) thought it probable that these are vestiges of dental papillae, but that the proposition is open to doubt. If they were dental papillae he conceded they may represent the vestiges of five deciduous teeth. Green (1937) concluded that the epithelial nodules "are detached and degenerating portions of what were at one time more prominent cusps. . . . In no sense can the epithelial bodies be considered as representing vestigial remains of a milk dentition." However Green did observe in the rostral part of the posterior portion of the dental lamina in five sucklings a nodule consisting of a mesodermal papilla overlain by a thin shell of dentine. These he designated $\underline{dv}$ and $\overline{dv}$ to indicate their positions in the upper and lower jaws, respectively. In specimen XXVIII B he found that "immediately in front of $\overline{dv}$ there is a definite indication of an aborted tooth rudiment"; this would correspond to the succeeding tooth $\overline{v}$. In an older specimen H. J., $\underline{dv}$ had disappeared but $\overline{dv}$ was present. In the place were $\underline{dv}$ had occurred the dental lamina had widened and become cupped on its deep aspect. Of this Green said "this I take to be the enamel organ of $\underline{v}$ which is later aborted." From these facts Green concluded "There is evidence of one milk tooth only in each jaw, this being in the premolar region." He gives the full dental formula of *Ornithorhyn-*

chus as i $\frac{0}{5}$ c $\frac{1}{1}$ pm $\frac{2}{3}$ m $\frac{3}{3}$ (v is pm 1). Kühne apparently accepts this as proof that there was a milk dentition in *Ornithorhynchus* and with a few strokes of his pen converts the monotreme to a marsupial by deeming the second premolar to be a molar and writes the dental formula pm $\frac{1}{1}$ m $\frac{4}{4}$, the premolar being the deciduous tooth.

It may be mentioned there is nothing novel about Kühne's thesis; Gregory in his 1947 paper described Green's work and concludes "There is evidence of one milk tooth in each jaw this being in the premolar region and recalling the condition in marsupials."

VIEW 3

A third, and widely accepted version of the affinities of the monotremes is based on D.M.S. Watson's (1916) opinion that the side walls of their brain cases are filled in, not by the alisphenoid bone as in Metatheria and Eutheria but by a rostral projection of the periotic bone. Hopson (1964) from a study of the tritylodont *Bienotherium* and Vandebroek (1964) from a study of a skull of a juvenile platypus independently found that this bone, presumed to be the periotic, forms in the membrana obturatoria of the sphenoparietal fissure. This membrane lies lateral to the cavum epitericum and forms its lateral wall; it is known that in some Metatheria and Eutheria that the lamina ascendens of the alisphenoid is ossified in this membrane. Hopson's interpretation of the condition of the side wall of the monotreme brain case is that the alisphenoid is a small ossification fused to the basicranium and that the lateral wall of the cavum epitericum is formed by a large anterior process of the periotic. Kermack (1963), however, considered that the anterior lamina of the periotic ossified within the wall of the neurocranium itself in the mammals *Eozostrodon, Trioracodon,* and *Triconodon* and in the tritylodontid *Oligokyphus*. Later Kermack (1967) acknowledged that Hopson's interpretation was the right one and expressed the opinion that groups of mammals having an anterior lamina of the periotic in the side wall of the brain case probably indicates real affinity. Hopson and Crompton (1969) expressed the concept thus: "The presence of a well-developed anterior lamina in the periotic of the nontherian orders of mammals, and its replacement by a broadly expanded alisphenoid in marsupials and placentals (the only therians in which the brain case is known) strongly suggest the manner in which the side wall of the braincase is constructed separates the known orders of mammals into two natural groups. . . . Thus the presence of an anterior periotic lamina in the Monotremata and Multituberculata clearly allies these orders with the Triconodontidae and Eozostrondontidae in which the anterior lamina is known." In further papers Hopson (1970) and Crompton and Jenkins (1973) put formal expression to the suggestion:

> Class Mammalia
> Subclass Prototheria
> Infraclass Ornithodelphia
> Infraclass Eotheria
> Infraclass Allotheria
> Subclass Theria
> Infraclass Trituberculata

This suggestion of division of the Mammalia into "two natural groups" seems to me to be premature and not very useful for a number of reasons. Nothing is known of the side wall in ancient marsupials and ancient eutherians, nor in Kuehneotheriidae, Amphidontidae, Spalacotheriidae, Amphitheriidae, Paurodontidae, Dryolestidae, and Aegialodontidae; Crompton and Jenkins themselves point out that they and others (Parrington, 1971; Crompton and Jenkins, 1968; Hopson and Crompton, 1969; Hopson, 1969) "do not believe that the marked differences between the braincase structure of therian (i.e., marsupial and placental) mammals from late Cretaceous and younger strata and that of nontherian mammals from the late Triassic indicate that there must also have been marked differences between the braincases of the immediate ancestors of the late Triassic therians and nontherians." In fact it would seem not unlikely that the early symmetrodont, *Kuehneotherium praecursorius*, currently classified as a "therian" will prove to have a brain case side wall similar to those of some of the "prototherians" since it has teeth similar to those of triconodonts (see especially Parrington, 1971, 1973; Crompton, 1974) which are classified as nontherian. I say some of the prototherians advisedly because there is no such thing, for example, as a monotreme condition of the side wall of the brain case. The literature is replete with references to this "condition," meaning that the bone consists of a rostral projection of the periotic. The periotic, however, plays no part whatever in filling in the sphenoparietal fissure in the echidna skull and it is of more than passing interest that in the two living representatives of what may well have been Mesozoic mammals the side walls of the brain case are filled in by bone of fundamentally different origin. Since the confusion has arisen from misinterpretation of defective skulls I will briefly indicate the history of what Bill Clemens (1970) aptly terms this bone of contention. In 1901 van Bemmelen diagnosed the bone as an alisphenoid. Doubtless he did so since it occupies a position identical to the position of the alisphenoid in other mammals and since the third or mandibular branch of the trigeminal nerve passes through a notch at the posterior part of the bone. Anteriorly van Bemmelen's alisphenoid in the echidna is fused with a prominent straplike bone, arising from the palatine which he named the temporal wing of the palatine. Gaupp (1908), however, did not believe that the huge membrana sphenoobturatoria (sphenoparietal membrane) which he found in the chondrocraniums and skulls of echidna pouch young could be filled in by an ossification from the ala temporalis, i.e., by an alisphenoid. He surmised that it was filled in by an ossification in the membrane itself: "Als

Alisphenoid bezeichnet van Bemmelen eine Knochenplatte, die zweifellos aus der Verknöcherung des Haupttheiles der Membrana spheno-obturatoria hervorgeht.'' He also considered that van Bemmelen's temporal wing of the palatine was formed in the same way and he summarized his conception of the ossification of the sphenoparietal membrane in these terms ''Ein aufsteigender Theil der Ala temporalis kommt bei *Echidna* nicht zur Entwickelung. Statt seiner entsteht eine Bindesgewebsplatte, Membrana spheno-obturatoria, die das Cavum epiptericum lateralwärts abschliesst, und aus deren später Verknöcherung das 'Temporalflügelchen des Palatinums' und das 'Alisphenoid' van Bemmelen's knöcherne Lamina spheno-obturatoria hervorgehen.'' However, he lacked the stages intermediate between his oldest pouch young and the adult condition to determine just how the sphenoparietal fissure was filled in. As it turns out his guess was partly right and partly wrong. Van Bemmelen also described a bone in the orbito-temporal region of the platypus skull as the ''Temporalflügelchen des Palatinums'' or ala temporalis palatini (see Appendix).

Eight years after publication of Gaupp's work, D.M.S. Watson (1916) had occasion to publish his findings on, inter alia, the ossification of the sphenoparietal membrane of the monotreme skull. His material consisted of a serially sectioned head, the incomplete skulls of two pouch young and the skulls of four young echidnas. The condition of the latter apparent from the illustrations given, was ruinous in the areas under consideration in that the membranes, and consequently any associated ossifications were not present. He also studied the serially sectioned heads of two platypus nestlings and the well-prepared skull of a subadult. He concluded from his examination of all these that the sphenoparietal fissure in the monotremes ''is not closed by an ossification of the ala temporalis, that is an alisphenoid, but by the gradual growth forward of an ossification which is always continuous with the anterior ossification of the otic capsule, and only very late in life acquires any connection with the ala temporalis.'' Furthermore he agreed that van Bemmelen's temporal wing of the palatine in the echidna was indeed a process of that bone but he strongly disagreed with the assertion that it was present in the platypus skull in these terms: ''In Fig. 4, Plate 32, of his paper a figure of the side of the skull of Platypus with the zygomatic arch removed, it is at once obvious that the area labeled plt. tmp. and supposed by him to be a 'temporal plate of the palatine,' is really entirely the ossified ala temporalis. . . .'' From 1916 to 1967 Watson's account was considered to be definitive, but almost simultaneously and independently, Giles MacIntyre (1967) and Griffiths (1968) found that the sphenoparietal fissure in echidnas was not filled-in in the way that Watson claimed was the case. MacIntyre states, speaking of Watson's interpretation: ''His specimens showed it to be continuous with the periotic, but not with the alisphenoid as it ossified. A series of seven skulls (American Museum of Natural History numbers 35679, 17355, 42176, 65831, 65843, 42258 and 1639) of *Tachyglossus* (echidna) shows that this is not quite correct. The separate membrane bone dorsal to the foramen for the mandibular nerve ossifies in a highly erratic manner, apparently sometimes from one center and sometimes from several asymmetrically arranged centers fusing finally comparatively late in life to all of the surrounding bones in no clearly definable sequence.... '' Griffiths, however, had seven freshly prepared intact skulls, three of which were from young, possible adult, two from suckling, and two from adult echidnas. In the skulls of the young echidnas, a substantial intrusion of bone could be seen invading the membrane in the sphenoparietal fissure. This he considered to be an alisphenoid since the intrusions appeared to originate in the ala temporalis region and were not connected in any way with the periotic. Griffiths took the view that van Bemmelen's interpretation was the right one, but with the reservation that no sutures, as drawn by van Bemmelen, could be detected. Subsequently Kühn (1971) published a beautifully illustrated monograph on the development and structure of the echidna skull but like Gaupp he lacked the stages to determine the mode of ossification of the membrana sphenoobturatoria except in one instance and this skull (63/98) suffered from the defects of Watson's skulls—a fact admitted by the author in the legend to his figure 34: "Die Lamina obturens (freiliegende Bindesgewebsverknöcherung in der Membrana spheno-obturatoria) ist wahrscheinlich bei der Präparation verloren gegangen.'' He dismisses Griffiths' diagnosis of the bony intrusion into the membrana sphenoobturatoria as incorrect and interprets the bone in these terms: "Die Abbildung 23 in Griffiths (1968) lässt vermuten, dass sie bei einzelnen Individuen von mehreren Zentren aus oder auch teilweise als Zuwachsknochen vom Orbitosphenoid her verknöchert.'' However it will be shown that it is not a matter of "einzelnen Individuen''; I have been in the fortunate position of being able to collect echidnas and platypus young and old, and to have had skulls prepared in my laboratory from freshly killed animals. This has yielded a series of 22 *Tachyglossus* skulls with the membrana sphenoobturatoria wholly or partially intact allowing study of the later development of the ala temporalis and of the way the sphenoparietal membrane is ossified, and a series of 15 platypus skulls.* In addition four skulls of *Zaglossus bruijnii* have been examined. Descriptions of the skulls follow:

Tachyglossus

1. CM 1399. A skull from a suckling 205 mm long. The roof was removed so that the ala temporalis could be studied. It consisted of an elongated, pear-shaped ossification, smaller end rostral, plastered along the side of the basisphenoid. Anterior to the ala temporalis the temporal wing of the palatine was detectable as a small nubbin of bone projecting almost vertically from the horizontal palatine. Incipient ossifications were detectable here and there on the membrana sphenoobturatoria.

**These specimens are housed in the collections of the Division of Wildlife Research, CSIRO, P.O. Box 84, Lyneham, 2602, Canberra and are available for study.

2. CM 3685. From inspection of this skull also with the roof removed, it is apparent that the ala temporalis still consists of a wing of bone rounded at the rostral end lying at the lateral margin of the basisphenoid. The ala temporalis is not visible when the skull is viewed from the side but immediately anterior to it is the prominent temporal wing of the palatine pointing outwards, forwards and upwards (Figs. 113 and 114). Its posterior margin is joined to the membrane of the parietal fissure. The base and distal projection of the temporal wing of the palatine serves as a support for cranial nerves II, III, IV, V(1 and 2), and VI emerging from the skull.

Isolated centers of ossification are apparent in the anterior parts of the sphenoparietal membrane and also in that membrane anterior to the squamosals, but quite separate from them, rounded discrete ossifications of different sizes can be seen (Fig. 115). The membrane between squamosal and periotic on the right side is ossified but not on the left.

3. CM 3686. This skull shows an advance on the previous one in that seen from the side the temporal wing of the palatine has elongated into a thin straplike

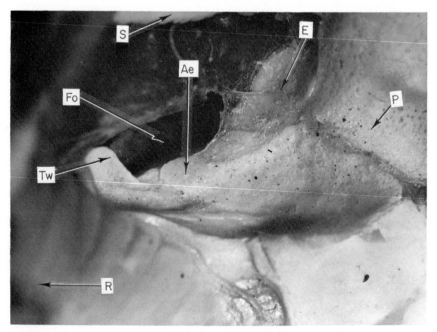

Figure 113. *Tachyglossus.* CM 3685; skull of a suckling 500 g in weight, length of skull 62 mm. Dorsal view of right ala temporalis and temporal wing of palatine (roof of skull removed). × 7.5. S, squamosal; E, ectopterygoid; P, periotic; Ae, anterior end of ala temporalis; Fo, foramen ovale; Tw, temporal wing of palatine.

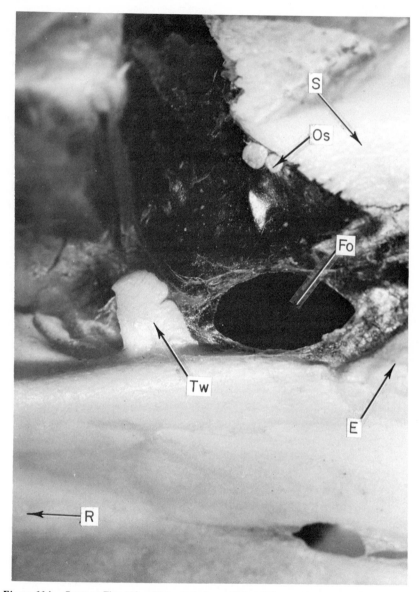

Figure 114. Same as Fig. 113, with ventral view of left palatine and its temporal wing, foramen ovale, and independently formed ossifications in the membrane of the sphenoparietal fissure. × 10.5. S, squamosal; Os, ossification in sphenoparietal membrane; Fo, foramen ovale; E, ectopterygoid; Tw, temporal wing of palatine; R, rostral.

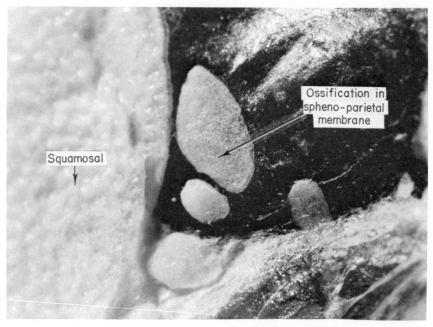

Figure 115. Ditto, ventral view of portion of sphenoparietal membrane between squamosal and ectopterygoid showing ossifications forming in the membrane. × 32.

bone pointing forwards and upwards to a ventrally directed process of the orbitosphenoid. Looking down onto the dorsal surface of the palatine wing (roof of skull removed) it is apparent that the ala temporalis has broadened somewhat but it still contributes nothing to the side wall of the brain case (Fig. 116).

4. CM 3683. The ala temporalis, or alisphenoid as it now may be called, is quite visible in this skull. It has grown forward, outward, and upward and its anterior margin has fused with and is investing the whole of the posterior part of the palatine wing (Fig. 117). Inside the skull it can also be seen that the alisphenoid is investing the internal surface of the palatine wing. At this stage it is apparent that the ala temporalis consists of a distal and a proximal moiety. The distal portion, which is contributing to ossification of the side wall of the skull, I consider to be a heterochronically retarded lamina ascendens of the alisphenoid (see Watson, 1916; Terry, 1917; Kesteven, 1918; Gregory and Noble, 1924, for discussions on the nature of the lamina ascendens and processus alaris; de Beer, 1930, for a discussion on the implications of heterochrony in ontogeny).

The left lamina ascendens was removed together with the adherent palatine wing; the two bones came away as an entity and the process of envelopment of the wing by the lamina ascendens could be seen taking place on both inner and outer surfaces.

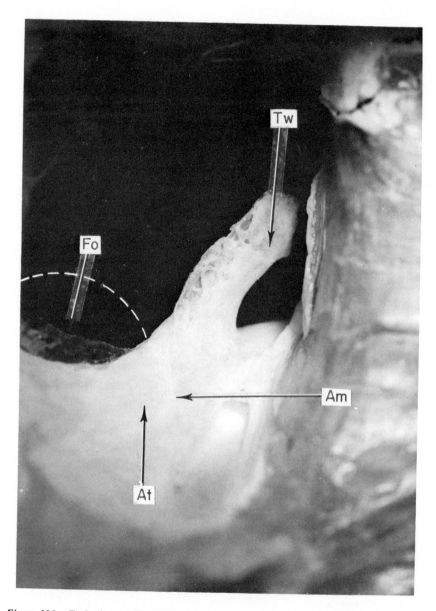

Figure 116. *Tachyglossus.* CM 3686; skull of a very young free-living echidna possibly only 7 months old, body weight 710 g, length of skull 77 mm. Dorsal view of left temporal wing of palatine and the ala temporalis. × 10.4. Tw, temporal wing of palatine; Am, anterior margin of ala temporalis; At, ala temporalis; Fo, foramen ovale.

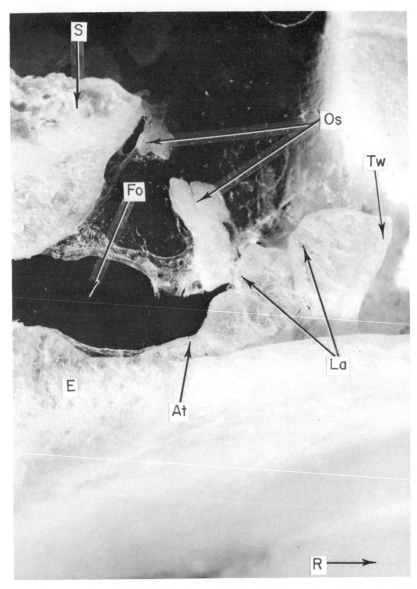

Figure 117. *Tachyglossus.* CM 3683. Skull of a young free-living echidna, body weight 810 g, length of skull 86 mm. Lateral view of right side foramen ovale and of the sphenoparietal fissure. ×
11.0. S, squamosal; Fo, foramen ovale; Os, ossification in sphenoparietal membrane; Tw, temporal wing of palatine; La, lamina ascendens; At, ala temporalis; E, ectopterygoid.

In the sphenoparietal membrane an ossification is present located just posterior to the alisphenoid, but independent of it, on the right side but not on the left. At the posterior end of the membrane one can discern independently formed discrete ossifications some of which have either fused with forwardly growing extensions of the periotic, or they have grown backwards to fuse with the anterior margin of the periotic. It is quite apparent, as MacIntyre found, that the sphenoparietal membrane can be ossified from "several asymmetrically arranged centers," but the process is not "highly erratic" as he supposed as will be shown below.

5. CM 3684, 3687, and 3692. In these skulls it can be seen that the process of investment of the palatine wing, except at its anterior margin, is practically complete: the alisphenoid has covered the wing both on its external and internal surfaces and has grown dorsally to unite with the ventrally directed process of the orbitosphenoid (Fig. 118). This union forms a tunnel—the foramen rotundum, for the emergence of the above mentioned cranial nerves. Thus the palatine wing serves in the earlier stages of development as a temporary strut supporting and

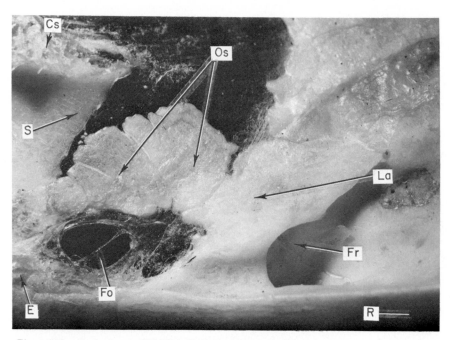

Figure 118. *Tachyglossus*. CM 3684. Skull of a subadult body weight 2060 g, length of skull 100 mm. Ventrolateral view of right side expanded lamina ascendens fused with orbitosphenoid and intramembranous ossifications. × 8. Cs, cut surface of squamosal; S, squamosal; Os, ossification in sphenoparietal membrane; La, lamina ascendens overlaying temporal wing of palatine; Fr, foramen rotundum; R, rostral; Fo, foramen ovale; E, ectopterygoid.

directing the exit of those nerves until the heterochronically retarded lamina ascendens of the alisphenoid grows out to take over its functions.

In all three of these skulls at about the middle of the sphenoparietal membrane are large ossifications dorsal to the foramen ovale and separate from the periotic. The anterior margins of these ossifications are in contact with, but separate from, the posterior margins of the lamina ascendens. The posterior border of the lamina ascendens exhibits backwardly directed processes which interdigitate with, and even overlay, processes on the ossifications on the membrane. Thus the alisphenoid makes a significant contribution to the structure of the side wall of the brain case.

The anteroventral border of the foramen ovale is formed by the alisphenoid, the dorsal and posterodorsal border by the membrane bone, and the posteroventral by the ectopterygoid (Fig. 119).

At the posterior end of the sphenoparietal membrane, on the left side only, of CM 3684, are two projections of bone pointing forward and apparently taking their origin from the periotic. If one did not know that these projections were the

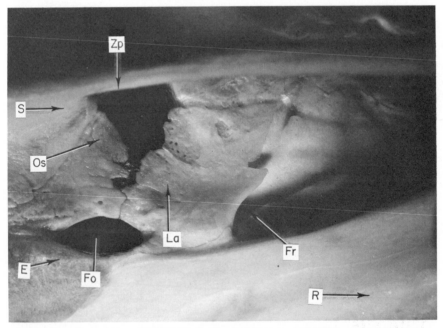

Figure 119. *Tachyglossus.* CM 3692. Skull of a subadult, body weight 2300 g. Ventrolateral view, right side, showing bones forming the foramen ovale and fusion of lamina ascendens with intramembranous ossifications. × 5.5 Zp, zygomatic process of squamosal; S, squamosal; Os, ossifications in sphenoparietal fissure; E, ectopterygoid; Fo, foramen ovale; La, lamina ascendens; R, rostral; Fr, foramen rotundum.

result of the growing together of many small discrete ossifications that finally unite with the periotic, one would be justified in concluding, as Watson did, that they are indeed forwardly directed outgrowths of the periotic.

6. CM 1394, 1396, 1398, 3676, 3680, 3690. These skulls are from large but subadult echidnas; in all of them the ossification of the sphenoparietal membrane has progressed beyond the condition seen in CM 3684, 3687, and 3692 but it is quite apparent that the alisphenoid has played a significant part in filling the side wall. In these seven skulls ossifications in contact with outgrowths from the alisphenoid occur dorsal to and form the dorsal margin of the foramen ovale. Thus in the ten subadult skulls available it has been found that the alisphenoid has played the same role in filling in the anterior part of the side wall.

In the nine skulls from adult echidnas the sphenoparietal fissure is completely ossified and gives no clue as to how this was achieved. However, contrary to the implications of MacIntyre's diagram (his Fig. 2D) it is quite apparent that the bones forming the margins of the foramen ovale are alisphenoid, ectopterygoid, and the membrane bone in the sphenoparietal fissure.

Zaglossus

In all four of these skulls of adult specimens the side wall appears as a smooth plate of bone identical to that of the adult *Tachyglossus* skull; the bones surrounding the foramen ovale are alisphenoid, ectopterygoid, and the membrane bone of the sphenoparietal fissure (Fig. 16). Furthermore van Bemmelen (1901) mentions that the relations of the temporal wing of the palatine in *Proechidna (Zaglossus)* are the same as in *Tachyglossus*.

Ornithorhynchus

1. CM 3706. This is the youngest skull in the series; the animal was only 33 cm long (measured shortly after death), weighed 315 g, still had teeth, and it was caught early in February which is about weaning time. It appears to be at the same stage of development as a skull examined by Vandebroek (1964). As Watson found, the ala temporalis is elongated, narrow, and does not contribute significantly to filling in the sphenoparietal gap. However, at its anterior end the ala temporalis exhibits a small lateral wing which forms the floor and part of the lateral wall of the foramen rotundum (Fig. 120). The side wall of the brain case is filled in by a forwardly growing flange of the periotic; it is completely united to the periotic posteriorly and exhibits the appearance of raying outwards and forwards from that bone but anteriorly the flange has not completely filled the membrana sphenoobturatora. Vandebroek's skull shows the same structure.

The foramen ovale is a hole bounded anteriorly and medially by the alisphenoid, posteriorly by the true periotic and laterally by the membrane bone in the sphenoparietal fissure (Fig. 120). In the light of this and of the above description of the foramen ovale in *Tachyglossus* the diagrams of egress of V3

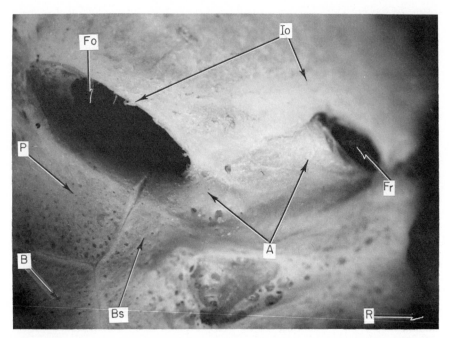

Figure 120. *Ornithorhynchus.* CM 3706. Skull of a juvenile, body weight 315 g, body length 33 cm. Dorsal view (roof of skull removed) of left foramen ovale showing long, narrow alisphenoid extending from periotic to foramen rotundum. Bones forming foramen ovale are alisphenoid, periotic, and anterior lamina of periotic. × 7.7. P, periotic; B, basioccipital; Bs, basisphenoid; R, rostral; A, alisphenoid; Fr, foramen rotundum; Io, intramembranous ossification; Fo, foramen ovale.

from the skull of a "monotreme" given by Hopson and Crompton (1969) and by Clemens (1970) are incorrect.

2. CM 3694–3708. The remaining 14 skulls in the series are from subadult and adult platypuses of both sexes. Inspection of these offers nothing to the elucidation of the nature of the bone in the sphenoparietal fissure; in fact the side walls of the brain cases look very like those of *Tachyglossus*—large expanses of thin smooth bone and no sutures, yet they are structurally fundamentally different. Thus it is not possible to speak of a monotreme condition of the side wall.

The foramina ovales are very different in the two genera also. That of *Tachyglossus* is displaced a long way rostral relative to that of *Ornithrohynchus* so that the true periotic plays no part in formation of the foramen in the former, the posterior border here being formed by the ectopterygoid. The rostral displacement in *Tachyglossus* is due to the presence of these enormous, robust, ectopterygoids preventing the possibility of egress of V3 from immediately anterior to the periotic. The relationship of the ectopterygoids to the grinding of the food has been discussed on p. 36.

The differences in the bones surrounding the foramen ovale in the two monotremes reminds one of the differences to be found in the marsupials and the placentals. Apparently these differences are not generally appreciated since in recent years nearly all authors who have had occasion to mention egress of the mandibular ramus of the trigeminal nerve from the monotreme skull refer to the foramen as pseudovale, pseudo-ovale, or even pseudoovale, as though there is something special about it that makes it different from or even nonhomologous with the foramen in the living Metatheria and Eutheria. That view implies that the foramen ovale in those two groups is an immutable entity but nothing could be further from the truth. Fifty-four years ago Gregory and Noble (1924) stated: "The mandibular branch (V3) in some marsupials and placentals passes quite behind the alisphenoid, in others it pierces it while in *Perameles obesula* its exit is at the front edge of the alisphenoid." In those instances where V3 emerges posterior to the alisphenoid the posterior margin of the foramen ovale is formed by the periotic; I have one other variation on the theme—in the marsupial *Vombatus ursinus* the anterior half of the foramen ovale is bordered by the alisphenoid, the posterior half by the squamosal. In view of this variety of bones surrounding the foramen ovale there is no reason to single out those of the monotremes as peculiar and especially as there are, apparently five holes for egress of V3 from the skull in *Kamptobaatar kuczynskii* (Kermack and Kielan-Jaworowska, 1971).* A further example of diversity of bones surrounding the foramen ovale within closely related groups is to be found in the mammals *Sinoconodon* and *Eozostrodon*. *Sinoconodon rigneyi* is classified by Crompton and Jenkins (1973) as a member of the family Eozostrodontidae along with *Megazostrodon, Erythrotherium*, and *Eozostrodon*. Simpson (1971) prefers to place *Sinoconodon* in a family of its own, the Sinoconodontidae, in the order Triconodonta in which he includes the family Eozostrodontidae. Mills (1971) places *Megazostrodon* and *Sinoconodon* in the family Sinoconodontidae thus excluding *Megazostrodon* from the Eozostrodontidae. Crompton (1974) says there is no justification for placing *Megazostrodon* in with the Sinoconodontidae. From all this it is quite apparent that all four genera are closely related yet V3 passed to the exterior through a hole in the alisphenoid of *Sinoconodon* (Patterson and Olson, 1961) and, apparently, through one in the periotic lamina of *Eozostrodon* (Kermack and Kielan-Jaworowska, 1971).

Another matter that may be mentioned in this context of alisphenoid and foramen ovale is a mysterious "membrane bone of uncertain homology" that

---

*Just how the authors decided that all these holes are for egress of V3 is not apparent. There are 14 foramina in the skull of *Tachyglossus* for distribution of the trigeminal nerve but only one for V3.

Giles MacIntyre (1967) discovered in the side wall of the brain case dorsal to the alisphenoid in some marsupials. It has been known for some time (de Beer, 1937) that the ascending lamina of the alisphenoid in the marsupials ossifies as a membrane bone in the huge sphenoparietal membrane; the latter is relatively as large as it is in *Tachyglossus*. In the 10-day old pouch young of *Megaleia rufa*, for example, the side wall of the brain case is largely membranous and the alisphenoid is hard to detect; by 100 days it can be seen that the anterodistal part of the lamina ascendens consists of very thin membrane bone and the proximal part of thicker bone; the rest of the side wall is still membrane although down growths of the parietal and back growths of the frontal and obitosphenoid are contributing to filling in the fissure (Fig. 121)—a condition not unlike that found in *Tachyglossus* when the lamina ascendens of tbe alisphenoid is ossifying and fusing with ingrowths of the surrounding bones. The difference, which appears to me to be of no great importance, is that in *Megaleia* the alisphenoid fills a large part of the sphenoparietal fissure whereas in *Tachyglossus* independent discrete ossifications contribute to filling it. When the distal part of the membrane is ossified in the marsupial the alisphenoid consists of a proximal part of

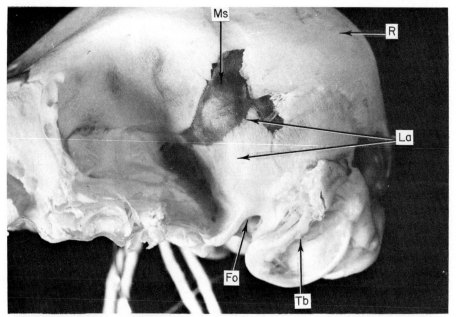

Figure 121. *Megaleia rufa.* Lateral view of posterior part of skull of 100-day-old pouch young. Malar arch has been removed. The objects dependent from ventral surface are threads attached to a label. × 3. Ms, membrane in sphenoparietal fissure; R, rostral; La, lamina ascendens of alisphenoid; Tb, tympanic bulla closing over; Fo, foramen ovale.

thick bone and a distal one of thin bone. An extreme condition is seen in the side wall of the brain case of *Trichosurus vulpecula*, the whole of which is of such thin bone that it is translucent; a very small part at the posterior end of the bone forming the anterior border of the foramen ovale is made up of thick opaque bone.

My final objection to the proposal to lump a polygot collection of mammals in the one taxon, Prototheria, is that it attempts to link highly evolved mammals like platypuses and echidnas possessing malleus, incus, stapes, and organ of Corti with mammals exhibiting malleus and incus still incorporated in the lower jaw.* George Gaylord Simpson (1971) has recently proposed a classification of the Mammalia which avoids a dichotomous arrangement: the Class to be divided into four subclasses, Prototheria (monotremes), Allotheria (multituberculates), Eotheria (triconodonts and docodonts), and Theria, the latter containing three infraclasses, Patriotheria (symmetrodonts and pantotheres), Metatheria, and Eutheria. I like very much his taxon Patriotheria of equal rank with Metatheria and Eutheria but again I see nothing useful in bringing another polyglot collection of mammals ranging from the most highly evolved on the planet to *Kuehneotherium* and other primitive mammals with incus and malleus still part of the lower jaw, into the one taxon—Theria. The qualification for admission to this taxon seems to be the candidate must have tribosphenic dentition or molars that might conceivably give rise to that kind of dentition.† This would appear to be a single-character classification since it is anticipated that the side wall of the brain case in early symmetrodonts and pantotheres could turn out to be like that of Eotheria; Parrington (1974b) has given an account of the dismal history of single-character classifications. It is implicit in the classification that Eutheria and Metatheria inherited tribosphenic dentition from common Late Jurassic ancestors. To date no Jurassic mammals with tribosphenic molars have been discovered, but several distinct lineages had independently achieved advanced protribosphenic dentition in those times. It is held that since the earliest marsupials and eutherians had similar tribosphenic dentitions they have descended from common ancestors with tribosphenic molars. Clemens (1968, 1971, 1977) makes plain that this is supposition and that tribosphenic molars could have developed independently in two or more lineages; indeed this appears to me to be quite likely since the evolution of the Mammalia, and according to Hopson and Crompton (1969) and Crompton (1972), of the Therapsida, is testimony to the fact that parallelism is "a strikingly pervasive phenomenon." Perhaps the most

---

*See Appendix.

†It would appear that Kühne believes that the molars of the platypus are derived from tribosphenic molars since he holds that the monotremes branched off the marsupial line after "the cladistic dichotomy of Placentals and Marsupials: hence not earlier than the Middle Cretaceous." This would be in the realm of the possible if Green's observation of prismatic enamel in platypus molars is confirmed (p. 57). After all the molars of aardvarks give no evidence of descent from tribosphenic molars.

striking instance is the independent and parallel evolution of malleus and incus. A consensus (see Hopson, 1969; Parrington, 1974b) including myself, but not Kühne, cheerfully entertains the notion that malleus and incus of monotremes evolved independently of those of Metatheria and Eutheria—by far a more complex feat of parallel evolution than say parallel evolution of tribosphenic molars from protribosphenic dentition. Again if one accepts along with a consensus holding that monotremes and marsupials have different origins one might say that the striking similarities of their ontogenies and of the caenogenetic modifications of their neonates and hatchlings are the result of parallel evolution since reptiles do not exhibit bilaminar blastocysts and minute specialized naked young so fantastically unlike their parents that they could be called larvae. However, Gregory believed and Kühne believes that these two examples of similarity (and many others) are evidence of affinity and of common ancestry of marsupials and monotremes. Sharman (1970) and Tyndale-Biscoe (1973), take it for granted that the Eutheria and the Metatheria had a common ancestor that was oviparous, and that their viviparity was independently acquired by parallel evolution. Lillegraven (1975) also takes it for granted that they had a common ancestor but he believes it was viviparous. Müller (1969) on the other hand arrived at the view, on ontogenetic grounds, that the Metatheria and Eutheria have descended from different but relatively similar ancestors.

Other examples of parallelism in the evolution of the Eutheria and Metatheria, although having nothing to do with their origins, support the notion of its all-pervasiveness: marsupials (Peramelidae) have evolved allantoic placentation independently of that in Eutheria; some marsupials have almost achieved the stirrup form of stapes (Segall, 1970) found in all Eutheria except pangolins; as is well-known eutherians and marsupials achieved molehood, bearhood, wolfhood, badgerhood, rathood and so on, independently (see Kirsch, 1977).

Apart from parallelism in the evolution of Metatheria and Eutheria there are differences that indicate differences of origin. Eutherians do not exhibit bilaminar blastocysts in their early ontogeny nor do they give birth to specialized larvae as marsupials and monotremes do, and, a source of wonderment to me, no eutherian exhibits epipubic bones not even relics during ontogeny. If the two groups had common ancestors, which presumably would have had epipubic bones, one would expect to find at least vestiges of those bones during development in some Eutheria even though some carnivorous marsupials have no, or reduced, epipubic bones (see Kirsch, 1977).

Thus in view of the facts that a common ancestor to the Eutheria and Metatheria, with tribosphenic dentition, has not yet been found, that several lineages of primitive mammals have evolved protribosphenic dentition by independent parallel evolution, that parallelism is all-pervasive in therapsid and mammalian evolution, and that the nature of the side wall in Patriotherians is unknown, it appears that Theria does not reflect close affinity. I suggest, therefore, that Eutheria, Metatheria, and Patriotheria be raised to the level of sub-classes:

Class Mammalia
Subclass Eotheria (triconodonts and docodonts)
Subclass Patriotheria (symmetrodonts and pantotheres)
Subclass Allotheria (multituberculates)
Subclass Prototheria (monotremes)
Subclass Metatheria (marsupials)
Subclass Eutheria (placentals)

Simpson's classification does away with any necessity to employ, to me, the revolting name of Atheria proposed by Kermack *et al.* (1973) instead of Pro-totheria, since in their opinion Prototheria is now irrevocably attached to the monotremes. Prototheria means first or primitive beasts, Atheria as far as I can see means "no beast" or "without beasts." Since echidnas and platypuses are very fine beasts indeed it would never do to include them in a taxon called Atheria. In any case Prototheria is not irrevocably attached to the monotremes alone since Cope (1888) wrote a paper called "The Multituberculate Monotremes" and Richard Semon (1899) grouped the monotremes and mul-tituberculates together as Prototheria. G. G. Simpson's classification also elimi-nates any necessity to resurrect another ghastly misnomer—Ornithodelphia.

The grouping of multituberculates with monotremes referred to above brings me to the fourth version of the affinities of the monotremes.

VIEW 4

Kielan-Jaworowska (1970) and Kermack and Kielan-Jaworowska (1971) have advanced the view, or rather revived it, that there is a close relationship between monotremes and multituberculates but the idea is not new: Broom (1914) dis-cussed the concepts of the affinities of multituberculates and the consensus, with the exception of Cope (1888), up to the year 1910, had it that upper Triassic and Cretaceous multituberculates were marsupials but Broom in 1910 (cited in Broom, 1914) doubted the relationship to marsupials stating "in the present state of our knowledge it seems wisest to leave the multituberculata as a distinct independent group with no very near affinities with the living monotremes, marsupials or Eutherians." In 1914, however, he inclined to the view that they are related to monotremes on the grounds that both groups exhibited uncoiled cochleas (*Ptilodus*), large coracoids, interclavicles (*Camptomus*), septomaxil-laries, small jugals (absent in *Tachyglossus*), union of maxillae to squamosals, and mandibles with small inflected borders. He also remarked that in "*Polymas-todon* there are only two large molariform teeth, and even in the Plagiaulacids only the last two in the lower jaw are multituberculate, and that in *Ornithorhyn-*

chus there are also only two molars retained." In his opinion the evidence indicates that a line of mammals arose from middle Jurassic herbivorous multituberculates which after considerable specialization and degeneration resulted in the monotremes.

Sometime later Kielan-Jaworowska (1969) published her discovery of epipubic bones in a multituberculate, *Kryptobaatar dashzevegi*. From consideration of the structure of the pelvis, despite the fact that marsupial bones are present, she concluded that the multituberculates are a side branch of the mammals not closely related to the monotremes, marsupials, and placentals. Some months later (Kielan-Jaworowska, 1970) she published a description of a beautifully preserved multituberculate skull and in that paper she concluded that there is a special close relationship of multituberculates to monotremes, the criterion of affinity being the structure of the side wall of the brain case. The skull in question was that of *Kamptobaatar kuczynskii* which exhibited well-preserved ectopterygoids, alisphenoids, and petrosals. Of the alisphenoid she remarks "there seems to be a suture between the alisphenoid and the petrosal," and "the alisphenoid is probably reduced to a comparatively small ventral element which does not contribute very much to the structure of the lateral wall of the braincase." Diagrammatic drawings of the lateral aspects of the braincase in *Kamptobaatar* and in *Ornithorhynchus* (the alisphenoid of which, as drawn, is quite unlike any I have seen in platypus skulls, see p. 323), are given and from this she concludes that the multituberculates are more closely allied to the monotremes than they are to Docodonta and Triconodonta. In a later paper Kermack and Kielan-Jaworowska (1971) give further information on the skull of *Kamptobaatar*. The middle ear is completely open, there is a crista parotica, and there is a posterior opening of the temporal fossa, all just as in the skulls of the monotremes. The discussion to their paper unfortunately includes slight inaccuracies and ambiguities:

a. The side walls of "therian" braincases are equated with those of late Cretaceous and living marsupials and eutherians, therefore all "therian" braincases are different from those of "nontherian" braincases.

b. It is assumed that the side wall of the echidna braincase is formed in the same way as the platypus side wall is formed, hence references to the (nonexistent) monotreme condition.

c. All foramina ovales are assumed to be holes in alisphenoids.

d. It is stated that "The ectopterygoid in multituberculates occupies a posterior position resembling that of the 'Echidna pterygoid' in monotremes." As we have seen the ectopterygoids in echidnas and platypuses do not resemble one another: in the platypus the ectopterygoids project out from the sides of the base of the skull rather like fragile, thin, membranous wings and the foramen ovale lies completely posterior to the ectopterygoids (Fig. 5). In tachyglossids the ectopterygoids are enormous, robust, closely applied to the ventral surface of the skull, and form part of the apparatus for the grinding of the food; the foramina ovales lie anterior to the ectopterygoids the whole length of which lies between the foramen ovale and the periotic (Fig. 16).

e. "Both multituberculates and monotremes lack a jugal." Broom (1914) describes the jugal of *Polymastodon taoensis* and gives a figure showing the jugal of *Ornithorhynchus* (see also p. 16).

However these are but minor matters; Kermack and Kielan-Jaworowska have brought exciting new light to bear on the theses of Cope and of Broom. Moreover the possibility that there are 2-5 foramina for the exit of V3 from the multituberculate skull argues that the trigeminal nerve was large as in monotremes and that perhaps the whole multituberculate brain was pervaded by a trigeminal aura as in the monotremes. It will be of great interest if some of the new finds also give information about the ear ossicles in multituberculates. Of these Professor Parrington (1974b) has recently stated "It is not at all likely that we shall ever discover the middle ear ossicles of say a docodont, a late triconodont or even a multituberculate (though this latter group might yield such a specimen as to show the middle ear structures).* There is no evidence, that, having evolved the squamoso-dentary hinge, these Mesozoic mammals evolved the mammalian ear." This is interesting since Broom (1914) infers that multituberculates had malleus and incus in the middle ear: "We might assume that the monotremes and multituberculates branched off independently from the Cynodont reptiles, and that all the resemblances are due to convergence; but against this is the extreme improbability of the articular and quadrate becoming converted into auditory ossicles of a similar type independently in two lines." I think Broom may have assumed the multituberculates had malleus and incus because the lower jaw exhibited a dentary only. In fact the similarity in structure of the multituberculate and marsupial dentaries was one of the factors leading to inclusion of the multituberculates in the Metatheria, so, presumably, it was assumed the ear ossicles were similar in the two groups. Anyway the fact that the lower jaw of the multituberculates consisted of dentary only is evidence that they had malleus and incus. If it turns out that the multituberculates had three ear ossicles it may indicate parallelism or perhaps relationship but if it were found that the incus was ankylosed to the malleus and that the latter had a long gracile process cemented to the tympanic it would, to an overt incorrigible typologist like myself, strengthen the argument that multituberculates and monotremes are related. It may be argued against the hypothesis that multituberculates and *Obdurodon insignis* have different kinds of teeth, but so do other specialized dead-end products of mammalian evolution: honey possums, numbats, wombats, aardvarks, elephants, and leopard seals.

*See Appendix.

APPENDIX

THE EAR OSSICLES OF TRICONODONTS AND MULTITUBERCULATES

Parrington (1946) and Hopson (1966) have suggested that the mammal-like reptiles had single-bone middle ears with a tympanic membrane located behind the quadrate and partially supported by that bone. Allin (1975), however, has published evidence that the stapes in advanced mammal-like reptiles was in contact with the quadrate and the articular, i.e., those bones, although still part of the lower jaw suspension, could be conducting sound waves to the cochlea. He put forward the hypothesis that a diverticulum from the pharynx in those reptiles extended to an area bounded by the angular (homologue of the mammalian tympanic bone) with its backwardly directed reflected lamina, and that the distal part of this diverticulum was thin and acted as a mandibular tympanic membrane. Crompton and Parker (1978) in a recent publication support this hypothesis and further suggest that a cartilaginous ear trumpet, similar to that found in monotremes, could have passed forward to reach the mandibular tympanic membrane thus conducting airborne sound waves to the membrane. Furthermore Crompton and Parker in that publication give an illustration showing the articular, quadrate, and stapes of the triconodont, *Eozostrodon,* to be in contact. That is, the bones could be conducting sound waves to the cochlea from a mandibular tympanic membrane considered to be present between the reflected lamina of the angular and the retro-articular process of the articular (homologue of the manubrium of the malleus). This chain of "ear ossicles" could be functional as a sound wave transmitter even though the quadrate and articular were still functioning as a suspension for the lower jaw alongside a small dentary-squamosol suspension. One might speculate that the act of chewing would "sound" thunderous to the bearer of such middle ears.

In view of all the above and the fact that Cretaceous multituberculates had dentary only in the lower jaw (p. 331) it would seem that the latter had incus and malleus in the middle ear.

OCCURRENCE OF EPIPUBIC BONES IN TRIASSIC AND JURASSIC MAMMALS

In a recent publication Parrington (1978) suggests that certain bones "found among the thousands of scraps from Pont Alun quarry" are the epipubic bones of *Eozostrodon*. The bones are asymmetrical, right and left forms having been found, and consist of a thin basal plate projecting a thin rod with a blunt end. Parrington points out that if these are epipubic bones and the rods projected anteriorly, then the basal plate would be equivalent to the expanded structure of the epipubes found among the monotremes and marsupials.

Almost simultaneously Henkel and Krebs (1977) reported that epipubic bones occurred in the pelvis of a Jurassic pantothere.

FOOD, MOVEMENTS, POPULATION SIZE, AND DISTRIBUTION OF THE PLATYPUS

T. Grant and F. N. Carrick (in press) have made an interim report of their observations on the food of platypuses living in a series of pools interrupted by riffle areas in the Shoalhaven River, New South Wales; the total length of the study area is 6.8 km of main river bed and 3.2 km of an adjoining creek; the main trapping area consists of two pools in the river divided by a riffle area—a total length of 1.8 km. These two pools were sampled at least bimonthly during 1976 and 1977. It was found that all year round the principal food (cheek-pouch contents) was benthic larvae of insects: Trichoptera, Diptera, Odonata, Ephemeroptera, and Megaloptera in that order of abundance. However, in winter 29% of the cheek pouches sampled contained nematomorph and decapod remains.

These authors in the same publication also report the latest and fascinating details of their trapping and marking program. Trapping carried out in 1977 outside the main study area (the 1.8 km of river bed mentioned above) confirmed the result reported on p. 47 that there is very little movement between the two pools of the main study area and trap sites upstream and downstream from them. In 1977 only four animals were found to have moved upstream out of that area; one had moved 2.6 km downstream and another was caught 2.4 km away in the adjoining creek. It was noted, however, that those animals caught outside the main area moved considerable distances in a short time: recapture of animals at sites over 1 km apart with only a few days between trap nights took place several times. It is considered likely, therefore, that some animals are more free-ranging than others. However 57% of the platypuses marked (total 111, sex ratio 1.4:1 in favor of females) in the whole study area (10 km of river bed) have been recaptured at least once and 26% recaptured more than once. One, a female, in the main study area has been caught 10 times since she was first marked in July 1973! Grant and Carrick also note that severe floods did not appear to displace some platypuses from the main study area.

An estimate of the size of the population in the main part of the study area was made using the Jolly-Seber method described by Caughley (1977). This is a stochastic method allowing for the immigration and emigration known to occur; the estimate was 18 ± 4 platypuses in the area at any one time.

Information has come to hand that has a bearing on the absence or scarcity of platypuses in the rivers of Cape York Peninsula (see p. 44). Harry Dick of Cooktown, Queensland tells me that he shot a 6-ft crocodile (*Crocodilus porosus*) at the junction of the Daintree River and Boollum Creek, and in its stomach he found a partly digested platypus. Mr. Dick is of the opinion that *C. porosus* is extending its range inland into areas where it was unknown 80 years ago. Perhaps predation by crocodiles is the reason for apparent absence of platypuses in the rivers of Queensland north of Cooktown.

OBSERVATIONS ON THE OSTEOLOGY OF A MIOCENE ORNITHORHYNCHID

M. Archer *et al.* (in press) have described fragments of fossilized bone bearing strong resemblances to an ornithorhyncid dentary and ilium. The specimens were detected by systematically screen-washing several tons of sand taken from deposits of Miocene age at Lake Palankarinna, the type locality of *Obdurodon insignis* (see Chapter 2). One fragment is the posterior part of a left dentary. The alveolar area is shallow in contrast to the wider and deeper alveolar region of the dentary of *Ornithorhynchus*. In the fossil the authors suggest that M_3 [corresponding to M_3 of *Ornithorhynchus,* Green's (1937) dental formula] is represented by one lingual alveolus that penetrates the mandibular canal and that M_2 is represented by four distinct alveoli—two lingual and two buccal, all four penetrating into the mandibular canal. M_1 is represented by a remnant of a posterobuccal alveolus. All these alveoli are very large and well formed in contrast to those seen in *Ornithorhynchus*. The mandibular canal and foramen are small relative to those of *Ornithorhynchus*.

A matter of interest is that the fossil dentary exhibits a small but distinct angular process. As mentioned in Chapter 1 this is discernible in some dentaries of *Ornithorhynchus* and not in others; a matter not mentioned before is that when present it can be quite sharp. (Specimens held in the collections of the CSIRO Division of Wildlife Research, Canberra.)

The other specimen, an isolated left ilium exhibits the anterior third of the acetabulum and is considered to be that of a juvenile ornithorhyncid since comparison with those of juvenile platypuses revealed significant likenesses; however, there was one marked difference: the distal head of the fossil shows little of the expansion of the head found in the ilium of *Ornithorhynchus* including juvenile forms of the latter.

Altogether this new evidence supports the suggestion of Woodburne and Tedford (1975) that a platypus-like mammal lived in Miocene times at Lake Palankarinna; in fact Archer *et al.* put it that the dentary and ilium are those of *Obdurodon insignis*. Since the ornithorhyncid tooth and the bones come from the same quarry at Lake Palankarinna, I think it is a safe bet that the bones are indeed those of *Obdurodon*. Discovery of a dentary with a tooth or two *in situ* will settle the matter and doubtless this will be only a matter of time in view of the systematic screening methods employed.

FOSSIL ZAGLOSSUS OF THE LATE CENOZOIC

P. F. Murray (1978) has examined fossil *Zaglossus* material held in Australian Museums in connection with his study of the fossil echidna found in Tasmania (p. 74). He inclines to the view that the specimen is *Zaglossus robustus*. This species, he demonstrates, differs anatomically and allometrically from *Z. bruijnii* in that it has a broad, blunt, relatively short beak less aquiline than that of *bruijnii*, and has stouter limb bones. From this Murray deduces that food and feeding behavior was probably similar to that of Tachyglossus. If that were so I should advance the notion that *robustus* had a tongue furnished with striated circular musculature and erectile tissue, quite unlike that of *bruijnii*, and lacking a groove with rows of keratinous teeth, i.e., it should be called *Tachyglossus robustus*. It is of interest that Murray stresses, on osteological grounds, the resemblance of *robustus* to *Tachyglossus*, and its difference from *Z. bruijnii*.

If *Z. robustus* is a species of *Tachyglossus* the question arises did a "real *Zaglossus*," i.e., a wormeater with a tongue specialized for worm-eating and fundamentally different from that of *Tachyglossus* ever exist in the region now known as Australia? A satisfactory answer to that will await discovery of fossil echidnas that had long, gracile, aquiline beaks.

SYNTHESIS OF ASCORBIC ACID IN MONOTREMES

E. C. Birney *et al.* (1978) have confirmed the observation of Dykhuizen and Richardson (p. 84) that the kidney, but not the liver, of *Tachyglossus* can synthesize ascorbic acid since the kidney contains the enzyme that catalyzes the oxidation of L-gulonolactone to L-ascorbic acid. The latter authors demonstrated, indirectly, the presence of a chain of reactions leading to the formation of ascorbic acid since their substrate was D-glucuronic acid: D-glucuronic acid $\rightarrow$ L-gulonic acid $\rightleftarrows$ L-gulonolactone $\rightarrow$ 3-keto-L-gulonolactone $\rightarrow$ L-ascorbic acid.

Birney *et al.* found L-gulonolactone oxidase in platypus kidney, but not in the liver, as in *Tachyglossus;* they also detected the presence of the oxidase in liver and kidney of the marsupials *Perameles nasuta, Isoodon macrourus,* of one specimen of *Macropus rufogriseus,* and of two specimens of *Thylogale thetis.* Its presence was detected in liver only in other specimens of *M. rufogriseus* and in one specimen of *T. thetis.*

The failure of Dykhuizen and Richardson to detect the synthetic system in the tissues of *Trichosurus vulpecula, Macropus giganteus* and of *Macropus eugenii* was possibly due to the presence of large amounts of ascorbic acid in the tissues leading to high blanks (personal communication).

MECHANISM OF PROTRUSION OF THE TONGUE OF TACHYGLOSSUS

H. Baggett (in press) has continued his studies of the mechanism of stiffening and protrusion of the tongue. These now show that the sinusoids of the erectile tissue (see p. 87) in the dorsal and ventral regions of the tongue unite in the midline anterior to a transverse sinus forming a large space lined with endothelium and containing arterial blood; striated fibers from the circular muscles are inserted into the wall of the sinus. Baggett surmises that contraction of the circular muscles compresses and distorts the large sinus so that the pressure of the blood contained therein is raised displacing blood into the dorsal and ventral sinusoids leading to rapid protrusion of the tongue. Retraction of the tongue brought about by contraction of the m. sternoglossi dilates the large sinus sucking blood back, from the sinusoids, in readiness for the next contraction of circular muscles.

TEMPERATURE REGULATION IN ZAGLOSSUS BRUIJNII

T. J. Dawson *et al.* (in press) have determined parameters of temperature regulation and metabolic rate in two specimens of *Zaglossus bruijnii* living in captivity at Taronga Park Zoo, Sydney. The specimens were of unknown sex and one of them was the heaviest on record—16.5 kg.

Between TA 15° and 30°C, deep body temperature, as predicted on p. 142, proved to be 32°C. However, SMR at those ambient temperatures was found to be only 17.85 kcal/kg$^{3/4}$/day. This very low value prompted the authors to determine the metabolic rates of captive *Tachyglossus aculeatus* using the same apparatus as for *Zaglossus.* These also proved to be very low, 20.22 kcal/kg$^{3/4}$/day whereas the values found by Schmidt-Nielsen *et al.* and Augee (Table 23) were 34.0 and 32.3, respectively. Discrepancies of this order are disturbing; apart from the doubts about taxonomic significance of SMR's expressed on p. 138, one wonders about the physical condition of the subjects in the experiments.

The fact that captive platypuses have an SMR of 45 kcal/kg$^{3/4}$/day, higher than that of many marsupials and eutherians, may be a reflection of the athletic life they lead—having to enter cold water daily and spend hours swimming and diving to get their food. Undoubtedly such animals would be in better condition than other kinds living a sedentary existence involving no exercise other than an occasional stroll to the feeding bowl.

Dawson *et al.* found indeed that conductance was low, even lower than that of the platypus—0.11 cal/cm^2/°C/hr for the platypus and 0.055 cal/cm^2/°C/hr for *Zaglossus* at 15°C ambient. The authors attribute this to the large body size of *Zaglossus* with its low surface area to body weight ratio and to the thickness of the fur which was 23 cm long on the dorsal surface.

At TA 30°C evaporation accounted for nearly 70% of the total heat loss, and, as might be expected from perusal of Fig. 45, the animal sweats at TA 30° with TB of 34.8°C.

THE ADRENAL CORTEX

The Binding of Adrenocortical and Sex Steroids to Plasma Proteins of Tachyglossus

C. Sernia has been studying the binding of steroids by plasma proteins in the echidna and he has kindly allowed me to summarize data from his thesis presented for the degree of Doctor of Philosophy, Monash University, Victoria Australia.

Plasma proteins analogous to the corticosteroid-binding globulin (CBG) found in man seem to be present in all terrestrial vertebrates investigated so far. Glucocorticoids bound to these proteins are metabolically inert while unbound and probably also albumin-bound steroids are available to corticosteroid-sensitive tissues. However, once again the echidna proved to be the odd man out: Sernia found that high-affinity binding of corticosteroids is absent in echidna plasma; of this circumstance he says it "seems especially appropriate in the echidna where adrenocortical secretion is very low and hence the maintenance of a reservoir of inert corticosteroid during times of most need would be a luxury it could ill afford." The corticosteroid binding properties of platypus plasma are, as yet, unknown.

In addition to CBG, some eutherian species have a sex-hormone-binding globulin (SHBG) that strongly binds to 17β-estradiol, testosterone, and 5α-dihydrotestosterone. In the echidna these steroids are bound by two proteins which may be clearly separated by electrophoresis in polyacrylamide gel. One SHBG migrates with the pre-albumin region and binds only estradiol while the second SHBG has a low electrophoretic mobility and a high affinity for dihydrotestosterone. These peculiarities are not shared with marsupials. From a selection of 16 marsupial species representing 7 families, Sernia found that 7 species had a single SHBG with electrophoretic properties like eutherian SHBG but with a low affinity for estradiol. A similar study of four reptilian species (two elapid snakes, one scincid lizard, one tortoise) revealed binding proteins with affinities for both adrenocortical and sex steroids. Interestingly, a minor estradiol-binding protein reminiscent of one of the echidna SHBG was detected in the pre-albumin fraction of plasma from the scincid lizard *Tiliqua scincoides*.

Aldosterone Secretion

C. Sernia and I. McDonald (personal communication) have recently determined plasma concentrations of aldosterone in conscious and in anesthetized echidnas. In the former the level was found to be 5.4 ± 1.3 pg/ml and in the latter 17.6 ± 3.8. Relative to those of metatherians and eutherians the levels in conscious echidnas are very low. It is suggested that the higher levels found in anesthetized echidnas are due to stress resulting in increased secretion of aldosterone by the adrenals. Intravenous infusion of ACTH (5 iu/kg/hr) in conscious animals likewise increased plasma aldosterone concentration but to a much greater degree: 53.8 ± 9.8 pg/ml.

The rate of secretion of aldosterone by the adrenals into the blood, as measured by isotope dilution techniques was also found to be quite low—5.0 ± 2.2 ng/kg/hr, compared with those for eutherians. These figures for plasma aldosterone concentrations and for secretion rates are in agreement with the evidence that adrenalectomy has little effect on the general well-being of unstressed echidnas (Chapter 6).

As in other mammals angiotensin II (p. 156) infused intravenously into echidnas at the rate of 0.1–0.5 μg/kg/hr increased plasma aldosterone levels to 24.1 ± 8.6 and 35.1 ± 10.9 pg/ml, respectively.

INSTRUMENTAL DISCRIMINATION—REVERSAL LEARNING IN ECHIDNAS

O. L. K. Buchmann and J. Rhodes (in press) have presented further evidence of the capacity of *Tachyglossus aculeatus,* in this instance *setosus,* to learn to make visual discriminations, to acquire a position habit, and to perform spatial habit reversal tasks, but with a difference: the echidnas were taught to operate a piece of apparatus and were tested by the use of repeated reversals of paired discrimination tasks.

The apparatus consisted essentially of a pair of treadles, capable of being operated simultaneously or independently, which when depressed (requiring a pressure of 25 g or more) activated a food dispenser situated some distance away from the treadles. This delivery of food was the reinforcement for the learning process.

The echidnas quickly learned to operate these treadles in order to get the reward of food and after mastering this procedure they had to learn to operate the correct treadle with the aid of spatial and visual-tactile cues. The results of the study confirm the findings described in Chapter 7 including the exhibition of one-trial learning and show that echidnas can achieve success in spatial-reversal tests employing operant rather than simple maze techniques. The authors also conclude that the overall performance of the echidnas was in no way inferior to those of eutherians.

ULTRASTRUCTURE AND PRECURSORS OF THE EGGSHELL OF THE MONOTREMES

Hughes (1977) has recently published electron micrographs of the shell of a near-term platypus egg whose long and short axes were 17 mm and 15 mm, respectively. From these it is apparent that the mucoid coat (identified by its ultrastructural properties) persists to full term. The open-lattice structure of the eggshell was attested by the fact that an erythrocyte was identified in the basal layer. A shell of an echidna egg exhibited similar ultrastructure.

Of great interest is Hughes' observation of presumptive shell membrane precursors in the cytoplasm of nonciliated secretory cells in the uterine endometrial glands of a platypus. Ultrastructurally these very much resemble those found in cells of the isthmus region of the avian oviduct. One wonders if the shell membrane precursors in marsupial reproductive tracts have like ultrastructure and if the ultrastructure of the freshly formed basal layer of the monotreme egg resembles that of the freshly-formed marsupial egg shell.

CARBOHYDRATES OF PLATYPUS MILK

M. Messer (personal communication) has carried out analyses for carbohydrates in two further samples of platypus milk. I took these from a total of four platypuses: one milked October 14, 1977 and a pooled sample from three milked December 7, 1977. These are earlier milks than those available for analysis previously and contained much larger amounts of carbohydrate than the sample taken late in the lactation season (p. 265). The October sample exhibited 3.5% hexose and 0.41% bound sialic acid w/w and the December sample 2.1% hexose and 0.37% sialic acid, whereas the late milk contained 1.7% total carbohydrate and practically no sialic acid.

It will be recalled that the milk of *Tachyglossus* contained sialic acid in the form of sialyl-lactose; it would appear likely, therefore, that the sialic acid in platypus milk will prove to be bound to lactose. As in late milk the present early samples contain difucosyllactose and practically no free lactose.

TEMPORAL WING OF THE PALATINE IN
ORNITHORHYNCHUS

Recently I acquired the skull,* of a juvenile platypus of body weight 470 g and length 29 cm, which exhibited quite clearly temporal wings of the palatines. It will be recalled that Watson (p. 314) disagreed with van Bemmelen's statement that the palatine bore an ala temporalis palatini. In this specimen, however, the anterior lamina of the periotic has not united to the alisphenoid, thus leaving a longitudinal gap communicating with the foramen rotundum anteriorly and the foramen ovale posteriorly. Just anterior to the alisphenoid on either side it can be seen that each palatine exhibits a rostrodorsally directed projection, quite sharp distally. It is apparent that the periotic lamina will grow down and fuse to the palatine wing and form the posterior margin of the foramen rotundum, thus obscuring its presence in later stages.

A temporal wing of the palatine would appear to be a characteristic of the monotreme skull but its presence in the skull of *Obdurodon,* if ever one is found, may be impossible to demonstrate since it cannot be detected in the adult skull of *Ornithorhynchus.* Finally it may be mentioned that van Bemmelen was right in stating that all three living genera had an ala temporalis palatini.

*Held in the collections of the Division of Wildlife Research, Canberra, and available for study.

References

Abbie, A. A. (1934). *Philos. Trans. R. Soc. London, Ser. B* **224**, 1–74.

Acher, R. (1971). *In* "Biochemical Evolution and the Origin of Life" (E. Schoffeniels, ed.), pp. 143–51. North-Holland Publ., Amsterdam.

Acher, R., Chauvet, J., and Chauvet, M. T. (1973). *Nature (London), New Biol.* **244**, 124–126.

Air, G. M., Thompson, E. O. P., Richardson, B. J., and Sharman, G. B. (1971). *Nature (London)* **229**, 391–394.

Aitken, L. M., and Johnstone, B. M. (1972). *J. Exp. Zool.* **180**, 245–250.

Aleksiuk, M., and Baldwin, J. (1973). *Can. J. Zool.* **51**, 17–19.

Alexander, G. (1904). *Denkschr. Med.-Naturwiss. Ges. Jena* **6**, Part 2, 3–118.

Allen, G. M. (1912). *Mem. Mus. Comp. Zool. Harv. Coll.* **40**, 253–307.

Allin, E. F. (1975). *J. Morphol.* **147**, 403–438.

Allison, T., and Goff, W. R. (1972). *Arch. Ital. Biol.* **110**, 195–216.

Allison, T., and Van Twyver, H. (1970). *Nat. Hist.*, N.Y. **79**, 56–65.

Allison, T., and Van Twyver, H. (1972). *Arch. Ital. Biol.* **110**, 185–194.

Allison, T., Van Twyver, H., and Goff, W. R. (1972). *Arch. Ital. Biol.* **110**, 145–184.

Allport, M. (1878). *Pap. Proc. R. Soc. Tasmania* 1878, 30–31.

Andres, K. H. (1970). *Taste Smell Vertebr., Ciba Found. Symp., 1969* pp. 177–196.

Arai, Y., Kubokura, A., Suzuki, Y., and Masuda, S. (1970). *Endocrinol. Jpn.* **17**, 441–446.

Archer, M., Plane, M. D., and Pledge, N. S. *Aust. Zool.,* in press.

Armit, W. E. (1878). *J. Linn. Soc. London* **14**, 411–413.

Arnold, J., and Shield, J. (1970). *J. Zool.* **160**, 391–404.

Atkins, A. M., and Schofield, G. C. (1971). *J. Anat.* **108**, 209.

Augee, M. L. (1969). Ph.D. Thesis, Monash, University, Clayton, Victoria, Australia.

Augee, M. L. (1975). *Koolewong* **4**, 4–7.

Augee, M. L. (1976). *J. Thermal Biol.* **1**, 181–184.

Augee, M. L., and Ealey, E. H. M. (1968). *J. Mammal.* **49**, 446–454.

Augee, M. L., and Grant, T. R. (1974). *Bull. Aust. Mammal Soc.* **3**, 18.

Augee, M. L., and McDonald, I. R. (1973). *J. Endocrinol.* **58**, 513–523.

Augee, M. L., Ealey, E. H. M., and Spencer, H. (1970). *J. Mammal.* **51**, 561–570.

Augee, M. L., Elsner, R. W., Gooden, B. A., and Wilson, P. R. (1971). *Respir. Physiol.* **11**, 327–334.

Augee, M. L., Ealey, E. H. M., and Price, I. P. (1975). *Aust. Wildl. Res.* **2**, 93–101.

Baggett, H. *Aust. Zool.,* in press.

Baird, J. A., Hales, J. R. S., and Lang, W. J. (1974). *J. Physiol. (London)* **236,** 539–548.

Baldwin, J. (1973). *Proc. Aust. Biochem. Soc.* **6,** 70.

Baldwin, J., and Aleksiuk, M. (1973). *Comp. Biochem. Physiol. B* **44,** 363–370.

Baldwin, J., and Temple-Smith, P. (1973). *Comp. Biochem. Physiol. B* **46,** 805–811.

Barer, R., Heller, H., and Lederis, K. (1963). *Proc. R. Soc. London* **158,** 388–416.

Barrett, C. (1942). *Wildlife* **4,** 350.

Barton, B. S. (1806). Reprint of letter to Monsieur Roume published in *Ann. Philos.* |N.S.| **6,** 349–354 (1823).

Basir, M. A. (1932). *J. Anat.* **66,** 628–649.

Bates, W. H. (1956). "Good Sight Without Glasses." Faber & Faber, London.

Benda, C. (1906). *Denkschr. Med.-Naturwiss. Ges. Jena* **6,** Part 2, 413–438.

Benedict, F. G., and Lee, R. C. (1938). *Carnegie Inst. Washington Publ.* **497,** 1–239.

Bennett, D. J., and Baldwin, J. (1972). *Proc. Aust. Biochem. Soc.* **5,** 6.

Bennett, G. (1835). *Trans. Zool. Soc. London* **1,** 229–258.

Bentley, P. J., and Schmidt-Nielsen, K. (1967). *Comp. Biochem. Physiol.* **20,** 285–290.

Bentley, P. J., Herreid, C. F., and Schmidt-Nielsen, K. (1967). *Am. J. Physiol.* **212,** 957–961.

Bergman, S. (1961). "My Father is a Cannibal." Robert Hale Ltd., London.

Berka, F. (1911). *Frankf. Z. Pathol.* **8,** 203–256.

Bick, Y. A. E., and Jackson, W. D. (1967a). *Am. Nat.* **101,** 79–86.

Bick, Y. A. E., and Jackson, W. D. (1967b). *Nature (London)* **214,** 600–601.

Bick, Y. A. E., and Sharman, G. B. (1975). *Cytobios* **14,** 17–28.

Bick, Y. A. E., Murtagh, C., and Sharman, G. B. (1973). *Cytobios* **7,** 233–243.

Biggers, J. D. (1966). *Symp. Zool. Soc. London* **15,** 251–277.

Birney, E. C., Jenness, R., and Hume, I. (1978). *Science,* in press.

Blandau, R. J. (1961). *In* "Sex and Internal Secretions" (W. C. Young, ed.), 3rd ed., Vol. 2, pp. 797–882. Williams & Wilkins, Baltimore, Maryland.

Blaxter, K. L., and Wainman, F. W. (1961). *J. Agric. Sci.* **56,** 81–90.

Bligh, J. (1961). *Nature (London)* **189,** 582–583.

Blumenbach, J. F. (1800). *Bull. Soc. Philomath. Paris* **2.**

Bohringer, R. C. (1976). *J. Anat.* **121,** 417.

Bohringer, R. C. (1977). *J. Anat.* **124,** 532.

Bohringer, R. C., and Rowe, M. J. (1977). *J. Comp. Neurol.* **174,** 1–14.

Bolliger, A., and Backhouse, T. C. (1960). *Proc. Zool. Soc. London* **135,** 91–97.

Bowler, J., Hope, G. S., Jennings, J. N., Singh, G., and Walker, D. (1976). *Quat. Res. (N.Y.)* **6,** 359–394.

Braden, A. W. H. (1952). *Aust. J. Sci., Ser. B* **5,** 460–471.

Bradley, S. R., and Hudson, J. W. (1974). *Comp. Biochem. Physiol. A* **48,** 55–60.

Braithwaite, L. W., and Miller, B. (1975). *Aust. Wildl. Res.* **2,** 47–61.

Brattstrom, B. H. (1973). *J. Mammal.* **54,** 50–70.

Braxton, D., and Knight, F. (1974). "Bush Families of Tidbinbilla." Australian Government Publishing Service, Canberra.

Bremer, J. L. (1904). *Am. J. Anat.* **3,** 67–73.

Bresslau, E. (1920). "The Mammary Apparatus of the Mammalia." Methuen, London.

Bridgewater, R. J., Haslewood, G. A. D., and Tammar, A. R. (1962). *Biochem. J.* **85**, 413–416.

Briggs, E. A. (1936). *Nature (London)* **138**, 762.

Brody, S. (1945). "Bioenergetics and Growth." Hafner, New York.

Broom, R. (1895). *Proc. Linn. Soc. N.S.W.* **10**, 576–577.

Broom, R. (1897). *J. Anat.* **31**, 513–515.

Broom, R. (1909). *Proc. Linn. Soc. N.S.W.* **34**, 195–214.

Broom, R. (1914). *Bull. Am. Mus. Nat. Hist.* **33**, 115–134.

Bruesch, S. R., and Arey, L. B. (1942). *J. Comp. Neurol.* **77**, 631–665.

Buchanan, G., and Fraser, E. A. (1918). *J. Anat.* **53**, 35–95.

Buchmann, O. L. K., and Rhodes, J. *Aust. Zool.,* in press.

Burrell, H. (1927). "The Platypus." Angus & Robertson, Sydney, Australia.

Cabrera, A. (1919). "Genera Mammalium Monotremata Marsupialia." Museo Nacional de Ciencias Naturales, Madrid.

Caldwell, W. H. (1884a). *Q. J. Microsc. Sci.* [N.S.] **24**, 655–658.

Caldwell, W. H. (1884b). Telegram "Monotremes oviparous, ovum meroblastic" (read in Montreal Sept. 2, 1884). Br. Assoc. Rep. Montreal Meet. 1884.

Caldwell, W. H. (1887). *Philos. Trans. R. Soc. London, Ser. B* **178**, 463–486.

Campbell, C. B. G., and Hayhow, W. R. (1971). *J. Comp. Neurol.* **143**, 119–136.

Campbell, C. B. G., and Hayhow, W. R. (1972). *J. Comp. Neurol.* **145**, 195–208.

Carmichael, G. G., and Krause, W. J. (1972). *J. Anat.* **111**, 209.

Carrick, F. N., and Cox, R. I. (1973). *J. Reprod. Fertil.* **32**, 338–339.

Carrick, F. N., Drinan, J. P., and Cox, R. I. (1975). *J. Reprod. Fertil.* **43**, 375–376.

Carrick, F. N., Hughes, R. L., Drinan, J. P., Cox, R. I., and Shorey, C. D. (1976). *Bull. Aust. Mammal Soc.* **3**, 22-23.

Caughley, G. (1977). "Analysis of Vertebrate Populations." Wiley, London.

Cave, A. J. E. (1970). *J. Zool.* **160**, 297–312.

Clausen, G., and Ersland, A. (1968). *Respir. Physiol.* **5**, 350–359.

Clemens, W. A. (1968). *Evolution* **22**, 1–18.

Clemens, W. A. (1970). *Annu. Rev. Ecol. Syst.* **1**, 357–390.

Clemens, W. A. (1971). *In* "Early Mammals" (D. M. Kermack and K. A. Kermack, eds.), pp. 165–180. Academic Press, New York.

Clemens, W. A. (1977). *In* "The Biology of the Marsupials" (B. Stonehouse and D. Gilmour, eds.), pp. 51–68. Macmillan, London.

Collett, R. (1885). *Proc. Sci. Meet., Zool. Soc. London* pp. 148–161.

Collins, D. (1802). "An Account of the English Colony in New South Wales." T. Cadell, Jr. and W. Davies, in The Strand, London.

Cooper, D. W., VandeBerg, J. L., Griffiths, M. E., and Ealey, E. H. M. (1973). *Aust. J. Biol. Sci.* **26**, 605–612.

Cope, E. D. (1888). *Am. Nat.* **22**, 258.

Cowie, A. T. (1972). *In* "Reproduction in Mammals" (C. R. Austin and R. V. Short, eds.), Vol. 3, pp. 106–143. Cambridge Univ. Press, London and New York.

Cowie, A. T., and Tindal, J. S. (1971). "The Physiology of Lactation," Monogr. Physiol. Soc., No. 22. Arnold, London.

CSIRO (1969). "The Echidna or Spiny Anteater." C.S.I.R.O. Film and Video Centre, Melbourne.

CSIRO (1974). "The Comparative Biology of Lactation." C.S.I.R.O. Film and Video Centre, Melbourne.

Crompton, A. W. (1972). *In* "Studies in Vertebrate Evolution" (K. A. Joysey and T. S. Kemp, eds.), pp. 231–251. Oliver & Boyd, Edinburgh.

Crompton, A. W. (1974). *Bull. Br. Mus. (Nat. Hist.), Geol.* **24**, 399–437.

Crompton, A. W., and Jenkins, F. A. (1968). *Biol. Rev. Cambridge Philos. Soc.* **43**, 427–458.

Crompton, A. W., and Jenkins, F. A. (1973). *Annu. Rev. Earth Planet. Sci.* **1**, 131–155.

Crompton, A. W., and Parker, P. (1978). *Am. Sci.* **66**, 192–201.

Crowther, A. B. (1879). *Pap. Proc. R. Soc. Tasmania* 1879, 96–99.

Cumpston, J. S. (1964). "Shipping Arrivals and Departures Sydney, 1788–1825." Publisbed by Author, Canberra, Australia.

Dabelow, A. (1957). *In* "Handbuch der mikroscopischen Anatomie des Menschen" (W. Möllendorff and W. Bargman, eds.), Vol. 3, pp. 277–439. Springer-Verlag, Berlin and New York.

Dalquest, W. W., and Werner, H. J. (1952). *Am. Midl. Nat.* **48**, 250–252.

Davy, J. (1840). *L'Institut* **8**, 441.

Dawson, T. J. (1973). *In* "Comparative Physiology of Thermoregulation" (G. C. Whittow, ed.), Vol. 3, pp. 1–46. Academic Press, New York.

Dawson, T. J., and Hulbert, A. J. (1970). *Am. J. Physiol.* **218**, 1233–1238.

Dawson, T. J., Fanning, D., and Bergin, T. J. *Aust. Zool.*, in press.

Dayhoff, M. O., ed. (1972). "Atlas of Protein Sequence and Structure." Natl. Biomed. Res. Found., Silver Spring, Maryland.

de Almeida, O., and de Fialho, B. A. (1924). *C.R. Seances Soc. Biol. Ses. Fil.* **91**, 1124–1125.

de Beer, G. R. (1926). "Anatomy, Histology, and Development of the Pituitary Body." Oliver & Boyd, London.

de Beer, G. R. (1930). "Embryology and Evolution." Oxford Univ. Press (Clarendon), London and New York.

de Beer, G. R. (1937). "The Development of the Vertebrate Skull." Oxford Univ. Press (Clarendon), London and New York.

de Beer, G. R., and Fell, W. A. (1936). *Trans. Zool. Soc. London* **23**, 1–43.

de Meijere, J. C. H. (1894). *Morphol. Jahrb.* **21**, 312–424.

Denker, A. (1901). *Denkschr. Med.-Naturwiss. Ges. Jena* **6**, Part 1, 637–662.

Denton, D. A., Reich, M., and Hird, F. (1963). *Science* **139**, 1225.

Devez, G. (1903). Thèses presentées à la Faculté de Sciences de Paris. Le Bigot Frères, Lille.

Diener, E., and Ealey, E. H. M. (1965). *Nature (London)* **208**, 950–953.

Diener, E., Wistar, R., and Ealey, E. H. M. (1967a). *Immunology* **13**, 329–337.

Diener, E., Ealey, E. H. M., and Legge, J. S. (1967b). *Immunology* **13**, 339–347.

Dillon, L. S. (1962). *J. Comp. Neurol.* **118**, 343–353.

Dodgson, S. J., Fisher, W. K., and Thompson, E. O. P. (1974). *Aust. J. Biol. Sci.* **27**, 111–115.

Doran, A. H. G. (1879). *Trans. Linn. Soc. London* **1**, 371–497.

Doran, G. A. (1973). *Anat. Anz.* **133**, 468–476.

Doran, G. A., and Baggett, H. (1970). *Anat. Rec.* **167**, 197–204.

Doran, G. A., and Baggett, H. (1972). *Anat. Rec.* **172**, 157–166.

Dowd, D. A. (1969). *Acta Anat.* **74**, 547–573.

Draper, E. A. (1971). Thesis presented for the degree of Doctor of Medicine in the University of Kansas, Lawrence.

Duke-Elder, W. S., ed. (1958). "System of Opthalmology." Kimpton, London.

Ebner, F. F. (1967). *J. Comp. Neurol.* **129**, 241–268.

Ebner, K. E., Denton, W. L., and Brodbeck. U. (1966). *Biochem. Biophys. Res. Commun.* **24**, 232–236.

Edmeades, R., and Baudinette, R. V. (1975). *Experientia* **31**, 935–936.

Eggeling, H. (1905). *Denkschr. Med.-Naturwiss. Ges. Jena* **7**, 299–332.

Fanning, F. D., and Dawson, T. J. (1977). *Bull. Aust. Mammal Soc.* **4**, 25.

Feakes, M. J., Hodgkin, E. P., Strahan, R., and Waring, H. (1950). *J. Exp. Biol.* **27**, 50–58.

Fernandez, C., and Schmidt. R. S. (1963). *J. Comp. Neurol.* **121**, 151–160.

Fewkes, J. W. (1877). *Bull. Essex Inst. Salem* **9**, 111–137.

Finck, A., and Sofouglu, M. (1966). *J. Aud. Res.* **6**, 313–319.

Fink, G., Smith, G. C., and Augee, M. L. (1975). *Cell Tissue Res.* **159**, 531–540.

Fink, R. P., and Heimer, L. (1967). *Brain Res.* **4**, 369–374.

Fisher, C., Ingram, W. R., and Ranson, S. W. (1938). "Diabetes Insipidus and the Neurohumoral Control of Water Balance: A Contribution to the Structure and Function of the Hypothalamico-hypophyseal System." Edwards, Ann Arbor, Michigan.

Fisher, W. K., and Thompson, E. O. P. (1976). *Aust. J. Biol. Sci.* **29**, 57–72.

Fitzhardinge, L. F. (1961). "Sydney's First Four Years." Angus & Robertson, Sydney, Australia.

Fleay, D. (1941). *Victorian Nat.* **57**, 214–215.

Fleay, D. (1944). *Victorian Nat.* **61**, 8–14, 29–37, 54–57, and 74–78.

Fleay, D. (1950). *Victorian Nat.* **67**, 81–87.

Flynn, T. T., and Hill, J. P. (1939). *Trans. Zool. Soc. London* **24**, 445–622.

Flynn, T. T., and Hill, J. P. (1947). *Trans. Zool. Soc. London* **26**, 1–151.

Folley, S. J. (1969). *J. Endocrinol.* **44**, ix–xx.

Fourie, S. (1963). *Nature (London)* **198**, 201.

Franz, V. (1934). *In* "Handbuch der vergleichenden Anatomie der Wirbeltiere" (L. Bolk *et al.*, eds.), Vol. 2, pp. 989–1292. Urban & Schwarzenberg, Berlin.

Fraser, R. D. B. (1969). *Sci. Am.* **221**, 87–96.

Frauca, H. (1969). *Walkabout* March, pp. 34–36.

Frisch, J., Oritsland, N. A., and Krog, J. (1974). *Comp. Biochem. Physiol. A* **47**, 403–410.

Gabe, M. (1970). *In* "Biology of the Reptilia" (C. Gans and T. Parsons, eds.), Vol. 3, pp. 263–318. Academic Press, New York.

Gates, G. R. (1973). M. Sc. Thesis, Monash University, Clayton, Victoria, Australia.

Gates, G. R., Saunders, J. C., Bock, G. R., Aitken, L. M., and Elliott, M. A. (1974). *J. Acoust. Soc. Am.* **56**, 152–156.

Gaupp, E. (1908). *Denkschr. Med.-Naturwiss, Ges. Jena* **6**, Part 2, 539–788.

Gegenbaur, C. (1886). "Zur Kenntniss der Mammarorgane der Monotremen." Engelmann, Leipzig.

Gersh, I. (1937). *Contrib. Embryol. Carnegie Inst.* **153**, 35–58.

Gervais, P. (1877–1878). "Ostéographie des Monotrèmes Vivants et Fossiles," Chapter 2. Librairie Scientifique et Maritime, Paris.

Gill, E. D. (1975). *Proc. R. Soc. Victoria* **87**, 215–234.

Gill, T. (1877). "Annual Record of Science and Industry for 1876." Harper, New York.

Gillespie, J. M., and Inglis, A. S. (1965). *Comp. Biochem. Physiol.* **15**, 175–185.

Gilmour, D. (1961). "Biochemistry of Insects." Academic Press, New York.

Glass, R. L., and Jenness, R. (1971). *Comp. Biochem. Physiol. B* **38**, 353–357.

Glass, R. L., Troolin, H. A., and Jenness, R. (1967). *Comp. Biochem. Physiol.* **22**, 415–425.

Goldby, F. (1939). *J. Anat.* **73**, 509–524.

Goodrich, E. S. (1958). "Studies on the Structure and Development of Vertebrates." Dover, New York.

Göppert, E. C. (1894). *Morphol. Jahrb.* **21**, 278–280.

Gottschalk, A., Bhargava, A. S., and Murty, V. L. N. (1966). "Glycoproteins," 1st ed., Part B. Elsevier, Amsterdam.

Grant, T. R. (1976). Ph.D. Thesis, University of New South Wales, Sydney, Australia.

Grant, T. R., and Carrick, F. N. *Aust. Zool.,* in press.

Grant, T. R., and Dawson, T. J. (1978a). *Physiol. Zool.* **51**, 1–6.

Grant, T. R., and Dawson, T. J. (1978b). *Physiol. Zool.* **51**.

Grant, T. R., and McBlain, R. B. (1976). *Bull. Aust. Mammal Soc.* **3**, 41–42.

Gray, A. A. (1908). *Proc. R. Soc. London. Ser. B* **80**, 507–530.

Green, A. A., and Redfield, A. C. (1933). *Biol. Bull. (Woods Hole, Mass.)* **64**, 44–52.

Green, H. L. H. H. (1930). *J. Anat.* **64**, 512–522.

Green, H. L. H. H. (1937). *Philos. Trans. R. Soc. London, Ser. B* **288**, 367–420.

Gregory, W. K. (1947). *Bull. Am. Mus. Nat. Hist.* **88**, 7–52.

Gregory, W. K., and Noble, G. K. (1924). *J. Morphol. Physiol.* **39**, 435–463.

Gresser, E. B., and Noback, C. V. (1935). *J. Morphol.* **58**, 279–284.

Griffiths, M. (1952). *J. Biol. Chem.* **197**, 399–407.

Griffiths, M. (1965a). *Comp. Biochem. Physiol.* **14**, 357–375.

Griffiths, M. (1965b). *Comp. Biochem. Physiol.* **16**, 383–392.

Griffiths, M. (1968). "Echidnas." Pergamon, Oxford.

Griffiths, M. (1972). *Aust. Nat. Hist.* **17**, 222–226.

Griffiths, M., and Simpson, K. S. (1966). *CSIRO Wildl. Res.* **11**, 137–143.

Griffiths, M., McIntosh, D. L., and Coles, R. E. A. (1969). *J. Zool.* **158**, 371–386.

Griffiths, M., McIntosh, D. L., and Leckie, R. M. C. (1972). *J. Zool.* **166**, 265–275.

Griffiths, M., Elliott, M. A., Leckie, R. M. C., and Schoefl, G. I. (1973). *J. Zool.* **169**, 255–279.

Haacke, W. (1885). *Proc. R. Soc. London, Ser. B* **38**, 72–74.

Halata, Z. (1972). *Z. Zellforsch. Mikrosk. Anat.* **125**, 108–120.

Hanström, B., and Wingstrand, K. G. (1951). *Lunds Univ. Arsskr., Avd. 2* [N.S.] **47**, No. 16.

Hart, J. S. (1962). *In* "Comparative Physiology of Temperature Regulation" (J.P. Hannon and E. Viereck, eds.), pp. 203–242. Arctic Aerosp. Lab., Fort Wainwright, Alaska.

Hashimoto, Y. *et al.* (1964). *Arch. Biochem. Biophys.* **104**, 282–291.

Hausmann, L. A. (1920). *Am. J. Anat.* **27**, 463–487.

Hayward, J. (1966). *J. Mammal.* **47**, 723–724.

Hearn, J. P. (1972). Ph.D. Thesis, Australian National University, Canberra.

Heath, C. J., and Jones, E. G. (1971). *J. Anat.* **109**, 253–270.

Heck, L. (1908). *Ges. Naturforch. Freunde, Berlin Sitzungsber.* pp. 187–189.

Henkel, S., and Krebs, B. (1977). *Umschau* **77**, 217–218.

Hill, C. J. (1933). *Trans. Zool. Soc. London* **21**, 413–443.

Hill, C. J. (1941). *Trans. Zool. Soc. London* **25**, 1–31.

Hill, J. P. (1910). *Q. J. Microsc. Sci.* [N.S.] **56**, 1–134.

Hill, J. P. (1933). *Trans. Zool. Soc. London* **21**, 443–476.

Hill, J. P., and de Beer, G. R. (1949). *Trans. Zool. Soc. London* **26**, 503–544.

Hill, J. P., and Gatenby, J. B. (1926). *Proc. Zool. Soc. London* **47**, 715–763.

Hill, J. P., and Hill, W. C. O. (1955). *Trans. Zool. Soc. London* **28**, 349–452.

Hines, M. (1929). *Philos. Trans. R. Soc. London, Ser. B* **217**, 155–287.

Hochstetter, F. (1896). *Denkschr. Med.-Naturwiss. Ges. Jena* **5**, 191–243.

Hodge, P., and Wakefield, N. (1959). *Victorian Nat.* **76,** 64–65.

Hollis, D. E., and Lyne, A. G. (1977). *Aust. J. Zool.* **25**, 207–223.

Home, E. (1802a). *Philos. Trans. R. Soc. London* pp. 67–83.

Home, E. (1802b). *Philos. Trans. R. Soc. London* pp. 348–364.

Hope, G. S. (1975). "Some Notes on the Non-forest vegetation and its Environment, Northern Flank of Mt. Albert Edward, Central District, Papua New Guinea." Department of Biogeography and Geomorphology, Australian National University (unpublished preliminary report).

Hope, G. S., and Hope, J. H. (1976). *In* "The Equatorial Glaciers of New Guinea" (G. S. Hope *et al.*, eds.), pp. 225–239. A. A. Balkema, Rotterdam.

Hope, J. H. (1976). *In* "The Equatorial Glaciers of New Guinea" (G. S. Hope *et al.*, eds.), pp. 207–224. A. A. Balkema, Rotterdam.

Hope, J. H. (1977). *In* "The Melanesian Environment" (J. Winslow, ed.), 21–27. Aust. Natl. Univ. Press, Canberra.

Hopper, K. E. (1971). Ph.D. Thesis, Aust. Natl. Univ., Canberra.

Hopper, K. E., and McKenzie, H. A. (1974). *Mol. Cell. Biochem.* **3**, 93–108.

Hopson, J. A. (1964). *Postilla* **87**, 1–30.

Hopson, J. A. (1966). *Am. Zool.* **6**, 437–450.

Hopson, J. A. (1969). *Ann. N.Y. Acad. Sci.* **167**, 199–216.

Hopson, J. A. (1970). *J. Mammal.* **51**, 1–9.

Hopson, J. A., and Crompton, A. W. (1969). *Evol. Biol.* **3,** 15–72.

Hüber, E. (1930). *Q. Rev. Biol.* **5**, 133–188.

Hudson, J. W. (1973). *In* "Comparative Physiology of Thermoregulation" (G. C. Whittow, ed.), Vol. 3, pp. 98–165. Academic Press, New York.

Hughes, R. L. (1962). *Aust. J. Zool.* **10**, 193–224.

Hughes, R. L. (1969). *J. Reprod. Fertil.* **19**, 387.

Hughes, R. L. (1974). *J. Reprod. Fertil.* **39**, 173–186.

Hughes, R. L. (1977). *In* "Reproduction and Evolution" (J. H. Calaby and C. H. Tyndale-Biscoe, eds.), pp. 281–291. Aust. Acad. Sci., Canberra.

Hughes, R. L., Thomson, J. A., and Owen, W. H. (1965). *Aust. J. Zool.* **13**, 383–406.

Hughes, R. L., Carrick, F. N., and Shorey, C. D. (1975). *J. Reprod. Fertil.* **43**, 374–375.

Hulbert, A. J., and Dawson, T. J. (1974). *Comp. Biochem. Physiol. A* **47,** 583–590.

Hydén, S., and Sperber, I. (1965). *In* "Physiology of Digestion in the Ruminants" (A. W. Dougherty, ed.), pp. 51–60. Butterworth, London.

Hyrtl, J. (1853). *Denkschr. Akad. Wiss. Wien.* **5**, 1–20.

Ingram, D. L. (1965). *Nature (London)* **207**, 415–416.

Insull, W., Hirsch, J., James, T., and Ahrens, E. H. (1959). *J. Clin. Invest.* **38**, 443–450.

Iredale, T., and Troughton, E. (1934). "A Checklist of Mammals Recorded from Australia," Mem. No. 6. Aust. Mus., Sydney.

Irving, L., Solandt, O., Solandt, D. Y., and Fisher, K. C. (1935). *J. Cell. Comp. Physiol.* **6**, 393–403.

Jacobsohn, D. (1961). *In*: "Milk: The Mammary Gland and its Secretion" (S. K. Kon and A. T. Cowie, eds.), Vol. 1, pp. 127–160. Academic Press, New York.

Jenkins, F. A. (1969). *Anat. Rec.* **164**, 173–184.

Jenkins, F. A. (1970a). *Science* **168**, 1473–1475.

Jenkins, F. A. (1970b). *Evolution* **24**, 230–252.

Jenkins, F. A. (1971). *J. Zool.* **165**, 303–315.

Jenkins, F. A. (1973). *Am. J. Anat.* **137**, 281–298.

Jenkins. F. A., and Parrington, F. R. (1976). *Philos. Trans. R. Soc. London, Ser. B* **273**, 387–431.

Johansen, K., Lenfant, C., and Grigg, G. C. (1966). *Comp. Biochem. Physiol.* **18**, 597–608.

Johnson, G. L. (1901). *Philos. Trans. R. Soc. London, Ser. B* **194**, 1–82.

Johnson, J. I. (1977). *In* "The Biology of Marsupials" (D. Hunsaker, II, ed.), Academic Press, New York.

Johnson, J. I., Rubel, E. W., and Hatton, G. I. (1974). *J. Comp. Neurol.* **158**, 81–108.

Johnstone, B. M., and Johnstone, J. R. (1965). *Int. Symp. Comp. Neurophysiol.*, Abstract No. 12A-2.

Jollie, M. (1962). "Chordate Morphology. Reinhold, New York.

Jones, F. W. (1923). "The Mammals of South Australia," Part I. Government Printer, Adelaide, Australia.

Jordan, S. M., and Morgan, E. H. (1969). *Comp. Biochem. Physiol.* **29**, 383–391.

Kaldor, I., and Ezekiel, E. (1962). *Nature (London)* **196**, 175.

Kayser, C. (1961). "The Physiology of Natural Hibernation." Pergamon, Oxford.

Keibel, F. (1904a). *Denkschr. Med.-Naturwiss. Ges. Jena* **6**, Part 2, 151–206.

Keibel, F. (1904b). *Denkschr. Med.-Naturwiss. Ges. Jena* **6**, Part 2, 207–228.

Kerbert, C. (1911). *Handel. Vlaam. Nat.-Geneeskd. Congr., 15th,* 1911 pp. 89–96.

Kerbert, C. (1913a). *Zool. Anz.* **42**, 162–167.

Kerbert, C. (1913b). *Bijdr. Dierkd* **19**, 167–184.

Kermack, K. A. (1963). *Philos. Trans. R. Soc. London, Ser. B* **246**, 83–103.

Kermack, K. A. (1967). *J. Linn. Soc. London, Zool.* **47**, 241–249.

Kermack, K. A., and Kielan-Jaworowska, Z. (1971). *In* "Early Mammals" (D. M. Kermack and K. A. Kermack, eds.), pp. 103–115. Academic Press, New York.

Kermack, K. A., Mussett, F., and Rigney, H. W. (1973). *J. Linn. Soc. London, Zool.* **53**, 87–175.

Kerry, K. (1969). *Comp. Biochem. Physiol.* **29**, 1015–1022.

Kersbaw, J. A. (1912). *Victoria Nat.* **19**, 102–106.

Kesteven, H. L. (1918). *J. Anat.* **52**, 449–466.

Kielan-Jaworowska, Z. (1969). *Nature (London)* **222**, 1091–1092.

Kielan-Jaworowska, Z. (1970). *Nature (London)* **226**, 974–976.

Kirsch, J. A. W. (1977). *In* "Biology of the Marsupials" (D. Hunsaker, ed.), pp. 1–50. Academic Press, New York.

Klaatsch, H. (1892). *Morphol. Jahrb.* **19**, 548–552.

Kleiber, M. (1961). "The Fire of Life." Wiley, New York.

Koelliker, A. (1901). "Die Medulla Oblongata und die Vierhügelgegend von Ornithorhynchus und Echidna." Engelmann, Leipzig.

Kohno, S. (1925). *Z. Gesamte Anat., Abt. 1* **77**, 419–479.

Kolmer, W. (1925). *Z. Wiss. Zool.* **125**, 448–482.

Kolmer, W. (1929). *Pflueger's Arch. Gesamte Physiol. Menschen Tiere* **221**, 319–320.

Korner, P. I., and Darian-Smith, I. (1954). *Aust. J. Exp. Biol. Med. Sci.* **32**, 499–510.

Krause, W. J. (1970). *Anat. Rec.* **167**, 473–488.

Krause, W. J. (1971a). *Am. J. Anat.* **132**, 147–166.

Krause, W. J. (1971b). *Acta Anat.* **80**, 435–448.

Krause, W. J. (1972). *Anat. Rec.* **172**, 603–622.

Krause, W. J. (1975). *Anat. Rec.* **181**, 251–266.

Krause, W. J., and Leeson, C. R. (1973). *Am. J. Anat.* **137**, 337–356.

Krause, W. J., and Leeson, C. R. (1974). *J. Morphol.* **142**, 285–300.

Kühn, H. (1971). *Abh. Senckenb. Naturforsch. Ges.* **528**.

Kühne, W. G. (1973). *Z. Morphol. Tiere* **75**, 59–64.

Kühne, W. G. (1977). *J. Nat. Hist.* **11**, 225–228.

Lake, J. S. (1966). *N.S.W. State Fish. Res. Bull.* **7**.

Lang, E. M. (1967). *Zolli, Bull. Zool. Gart. Basel* **19**, 5–8.

Larsell, O., McCrady, E., and Zimmermann, A. A. (1935). *J. Comp. Neurol.* **63**, 95–118.

Lashley, K. S. (1939). *J. Comp. Neurol.* **70**, 45–67.

Laurie, E. (1952). *Bull. Br. Mus. (Nat. Hist.), Zool.* **1**, 273–274.

Lee, I. (1920). "Captain Bligh's Second Voyage to the South Sea." Longmans, Green, New York.

Lende, R. A. (1963). *Science* **141**, 730–732.

Lende, R. A. (1964). *J. Neurophysiol.* **27**, 37–48.

Lende, R. A. (1969). *Ann. N.Y. Acad. Sci.* **167**, 262–276.

Lende, R. A., and Sadler, R. M. (1967). *Brain Res.* **5**, 390–405.

Lessertisseur, J., and Sigogneau, D. (1965). *Mammalia* **29**, 95–168.

Lew, J. Y., Heidelberger, M., and Griffiths, M. (1975). *Int. J. Pept. Protein Res.* **7**, 289–293.

Lewis, J. A., Phillips, L. L., and Hann, C. (1968). *Comp. Biochem. Physiol.* **25**, 1129–1135.

Leydig, F. (1859). *Arch. Anat. Physiol. Gahrgang 1859 Leipzig* pp. 677–747.

Lillegraven, J. A. (1969). "Latest Cretaceous Mammals of Upper Part of Edmonton Formation of Alberta, Canada, and Review of Marsupial-Placental Dichotomy in Mammalian Evolution." Univ. of Kansas Publ., Lawrence.

Lillegraven, J. A. (1975). *Evolution* **29**, 707–722.

Linzell, J. A. (1955). *J. Physiol. (London)* **130**, 257–267.

Linzell, J. A., and Peaker, M. (1971). *J. Physiol. (London)* **216**, 717–734.

Lombart, C., and Winzler, R. J. (1972). *Biochem. J.* **128**, 975–977.

Luckett, W. P. (1976). *Anat. Rec.* **184**, 466.

Lyne, A. G. (1974). *Aust. J. Zool.* **22**, 303–309.

Lyne, A. G., and Hollis, D. E. (1976). *Aust. J. Zool.* **24**, 361–382.

McCrady, E. (1938). "The Embryology of the Opossum," Am. Anat. Mem. No. 16. Wistar Inst. Anat. Biol., Philadelphia, Pennsylvania.

McDonald, I. R., and Augee, M. L. (1968). *Comp. Biochem. Physiol.* **27**, 669–678.

McIntyre, A. K., and Kenins, P. (1974). *Proc. Int. Union Physiol. Sci.* **11**, 237.

MacIntyre, G. (1967). *Evolution* **21**, 834–841.

McKenna, M. C. (1961). *Am. Mus. Novit.* **2066**, 1–7.

MacMillen, R. E., and Nelson, J. E. (1969). *Am. J. Physiol.* **217**, 1246–1251.

McMurchie, E. J., and Raison, J. K. (1975). *J. Therm. Biol.* **1**, 113–118.

Mahoney, J. A., and Ride, W. D. L. (1975). *West. Aust. Mus., Spec. Publ.* **6**.

Marchalonis, J. J., Ealey, E. H. M., and Diener, E. (1969). *Aust. J. Exp. Biol. Med. Sci.* **47**, 367–380.

Marston, H. R. (1926). *Aust. J. Exp. Biol. Med. Sci.* **3**, 217–220.

Martin, C. J. (1903). *Philos. Trans. R. Soc. London Ser. B* **195**, 1–37.

Martin, R. D. (1968). *Z. Tierpsychol.* **25**, 409–532.

Maule, L. (1832). *Proc. Zool. Soc. London* pp. 145–146.

Maurer, F. (1892). *Morphol. Jahrb.* **18** (quoted by Römer, 1898).

Maurer, F. (1899). *Denkschr. Med.-Naturwiss. Ges. Jena* **6**, Part 1, 403–444.

Meckel, J. F. (1827). *Meckel's Arch.* pp. 23–27.

Meckelio, I. F. (1826). "Ornithorhynchi paradoxi Descriptio Anatomica." Gerhardum Fleischerum, Lipsiae.

Merchant, J. C., and Sharman, G. B. (1966). *Aust. J. Zool.* **14**, 593–609.

Messer, M. (1974). *Biochem. J.* **139**, 415–420.

Messer, M., and Kerry, K. R. (1973). *Science* **180**, 201–202.

Miller, J. J., and Nossal, G. J. V. (1964). *J. Exp. Med.* **120**, 1075.

Miller, M. R., and Lagios, M. D. (1970). *In* "Biology of the Reptilia" (C. Gans and T. Parsons, eds.), Vol. 3, pp. 319–346. Academic Press, New York.

Mills, J. R. E. (1971). *In* "Early Mammals" (D. M. Kermack and K. H. Kermack, eds.), pp. 29–63. Academic Press, New York.

Montagna, W., and Ellis, R. A. (1960). *Anat. Rec.* **137**, 271–273.

Morrill, C. C. (1938). *Cornell. Vet.* **28**, 196–210.

Moss, M. L. (1969). *Am. Mus. Novit.* **2360**, 1–39.

Moss, M. L., and Kermack, K. A. (1967). *J. Dent. Res.* **46**, 745–747.

Müller, F. (1969). *Acta Anat.* **74**, 297–404.

Muntz, W. R. A., and Sutherland, N. S. (1963). *J. Comp. Neurol.* **122**, 69–77.

Murray, P. F. (1976). *Bull. Aust. Mammal Soc.* **3**, 61.

Murray, P. (1978). *Papers and Proc. Roy. Soc. Tasmania* **112**, 39–68.

Murtagh, C. E., and Sharman, G. B. (1977). *In* "Reproduction and Evolution" (J. H. Calaby and C. H. Tyndale-Biscoe, eds.), pp. 61–66. Aust. Acad. Sci., Canberra.

Narath, A. (1896). *Denkschr. Med.-Naturwiss. Ges. Jena* **5**, 247–274.

Nauta, W. J. H., and Gygax, P. A. (1954). *Stain Technol.* **29**, 91–93.

Neumeister, R. (1894). *Denkschr. Med.-Naturwiss. Ges. Jena* **5**, 59–74.

O'Day, K. (1938). *Med. J. Aust.* **1**, 326–328.

O'Day, K. (1952). *Trans. Ophthalmol. Soc. Aust.* **12**, 95–104.

Oppel, A. (1896). *Denkschr. Med.-Naturwiss. Ges. Jena* **5**, 275–300.

Oppel, A. (1897). *Denkschr. Med.-Naturwiss. Ges. Jena* **5**, 403–433.

Oppel, A. (1899). *Denkschr. Med.-Naturwiss. Ges. Jena* **7**, 105–172.

Owen, R. (1839–1847). "Monotremata" in "Cyclopaedia of Anatomy and Physiology," Vol. 3, Longman, Brown, Green, Longmans, and Roberts, London.

Owen, R. (1845). *Proc. Zool. Soc. London* 1845, 80–82.

Owen, R. (1865). Philos. Trans. R. Soc. London **155**, 671–686.

Owen, R. (1868). "On the Anatomy of Vertebrates," Vol. 3. Longmans, Green, New York.

Parer, J. T., and Hodson, W. A. (1974). *Respir. Physiol.* **21**, 307–316.

Parer, J. T., and Metcalfe, J. (1967a). *Respir. Physiol.* **3**, 136–142.

Parer, J. T., and Metcalfe, J. (1967b). *Respir. Physiol.* **3**, 143–150.

Parer, J. T., and Metcalfe, J. (1967c). *Respir. Physiol.* **3**, 151–159.

Parker, W. K. (1868). "A Monograph on the Structure and Development of the Shoulder Girdle in the Vertebrate." Ray Society, London.

Parrington, F. R. (1946). *Proc. Zool. Soc. London* **116**, 181–187.

Parrington, F. R. (1971). *Philos. Trans. R. Soc. London, Ser. B* **261**, 231–272.

Parrington, F. R. (1973). *J. Linn. Soc. London, Zool.* **52**, 85–95.

Parrington, F. R. (1974a). *J. Nat. Hist.* **8**, 231–233.

Parrington, F. R. (1974b), *J. Nat. Hist.* **8**, 421–426.

Parrington, F. R. (1978). *Philos. Trans. R. Soc. London, Ser. B*. **282**, 177–204.

Parsons, R. S., Atwood, J., Guiler, E. R., and Heddle, R. W. L. (1971). *Comp. Biochem. Physiol. B* **39**, 203–208.

Patterson, B., and Olson, E. C. (1961). *In* "International Colloquium on the Evolution of Lower and Non-specialized Mammals" (G. Vanderbroek, ed.), pp. 129–191. Paleis der Academien, Brussels.

Péron, F., and de Freycinet, L. (1807–1816). "Voyages de découvertes aux Terres Australes," 2 vols. Atlas, Paris.

Peters, W. C. H., and Doria, G. (1876). *Ann. Mus. Civ. Stor. Nat. Giacomo Doria* **9**, 183–187.

Pinkus, F. (1906). *Denkschr. Med.-Naturwiss. Ges. Jena* **6**, Part 2, 459–480.

Pizzey, G. (1975). Article in Melbourne Herald, 21.6.75.

Pocock, R. I. (1912). *Field* **3129**, 1231.

Poole, D. F. G. (1956). *Q. J. Microsc. Sci.* [N.S.] **97**, 303–312.

Poulton, E. B. (1885). *J. Physiol. (London)* **5**, xv–xvi.

Poulton, E. B. (1894). *Q. J. Microsc. Sci.* [N.S.] **36**, 143–200.

Pridmore, P. (1970). B. Sc. Honours Thesis, Monash University, Clayton, Victoria, Australia.

Prince, J. H. (1956). "Comparative Anatomy of the Eye." Thomas, Springfield, Illinois.

Pritchard, U. (1881). *Philos. Trans. R. Soc. London* **172**, 267–282.

Pucek, M. (1968). *In* "Venomous Animals and their Venoms" (W. Bücherl, V. Deulofen, and E. E. Buckley, eds.), Vol. 1, pp. 43–50. Academic Press, New York.

Racey, P. A., and Potts, D. M. (1970). *J. Reprod. Fertil.* **22**, 57–63.

Raison, J. K., and McMurchie, E. J. (1974). *Biochim. Biophys. Acta* **363**, 135–140.

Ramsay, E. P. (1877). *Proc. Linn. Soc. N.S.W.* **2**, 31–33.

Ramsay, E. P. (1878). *Proc. Linn. Soc. N.S.W.* **3**, 244.

Rasmussen, A. T. (1938). *Endocrinology* **23**, 263–278.

Reid, I. A. (1971). *Comp. Biochem. Physiol. A* **40**, 249–255.

Renfree, M. B. (1972). Ph.D. Thesis, Aust. Natl. Univ., Canberra.

Renfree, M. B. (1973). *Dev. Biol.* **32**, 41–49.

Renfree, M. B. (1975). *J. Reprod. Fertil.* **42**, 163–166.

Richardson, K. C. (1949). *Proc. R. Soc. London, Ser. B* **136**, 30–45.

Rienits, R., and Rienits, T. (1963). "Early Artists of Australia." Angus & Robertson, Sydney, Australia.

Roberts, A. C., Makey, D., and Seal, U. S. (1966). *J. Biol. Chem.* **241**, 4907–4913.

Roberts, R. C., and Seal, U. S. (1965). *Comp. Biochem. Physiol.* **16**, 327–331.

Robinson, K. W. (1954). *Aust. J. Biol. Sci.* **7**, 348–360.

Römer, F. (1898). *Denkschr. Med.-Naturwiss. Ges. Jena* **6**, Part 1, 189–241.

Rothschild, W. (1905). *Novit. Zool.* **12**, 305–306.

Rothschild, W. (1913). *Novit. Zool.* **20**, 188–191.

Rudin, W. (1974). *Acta Anat.* **89**, 481–515.

Saban, R. (1969a). *C.R. Hebd. Seances Acad. Sci* **268**, 2351–2354.

Saban, R. (1969b). *C.R. Hebd. Seances Acad. Sci.* **268**, 2779–2782.

Saint Girons, H. (1970). *In* "Biology of the Reptilia" (C. Gans and T. S. Parsons, eds.), Vol. 3, pp. 73–91. Academic Press, New York.

Saraiva, P., and Magalhaes-Castro, B. (1975). *Brain Res.* **90**, 181–193.

Saunders, J. C., Chen, C., and Pridmore, P. A. (1971a). *Anim. Behav.* **19**, 552–553.

Saunders, J. C., Teague, J., Slonim, D., and Pridmore, P. A. (1971b). *Aust. J. Psychol.* **23**, 47–51.

Schmalbeck, J., and Rohr, H. (1967). *Z. Zellforsch. Mikrosk. Anat.* **80**, 329–344.

Schmidt, R. S. (1963). *Comp. Biochem. Physiol.* **10**, 83–87.

Schmidt, R. S., and Fernandez, C. (1963). *J. Exp. Zool.* **153**, 227–236.

Schmidt-Nielsen, K., Dawson, T. J., and Crawford, E. C. (1966). *J. Cell. Comp. Physiol.* **67**, 63–71.

Schodde, R., van Tets, G. F., Champion, C. R., and Hope, G. S. (1975). *Emu* **75**, 65–72.

Schofield, G. C., and Cahill, R. N. P. (1969). *J. Anat.* **105**, 447–456.

Scholander, P. F., Irving, L., and Grinnell, S. W. (1943). *J. Cell. Comp. Biol.* **21**, 53–63.

Scholander, P. F., Walters, V., Hock, R., and Irving, L. (1950). *Biol. Bull. (Woods Hole, Mass.)* **99**, 225–236.

Schuknecht, H. F. (1970). *In* "Contributions to Sensory Physiology" (W. D. Neff, ed.), Vol. 4, pp. 75–93. Academic Press, New York.

Schultz, W. (1967). *Z. Anat. Entwicklungsgesch.* **126**, 303–319.

Schuster, E. (1910). *Proc. R. Soc. London, Ser. B* **82**, 113–123.

Segall, W. (1970). *Fieldiana, Zool.* **51**, 169–205.

Selenka, E. (1887). "Das Opossum." C. W. Kreidels Verlag, Wiesbaden.

Semon, R. (1894a). *Denkschr. Med.-Naturwiss. Ges. Jena* **5**, 3–15.

Semon, R. (1894b). *Denkschr. Med.-Naturwiss. Ges. Jena* **5**, 19–58.

Semon, R. (1894c). *Denkschr. Med.-Naturwiss. Ges. Jena* **5**, 61–74.

Semon, R. (1899). "In the Australian Bush." Macmillan, London.

Sernia, C. (1977). Ph.D. Thesis, Monash University, Clayton, Victoria, Australia.

Sernia, C., and McDonald, I. R. (1977a). *J. Endocrinol.* **72**, 41–52.

Sernia, C., and McDonald, I. R. (1977b). *J. Endocrinol.* **75**, 261–269.

Sernia, C., and McDonald, I. R. (1977c). *Bull. Aust. Mammal Soc.* **4**, 35–36.

Setchell, B. P. (1977). *In* "The Biology of the Marsupials" (B. Stonehouse and D. P. Gilmore, eds.), pp. 411–457. Macmillan, London.

Seydel, O. (1899). *Denkschr. Med.-Naturwiss. Ges. Jena* **6**, Part 1, 445–532.

Sharman, G. B. (1961). *Proc. Zool. Soc. London* **137**, 197–220.

Sharman, G. B. (1962). *J. Endocrinol.* **25**, 375–385.

Sharman, G. B. (1970). *Science* **167**, 1221–1228.

Shaw, G. (1792). "The Naturalist's Miscellany," Vol. 3. London.

Shaw, G. (1799). "The Naturalist's Miscellany," Vol. 10. London.

Simpson, G. G. (1928). "A Catalogue of the Mesozoic Mammalia in the Geological Department of the British Museum." British Museum, London.

Simpson, G. G. (1929). *Am. Mus. Novit.* **390**, 1–15.

Simpson, G. G. (1938). *Am. Mus. Novit.* **978**, 1–15.

Simpson, G. G. (1971). *In* "Early Mammals" (K. Kermack and D. M. Kermack, eds.), pp. 181–198. Academic Press, New York.

Smith, C. A., and Takasaka, T. (1971). *In* "Contributions to Sensory Physiology" (W. D. Neff, ed.), Vol. 5, pp. 129–178. Academic Press, New York.

Smith, G. E. (1902). *R. Coll. Surg. Mus. Cat. Physiol. Ser.* **2**, 133–157.

Smyth, D. M. (1973). *Comp. Biochem. Physiol. A* **45**, 705–715.

Spencer, B., and Sweet, G. (1899). *Q. J. Microsc. Sci.* [N.S.] **41**, 549–558.

Stirton, R. A., Tedford, R. H., and Woodburne, M. O. (1967). *Rec. South Aust. Mus.* **15**, 427–462.

Stirton, R. A., Tedford, R. H., and Woodburne, M. O. (1968). *Univ. Calif., Publ. Geol. Sci.* **77**, 1–30.

Strahan, R., and Thomas, D. E. (1975). *Aust. Zool.* **18**, 165–178.

Strother, W. F. (1967). *J. Aud. Res.* **7**, 145–155.

Sturkie, P. (1965). "Avian Physiology." Cornell Univ. Press (Comstock), Ithaca, New York.

Sutherland, N. S., and Mackintosh, N. J. (1971). "Mechanisms of Animal Discrimination Learning." Academic Press, New York.

Switzer, R. C., and Johnson, J. I. (1977). *Acta Anat.* **9**, 36–42.

Temple, P. (1962). "Nawok! The New Zealand Expedition to New Guinea's Highest Mountains." J. M. Dent and Sons, London.

Temple-Smith, P. D. (1973). Ph.D. Thesis, Aust. Nat. Univ., Canberra.

Terry, R. (1917). *J. Morphol.* **29**, 381–436.

Tettamanti, G., and Pigman, W. (1968). *Arch. Biochem. Biophys.* **124**, 41–50.

Thomas, O. (1885). *Proc. Zool. Soc. London* 1885, 329–339.

Thomas. O. (1906). *Proc. Zool. Soc. London* 1906, 468–478.

Thomas, O. (1907a). *Ann. Mag. Nat. Hist.* [7] **20**, 293–294.

Thomas, O. (1907b). *Ann. Mag. Nat. Hist.* [7] **20**, 498–499.

Thomas, O. (1923). *Ann. Mag. Nat. Hist.* [9] **11**, 170–178.

Thomas, O., and Rothschild, W. (1922). *Ann. Mag. Nat. Hist.* [9] **10**, 129–131.

Thompson, E. O. P., Fisher, W. K., and Whittaker, R. G. (1973). *Aust. J. Biol. Sci.* **26**, 1327–1335.

Tobin, G. (1792). Journal. Ms A562-3 in Mitchell Library, Sydney.

Toldt, K. (1906). *K. K. Naturhist. Hofmus., Ann.* **21**, 1–21.

Tucker, V. (1968). *Comp. Biochem. Physiol.* **24**, 307–310.

Turner, C. W. (1952). "The Mammary Gland." Lucas Brothers, Columbia, Missouri.

Tyndale-Biscoe, C. H. (1973). "The Life of Marsupials." Arnold, London.

van Bemmelen, J. F. (1901). *Denskchr. Med.-Naturwiss. Ges. Jena* **6**, Part 1, 729–798.

Vandebroek, G. (1964). *Ann. Soc. R. Zool. Belg.* **94**, 117–160.

Van Deusen, H. M. (1971). *Fauna* **2**, 12–19.

Van Deusen, H. M., and George, G. G. (1969). *Am. Mus. Novit.* **2383**, 1–23.

van Rijnberk, G. (1913). *Bijd Dierkd.* **19**, 187–196.

Verreaux, J. (1848). *Revue Zoologique* **11**, 127–134.

Waite, E. (1896). *Proc. Linn. Soc. N.S.W.* **21**, 500–502.

Walls, G. L. (1942). "The Vertebrate Eye," Bull. No. 19. Cranbrook Inst. Sci., Michigan.

Waring, H., Moir, R. J., and Tyndale-Biscoe, C. H. (1966). *Adv. Comp. Physiol. Biochem.* **2**, 237–376.

Watkins, W. M., and Hassid, W. Z. (1962). *J. Biol. Chem.* **237**, 1432.

Watson, C. R. R., Provis, J. M., and Bohringer, R. C. (1977). *J. Anat.* **124**, 533.

Watson, D. M. S. (1916). *Philos. Trans. R. Soc. London, Ser. B* **207**, 311–374.

Watson, D. M. S. (1918). *J. Anat.* **52**, 1–63.

Weber, M. (1888). *Bijdr. Dierkd. Feestnummer,* 1–8.

Weiss, M. (1973). *J. Endocrinol.* **58**, 251–262.

Weiss, M., and McDonald, I. R. (1965). *J. Endocrinol.* **33**, 203–210.

Weiss, M., and McDonald, I. R. (1966). *J. Endocrinol.* **35**, 207–208.

Welker, W., and Lende, R. A. (1978). *In* "Comparative Neurology of the Telencephalon" (S. D. E. Ebbesson and H. Vanegas, eds.), Plenum, New York (in press).

Wells, R. T. (1973). Ph.D. Thesis, University of Adelaide, Adelaide.

Westling, C. (1889). *Bih. Sven. Vetenskapsakad. Handl.* **15**, 3–71.

Wetherall, J. D., and Turner, K. J. (1972). *Aust. J. Exp. Biol. Med. Sci.* **50**, 79–95.

Wever, E. G., Vernon, J. A., Peterson, E. A., and Crowley, D. E. (1963). *Proc. Natl. Acad. Sci. U.S.A.* **50**, 806–811.

Whitley, G. P. (1975). "More Early History of Australian Zoology." R. Zool. Soc. N.S.W., Sydney, Australia.

Whittaker, R. G., and Thompson, E. O. P. (1974). *Aust. J. Biol. Sci.* **27**, 591–605.

Whittaker, R. G., and Thompson, E. O. P. (1975). *Aust. J. Biol. Sci.* **28**, 353–365.

Whittaker, R. G., Fisher, W. K., and Thompson, E. O. P. (1972). *Aust. J. Biol. Sci.* **25**, 989–1004.

Whittaker, R. G., Fisher, W. K., and Thompson, E. O. P. (1973). *Aust. J. Biol. Sci.* **26**, 877–888.

Wilson, J. T., and Hill, J. P. (1907). *Q. J. Microsc. Sci.* [N.S.] **51**, 137–165.

Wilson, J. T., and Hill, J. P. (1908). *Philos. Trans. R. Soc. London, Ser. B* **199**, 31–168.

Wilson, J. T., and Hill, J. P. (1915). *Q. J. Microsc. Sci.* [N.S.] **61**, 15–25.

Wilson, J. T., and Martin, C. J. (1893a). *In* "The Macleay Memorial Volume" (J. J. Fletcher, ed.), pp. 179–189. Linn. Soc. N.S.W., Sydney, Australia.

Wilson, J. T., and Martin, C. J. (1893b). *In* "The Macleay Memorial Volume" (J. J. Fletcher, ed.), pp. 190–200. Linn. Soc. N.S.W., Sydney, Australia.

Wood, B. G., Washburn, L. L., Mukerjee, A. S., and Banerjee, M. R. (1975). *J. Endocrinol.* **65**, 1–6.

Woodburne, M. O., and Clemens, W. A. (1978). In preparation.

Woodburne, M. O., and Tedford, R. H. (1975). *Am. Mus. Novit.* **2588**, 1–11.

Wooding, F. B. P. (1977). *In* "Comparative Aspects of Lactation" (M. Peaker, ed.), pp. 1–41. Symposia of The Zoological Society of London, Vol. 41, Academic Press, New York.

Wright, A., Jones, I. C., and Phillips, J. G. (1957). *J. Endocrinol.* **15**, 100–107.

Zarnik, B. (1910). "Vergleichende Studien über den Bau der Niere von Echidna und der Reptilienniere." Fischer, Jena.

Ziehen, T. (1908). *Denkschr. Med.-Naturwiss. Ges. Jena* **6**, Part 2, 789–921.

INDEX